Ecological Assessment of Natural Resources

Ecological Assessment of
Natural Resources

Edited by **Alfred Muller**

SYRAWOOD
PUBLISHING HOUSE

New York

Published by Syrawood Publishing House,
750 Third Avenue, 9th Floor,
New York, NY 10017, USA
www.syrawoodpublishinghouse.com

Ecological Assessment of Natural Resources
Edited by Alfred Muller

Contents

Preface

It is often said that books are a boon to mankind. They document every progress and pass on the knowledge from one generation to the other. They play a crucial role in our lives. Thus I was both excited and nervous while editing this book. I was pleased by the thought of being able to make a mark but I was also nervous to do it right because the future of students depends upon it. Hence, I took a few months to research further into the discipline, revise my knowledge and also explore some more aspects. Post this process, I began with the editing of this book.

Ecological monitoring of natural resources has become vital for assessing existing reserves of natural resources as well as for prevention of environmental degradation and contamination. This book compiles researches and case studies related to ecological assessment of resources and related topics such as management of various water resources, sustainable land management and agricultural policies, evaluation of industrial pollutants, etc. It aims to provide a comprehensive overview of the prominent concepts and methods of ecological assessment from different parts of the world. Through this book, an attempt has been made to further enlighten the readers about the emerging aspects in this field.

I thank my publisher with all my heart for considering me worthy of this unparalleled opportunity and for showing unwavering faith in my skills. I would also like to thank the editorial team who worked closely with me at every step and contributed immensely towards the successful completion of this book. Last but not the least, I wish to thank my friends and colleagues for their support.

Editor

Experts Valuating the Climate Change Policies in Greece: Self-Interested Versus Ethically Motivated Values

Vasileios Markantonis[1,2] & Kostas Bithas[3]

[1] Joint Research Center of the European Commission (JRC), Ispra, Italy

[2] Research Institute of Urban Environment and Human Resources (UEHR), Panteion University, Athens, Greece

[3] Faculty of Economics and Regional Development, Panteion University, Athens, Greece

Correspondence: Vasileios Markantonis, Joint Research Center of the European Commission (JRC), Ispra, Italy. E-mail: vasileios.markantonis@jrc.ec.europa.eu

Abstract

Understanding and estimating the climate change costs has been in focus of the scientific community in the last years, whereas several studies are dealing with this challenging issue. In this context, the present paper aims at valuating the climate change mitigation and adaptation measures in Greece. To achieve that, we carried out a contingent valuation survey. In addition it explores the coexistence of ethically motivated values and self-interested values held by Greek climate experts, a condition in economics underlying the existence of the "Bergson-Tintner-Samuelson (BTS) value formulation effect". This is an experimental attempt in recent valuation literature and carries significant implications for the valuation issue and its policy implications. The results indicate that ethically motivated values of crucial environmental functions such as climate far exceed the individualistic ones. Furthermore, the coexistence of public ethically-based values alongside self-interested ones supports earlier findings in the literature and indicates that solely self-interested individual values do not reflect the real welfare contribution of crucial environmental functions and, therefore, should not form the exclusive guide for environmental policy.

Keywords: climate change costs, environmental valuation, mitigation and adaptation, experts' preferences

1. Introduction

According to the IPCC 5th Assessment Report, Greece, as a geographical part of the Mediterranean, appears to be an area highly vulnerable to the effects of climate change. In particular, the Report marks out Southern Greece as one of the regions potentially most affected, an evolution expressed mainly through increased incidence of heat waves and droughts (IPCC, 2013). The severity of those climate change impacts indicates the need for comprehensive mitigation and adaptation policy measures in Greece. In that context, the objective of this paper is twofold: First, to estimate the economic cost of climate change in Greece by applying a stated preferences survey to national climate experts. In designing our valuation framework we consider climate change as an indicative example of crucial environmental function. Second, the paper investigates the existence of the "Bergson-Tintner-Samuelson (BTS) value formulation effect" for climate change costs. This theory suggests that people hold two categories of values for public environmental goods: those motivated by private preferences and those driven by social preferences (Ami et al., 2014). In that respect the study investigates the dual values individual hold for crucial environmental functions. To our knowledge, it is the first time in recent literature that the BTS effect, which has significant implications for the valuation problem itself and for its policy relevance, has been investigated in connection with the valuation of environmental goods and services.

Our study carries out a Contingent Valuation (CV) survey (Spash & Hanley, 1995) following widely applied methodology. We commence by discussing an essential strategic decision for the study, namely, the selection of participants in the survey. At this point, by defining "climate change experts" we include those individuals who have a proven professional knowledge and expertise in this scientific field. Our choice was to survey exclusively climate change experts so as to eliminate potential measurement biases and have the study focus on the investigation of two kinds of values held simultaneously by experts: self-interested and ethically motivated ones. The relative elimination of measurement biases is a necessary condition for placing emphasis on the evaluation of the moral incentives of preferences. An environmental policy that is based solely on self-interested values is ineffective when the environmental asset at hand inspires ethically motivated values as well. In this context, the ethically motivated values and preferences could be measured and taken into account by environmental policy.

On the other hand, the lack of sufficient knowledge and personal experience of "average" citizens in valuing complex environmental goods and services has become the subject of debate (Vatn & Bromley, 1994). As Stevens et al. (1991) assert, "many respondents probably did not give the valuation question much thought". Those limitations result in various estimation biases (ordering and embedding effects, information biases, etc.) that have been addressed extensively in the relevant literature (Bishop & Heberlein, 1979; Kahneman et al., 1990; Knetsch, 1995; Bergstrom et al., 1990; MacDonald & Bowker, 1994; Brown & Gregory, 1999; Kolstad & Guzman, 1999). We consider eliminating the several existing biases to the maximum possible a necessary condition for tracing the co-existence of ethically motivated and self-interested values. This condition is quite crucial when valuating critical environmnental issues, such as climate change, that define the prospects of sustainability (Bithas, 2008). We consider as critical environmental functions those that provide a biologically 'safe operating space' for humanity (Rockstrom et al., 2009; Karr, 1991; Bithas, 2011). Clearly, critical environmental functions strongly concern the well-being of future generations. That characteristic may result in an inherent limitation of the ability of average citizens to value such functions and services. In contrast, experts are better informed on the scientific and technical aspects of crucial environment issues. Experts can, in this case, potentially offer a relatively more valid and authoritative evaluation, which could overcome some of the possible biases that arise in CV surveys of the general population.

Thus, limiting the survey to experts was a strategic methodological decision guided by the following characteristics of the climate change phenomenon:

- It concerns a crucial environmental function that defines the evolution of the earth's biosphere.

- It significantly affects the functioning of the economy and society.

- There are strong uncertainties regarding the function of the climate, climate change trends, and the impacts of climate change on society and the economy.

- Average citizens cannot understand the climate change phenomenon in detail because they lack previous experience and their opinions are influenced by current mass media trends.

- Although it directly concerns the present generation, it will profoundly determine the potentials of future generations due to its long-term impacts.

For multidimensional phenomena, like climate change, whose properties elude the perception of ordinary individuals, experts can offer a substantial contribution in shaping public policies. Experts can lead to the shaping of public choices and guide informed public decision making. Plato establishes the same argument in the Kriton dialogue (Woodruff, 2010; Peterson, 2011) where a democratic society acts and decides on the basis of knowledge, experience, and valid information which alter the knowledge of the average citizens; otherwise, the decision-makers may justify their decisions by invoking the essential ignorance of the average citizens. In the study of Van Houtven et al. (2014) experts' elicitations where combined with stated preferences methods to estimate ecosystem services values occurring from lake water quality improvement.

In any case, we do not assume that a valuation based on the experts' preferences is representative of the general population. However, we suggest this valuation as representative of the 'informed citizens' thereby overcoming the inherent limitation of the average citizens.

Furthermore, the present paper investigates the dual identity of the experts as self-interested individuals and as citizens motivated by ethical incentives, who evaluate crucial environmental goods. In that context, our paper analyses the Bergson-Tintner-Samuelson (BTS) value formulation effect according to which people hold two kinds of values for public environmental goods: values motivated by private preferences and values inspired by social preferences founded on moral duty and social commitment (Van Kooten & Schmitz, 1992; Bergstrom et al., 1986; Bergson, 1938; Samuelson, 1977, 1981; Tintner, 1946). As citizens, respondents base their decisions on social welfare rather than their personal self-interested preferences when contributing to public goods (Nyborg, 2000; Sugden, 2005).

2. The Methodological Framework

2.1 The Cost of Climate Change

The direct valuation of the impacts of climate change is a difficult and uncertain process. Strong criticism has been leveled against efforts towards directly estimating climate change impacts at a macro level, such as the Stern

Review (Stern, 2007). The Stern Review has provoked a debate (Weitzman, 2008; Nordhaus, 2008) with a focus on the selected discount rate and the valuation of future climate impacts. With regard to the latter, criticism focused on the underestimation of the non-market impacts on environmental functions and services (Sterner & Persson, 2007; Neumayer, 2007). Moreover, the Stern Review has been criticized due to its advocating applying conventional methods of economic analysis to a problem for which they are unsuitable (Spash, 2007). In addition, Toman (2006) questions the empirical capacities and the implicit assumptions of climate change economics and develops a way for integrating economic analysis and public dialogue. Tol (2005) has used the results of twenty-eight published studies combined to form a probability density function for estimating the external costs of the carbon emissions and the uncertainties those estimates entail. In a later study (Tol, 2009), he reviewed the literature on the economic impacts of climate change, analyzing fourteen estimates of the total damage cost of climate change. In that study he perceives climate change as an externality that is unprecedentedly large, complex, and uncertain, where the most important among the non-estimated impacts include the indirect effects of climate change on economic development; large-scale biodiversity loss; low-probability; high-impact scenarios; climate change impact on violent conflict; and the impacts climate change will have beyond 2100.

Our approach avoids directly estimating climate change costs in Greece. Instead, it follows the alternative line of estimating the mitigation and adaptation costs required when dealing with climate change impacts at the national level. Taking into account the long-term duration of the climate change impacts, in tandem with the delay in implementing relevant policy measures, the present study estimates the national mitigation and adaptation costs at two time periods: at the present one and in 25 years, assuming that the current status of the relevant policies remains unchanged ("business as usual" policies). The specific adaptation and mitigation measure selected for this study were based on those indicated in the Stern Review (Stern, 2007), which was the prominent document, at the time of conducting the survey, setting the basis for the climate change costs.

The CV method is based on the assumption of standard economics that environmental assets and services can be treated identically to marketed goods for the purposes of valuation (Spash & Hanley, 1995; Braden & Kolstad, 1991). CV creates the conditions of a hypothetical market for non-marketed environmental assets. Although the results of the hypothetical markets have certain intrinsic shortcomings, they are accepted as a feasible valuation in the absence of more realistic estimates. In this context, Parks and Cowdy (2013) analyze the importance and challenges of traditional welfare methods, including stated preference, to estimate social values of crucial environmental functions, such as climate change.

Willingness-to-Pay (WTP) and Willingness-to-Accept (WTA) methods are expected to provide similar values. However, many differences do arise in practice either because of the presence of 'irrational choices' by respondents or because respondents may be unable or unwilling to offer meaningful valuations. Simultaneously, the income, wealth, and endowment effects may strongly influence the elicited values (Stevens et al., 1991; Whitehead & Blomquist, 1991; Bergstrom et al., 1990; Coursey et al., 1987; Adamowicz et al., 1993). Although methodological reasons do exist for making a choice between WTP and WTA, the current literature is largely dominated by the WTP method.

With regard to the economics of climate change, several studies have used CV and other stated preferences methods to value the impacts of climate change on non-market goods. More specifically, Layton and Gardner (2000) used the conjoint analysis method to analyze the willingness to pay in order to prevent a long-term shift in the forest line caused by climate change. Berrens et al. (2004) applied a CV study to estimate American family willingness to pay in order to support the Kyoto Protocol. Fleischer and Sternberg (2005) used the choice modelling method to value the economic impact of global climate change on Mediterranean rangeland ecosystems. Riera et al. (2007) used CV to elicit values of climate change effects on Mediterranean shrublands and to estimate the costs of climate change mitigation programs. Moreover, Cameron (2004) applied a discrete-choice, stated preference method to estimate individual willingness to pay for climate change mitigation aspects.

2.2 Self-Interested and Ethically Motivated, Social Values: The Bergson-Tintner-Samuelson Values Formulation

In several cases, the extensive application of WTP has led to findings which, if one goes by economic theory, could not have been expected (Spash & Hanley, 1995; Spash, 2000; Stevens et al., 1991; Whitehead & Blomquist, 1991). The phenomenon that is most difficult to explain, in the context of standard welfare economics, is the coexistence of positive bids with lexicographic preferences. The standard approach predicts that an individual with lexicographic preferences accepts an absolute right for environmental services and must, therefore, refuse all monetary trade-offs. In that context, WTA tends to infinity while WTP approaches the individual's entire budget. In practice, individuals with lexicographic preferences for environmental services usually define WTP as zero, in protest against the monetary valuation. This is a stance expected in the context of standard theory. However, recent studies have identified positive 'moderate' bids in WTP by citizens who hold lexicographic preferences and hence attribute rights of protection to environmental functions and assets. That finding can hardly be explained in the context of standard welfare economics (Spash, 2000; Spash & Hanley, 1995; Stevens et al., 1991).

In order to fathom the coexistence of ethically motivated and self-interested values, certain explanations have been developed. Spash accepts that rights-based respondents may express positive WTP values motivated by a risk-averse instinct for environmental functions (Spash, 2000). Stevens et al. (1991) suggest that positive bids can coexist with lexicographic preferences which are inspired by moral considerations. In this context, positive bids may express a desire for conservation, a 'good cause', irrespective of the value of the environmental function itself.

A systematic explanation can be traced through the Bergson-Tintner-Samuelson (BTS) value formulation (Bergson, 1938; Samuelson, 1977, 1981; Tintner, 1946). As adapted by Kohn (1993) and Stevens et al. (1993) for the valuation of environmental goods, the BTS value formulation suggests that people hold two categories of values for public environmental goods and services: values motivated by private preferences and values inspired by social preferences founded on moral duty and social commitment (Van Kooten & Schmitz, 1992).

In that context, the present study aims at examining the existence of the BTS values formulation for climate change adaptation and mitigation measures and, hence, for climate impacts. Specifically the Greek climate experts were asked to provide two values:

a) The individual WTP for financing mitigation and adaptation measures. Here, the experts are expected to act as conventional individuals who express their WTP for the preservation or restoration of environmental functions (protecting climate and avoiding climate change impacts). Within this valuation framework there is an explicit trade-off between income and protection of the environmental functions and services.

b) The percentage of Greek GDP that should be allocated to mitigation and adaptation policies. That is the amount of public funding that should be allocated towards the preservation of the public good of climate and the avoidance of the corresponding 'public bads' (climate change impacts). In that second valuation, experts are expected to act as citizens and members of a society with an ethical commitment (Kohn, 1993; Stevens et al., 1993). Two separate valuations were requested: the present-day one and one in 25 years' time, under the assumption that the current status of climate change policy continues in effect. In both valuations, the GDP percentage is defined as an annual financial contribution for those adaptation and mitigation measures that are necessary in order to maintain the current level of social welfare and economic activities (2009 estimates).

The survey also contained a set of questions to evaluate opinions of the criticality of climate change at present and in the future. Eventually, forty-one national experts with a vast working experience in climate change were identified. They were affiliated with insitutions including research institutes, universities, non-governmental environmental organisations, public administration and environmental consultancies. This is a small sample in absolute terms. Nevertheless, it does contain all of the Greek experts in the relevant field. Eventually, 30 out of those 41 experts (73%) agreed to participate in the survey, providing results concerning their preferences and values on the complex phenomenon of climate change. Data were collected by in-person interviews between June and November 2009.

3. Findings and Analysis

For the investigation of the Bergon-Tinter-Samuelson value formulation effect, that offers evidence of the co-existence of private and public preferences for climate change adaptation and mitigation measures, we should compare the individual WTP with the public contribution (percentance of GDP). For a meaningful comparison the public contribution will be converted to the household level. The experts indicate that an amount equal to 1.71% of GDP is the necessary public contribution. That is estimated at 3,881,700,000 euros, based on a GDP of 227 billion euros (Eurostat estimate for 2009). This is equivalent to a public contribution of 1358.60 euros for each of the 2,857,000 Greek households. On the other hand, the average individual WTP comes to 306.11 euros at the household level (Table 1). As a result, the public contribution at the household level is 4.4 times the individual contribution. It is evident that the disparity between individual (WTP) and public contributions is extremely wide.

Table 1. Willingness-to-pay estimates of a national annual household fee for mitigation and adaptation costs

If Yes, how much would you be willing to pay and how would you allocate that payment to the specific mitigation and adaptation measures?	Mean WTP (€)	S.D.	%
Mitigation Total	**209,17**	**338,14**	*68.3%*
1 Improved energy efficiency	50,64	96.36	*16.5%*
2 Promotion of new low-carbon technologies	58,82	136,92	*19.2%*
3 Investment in renewable energy sources	74,31	117,19	*24.3%*
4 Reduction of non-CO_2 emissions (land use, farming, stockbreeding)	25,01	33,82	*8.2%*
5 Other	0,50	1,33	*0.2%*
Adaptation Total	**96,94**	**119,57**	*31.7%*
1. Restructuring the economy (new cultivations, eco-tourism, etc.)	19,14	29,19	*6.3%*
2. Infrastructure for protection from natural hazards (floods, forest fires, etc.)	16,32	20,06	*5.3%*
3. Infrastructure for the creation of new water reservoirs	18,65	24,91	*6.1%*
4. Protection of the natural environment (rivers, threatened species, marine ecosystems, etc.)	15,54	18,31	*5.1%*
5. Establishment of new institutional, regulatory and management structures	17,49	26,56	*5.7%*
6. Reform of the healthcare system	6,86	10,43	*2.2%*
7. Other (Climate change refugees)	3,79	11,96	*1.2%*
TOTAL WTP for mitigation and adaptation	***306,11***	***445,46***	***100.0%***

This is strong evidence for the existence of a dual valuation described by to the BTS value formulation effect. The existence of the BTS effect is further supported by the data of Tables 2 and 3. Table 2 indicates that, in the WTP question, all the negative responses were the outcome of ethical reasoning. The 'zero protest bidders', according to Table 3, evaluate as significant the climate change impacts and, hence, the corresponding public good. According to Table 4, the majority of 'zero protest bidders' proposes a public contribution between 0.5-1.5% of GDP today and 0.75%-3% in 25 years' time.

Table 2. Acceptance of the individual valuation scenario

Would you be willing to pay a 'national annual household fee' for the mitigation and adaptation of the climate change?	n	%
No	12	40.0%
a. Zero Protest Bids	12	
Justification of zero protest bids replies		
I don't believe in this approach	6	
It is an international issue and should be treated globally	3	
The state should pay for this	3	
b. Zero bids	0	
YES	*18*	*60.0%*

Table 3. Ranking the importance of climate change impacts at the national level

In your opinion, how important are the impacts of climate change in social, economic, and environmental terms at the national level?

	Protest Zero bids respondents (n=12)				All respondents (n=30)			
Importance	*Present*		*In 25 years*		*Present*		*In 25 years*	
	n	%	n	%	n	%	n	%
Not at all important	0	0.0%	0	0.0%	0	0.0%	0	0.0%
Somewhat important	2	16.7%	0	0.0%	6	20.0%	0	0.0%
Important	4	33.3%	1	8.3%	11	36.7%	3	10.0%
Rather important	4	33.3%	7	58.3%	9	30.0%	14	46.7%
Very important	2	16.7%	3	25.0%	4	13.3%	12	40.0%
I don't know	0	0.0%	1	8.4%	0	0.0%	1	3.3%
Total	12	100%	12	100%	30	100%	30	100%

Table 4. GDP contribution for mitigation and adaptation measures

What percentage of national GDP should the Greek state allocate for mitigation and adaptation of the climate change?

	Protest Zero bids respondents (n=12)				All respondents (n=30)			
GDP percentage	National GDP at present		National GDP in 25 years		National GDP at present		National GDP in 25 years	
	n	%	n	%	n	%	n	%
No contribution	0	0.0%	1	8.3%	0	0.0%	1	3.3%
0.01%	0	0.0%	0	0.0%	0	0.0%	0	0.0%
0.05%	0	0.0%	0	0.0%	0	0.0%	0	0.0%
0.1%	1	8.3%	0	0.0%	2	6.7%	0	0.0%
0.25%	0	0.0%	0	0.0%	0	0,0%	0	0.0%
0.5%	4	33.3%	0	0.0%	7	23.3%	0	0.0%
0.75%	1	8.3%	1	8.3%	2	6.7%	2	6.7%
1%	2	16.7%	3	25.0%	4	13.3%	5	16.7%
1.5%	1	8.3%	0	0.0%	3	10.0%	1	3.3%
2%	0	0.0%	2	16.7%	4	13.3%	3	10.0%
2.5%	0	0.0%	1	8.3%	3	10.0%	3	10.0%
3%	1	8.3%	1	8.3%	1	3.3%	4	13.3%
4%	1	8.3%	0	0.0%	1	3.3%	6	20.0%
5%	1	8.3%	3	25.0%	3	10.0%	5	16.7%
Total	12	100%	12	100%	30	100%	30	100%

60% of the experts participating in the survey are willing to pay a significant sum to confront climate change impacts (Table 2). However, that sum amounts to a small percentage of the aggregate public contribution proposed by the same experts. Thus, the experts in question are explicitly accepting two distinct valuations of mitigation and adaptation measures. They attempt an individual valuation but, at the same time, they define a public valuation which by far exceeds the individual one. It appears that what the experts are proposing is that climate change mitigation and adaptation measures should both receive public funding that reflects their public value.

The criticality of 'climate' environmental functions (and, hence, of climate change impacts) as a public asset is reflected in Table 3: 80% of the experts consider the phenomenon to be important/very important today and 97% that it will be important/very important in 25 years' time. The increasing criticality of the phenomenon in the future is accompanied by a substantial increase in the public valuation for its confrontation. Table 5 indicates that the public contribution should reach 2.75% of GDP in 25 years whereas it is currently estimated at 1.71% of GDP.

Table 5. GDP allocation for present and future mitigation and adaptation costs

What percentage of the National GDP should the Greek state provide for mitigation and adaptation of the climate change?	Mean GDP	S.D.	%
Present context			
Mitigation Total	**1.13%**	**1.12%**	*66.1%*
1. Reduction in the demand for carbon-intensive goods and services	0.19%	0.21%	*11.1%*
2. Improved energy efficiency	0.28%	0.38%	*16.4%*
3. Promotion of new, low-carbon technologies	0.20%	0.16%	*11.7%*
4. Investment in renewable energy sources	0.35%	0.49%	*20.5%*
5. Reduction of non-CO_2 emissions (land use, farming etc)	0.12%	0.14%	*7.0%*
6. Other (Research – Education)	0.02%	0.05%	*1.2%*
Adaptation Total	**0.58%**	**0.68%**	*33.9%*
1. Restructuring the economy (new cultivations, eco-tourism, etc.)	0.12%	0.14%	*7.0%*
2. Infrastructure for protection from natural hazards (floods, droughts, wildfires, etc.)	0.10%	0.12%	*5.8%*
3. Infrastructure for the creation of new water reservoirs	0.11%	0.13%	*6.4%*
4. Protection of the natural environment (rivers, threatened species, marine ecosystems, etc.)	0.12%	0.17%	*7.0%*
5. Establishment of new institutional, regulatory and management structures	0.09%	0.11%	*5.3%*
6. Reform of the healthcare system	0.05%	0.08%	*2.9%*
7. Other (Climate change refugees)	0.01%	0.04%	*0.6%*
TOTAL GDP for mitigation and adaptation	**1.71%**	**1.45%**	*100.0%*
Future context (25 years from now, provided that future policy is identical with present policy)			
Mitigation Total	**1.45%**	**0.97%**	*52.7%*
1. Reduction in demand for carbon-intensive goods and servic	0.21%	0.21%	*7.6%*
2. Improved energy efficiency	0.30%	0.21%	*10.9%*
3. Promotion of new, low-carbon technologies	0.33%	0.34%	*12.0%*
4. Investment in renewable energy sources	0.44%	0.41%	*16.0%*
5. Reduction of non-CO_2 emissions (land use, farming, stockbreeding)	0.14%	0.12%	*5.1%*
6. Other (Research – Education)	0.01%	0.04%	*0.4%*
Adaptation Total	**1.30%**	**0.88%**	*47.3%*
1. Restructuring the economy (new cultivations, eco-tourism, etc.)	0.17%	0.14%	*6.2%*
2. Infrastructure for protection from natural hazards (floods, forest fires, etc.)	0.29%	0.21%	*10.5%*
3. Infrastructure for the creation of new water reservoirs	0.31%	0.28%	*11.3%*
4. Protection of the natural environment (rivers, threatened species, marine ecosystems, etc.)	0.23%	0.19%	*8.4%*
5. Establishment of new institutional, regulatory, and management structures	0.16%	0.16%	*5.8%*
6. Reform of the healthcare system	0.11%	0.09%	*4.0%*
7. Other (Climate change refugees)	0.02%	0.06%	*0.7%*
TOTAL GDP for mitigation and adaptation	*2.75%*	*1.55%*	*100.0%*

Another finding, which is probably the opposite of what is intuitively expected, strongly supports the criticality of climate as well as its public nature and the existence of ethical motives in the valuation. Table 1 indicates that 68.3% of the individual contributions is allocated to mitigation policies and only 31.7% to adaptation measures. Similarly, according to Table 5, 1.13% of GDP should fund mitigation policies and 0.58% should go towards adaptation strategies. One might expect that both individual and public contributions would be oriented towards adaptation measures that protect the national socio-economic system and national welfare and, hence, the welfare of the experts surveyed. On the contrary, the findings of the research indicate that Greek experts regard as more valuable the national mitigation measures that control the Greek greenhouse emissions and, consequently, prevent deterioration in the global climate. Taking into account that Greece is a small contributor to climate change gases [Greek emissions, were 122.543 Tg in 2009, in other words, 2.7% of the EU total of 4614.526 Tg (EEA, 2012)], it appears likely that an ethical motivation is behind the allocation of individual and public contributions between mitigation and adaptation measures: instead of protecting themselves from climate change impacts, the experts consider it more valuable to prevent the worldwide public good from deteriorating further.

In conclusion, the existence of the BTS value formulation effect is supported by the findings of the survey with climate experts. The analysis strongly indicates the existence of ethically motivated values for the climate's crucial environmental functions and services. Those ethical values co-exist with individual values which, however, are substantially lower than the corresponding ethically motivated values. In the following section we attempt to explain the phenomenon of the co-existence of ethically motivated values alongside self-interested ones.

4. Discussion and Concluding Remarks

The BTS value formulation effect is confirmed by the results of our survey. This confirmation is important since it is expected that experts are better informed in comparison to average citizens. The experts' valuations are less affected by measurement biases (Harris et al., 1989). The elimination of measurement biases is important in a study that seeks to place emphasis on the co-existence of ethical motivated and self-interested preferences through the identification of the BTS effect. Our results indicate that the individual values stated by the experts are much lower than the public values that those experts define for the same goods. Based on these findings we essay to explain the origins of ethical values towards the protection of crucial environmental functions and services.

The experts consider that the impact of climate change represents a major threat to public welfare. Therefore, climate change impacts must be confronted with appropriate mitigation and adaptation policies. Experts, acting as self-interested individuals, are willing to pay a significant amount for the appropriate policies. This amount emerges as a 'fair share' contribution for the protection of a significant public good. However, applying such a survey only to experts leads to methodological shortcomings and limitations. In order to estimate comprehensively and reveal the preferences of the society this survey should be redesigned and applied to a representative sample of the total population.

On the other hand, the ethically motivated valuation indicates that individual contributions lag behind the amount necessary to protect society's welfare from climate change impacts. Climate policy is a public initiative and requires public financial contribution. Climate functions and services and, therefore, climate policies contribute to the welfare of society in a significant way. Furthermore, climate functions are an important factor in the wellbeing of future generations. The preservation of such crucial public environmental functions should receive public funds. Individuals could contribute to these funding schemes by a 'fair share' which is determined by their 'time span' and 'space span' effects (Bithas, 2011). The time span effect is determined by the expected lifetime of an individual plus a period covering the lifetime of direct descendants. The space span effect is defined by the geographical area that is related functionally to the welfare of an individual. The area within individuals perceive their welfare is extremely limited in comparison to the geographical dimensions of climate change impacts which cover the ecosystem of the Earth in its entirety.

The co-existence of ethically motivated values with self-interested ones may well be explained by the institutional context of western societies, in which the 'marketization' of every sphere of social life is extensively promoted. Traditional public spheres of life, actions, and goods are constantly pressed towards 'marketization'. In that institutional setting, the so-called 'self-interested rational consumer' is the dominant prototype. Adhering to that contemporary social prototype, individuals are willing to pay to preserve public environmental goods and environmental functions. On the other hand, the contemporary ideological prototype cannot eliminate the fundamental properties of crucial environmental goods and functions, which are irrevocably public and directly concern the welfare potential of the current and future generations. Under those conditions and if they are to be effective, the appropriate preservation schemes inevitably acquire public dimensions. Although individual valuations can be substantial, they cannot be but a 'fair share' of the aggregate social value of crucial public environment assets and functions. Therefore, individual valuations are not the appropriate guide for effective environmental policies.

The empirical findings of our study support a prototype citizen who, in the case of climate change, acts according to a rationale not predicted by economics. Informed citizens hold ethical social commitments that result in attributing significant public values that far exceed the self-interested values. Strategies for crucial environmental functions should recognize the existence of social values and include them in the policy-making process. Crucial environmental functions determine the welfare of all future generations and, therefore, may be evaluated in a way that overcomes the individualistic, egoistic prototype. When they can be estimated, social values offer an indicator for climate change policies.

References

Adamowicz, W. L., Bhardwaj, V., & Macnab, B. (1993). Experiments on the difference between willingness to pay and willingness to accept. *Land Economics, 69*, 416-427. http://dx.doi.org/10.2307/3146458

Ami, D., Aprahamian, F., Chanel, O., Joulé, R. V., & Luchini, S. (2014). Willingness to pay of committed citizens: A field experiment. *Ecological Economics*. http://dx.doi.org/10.1016/j.ecolecon.2014.04.014

Bergson, A. (1938). A reformulation of Certain Aspects of Welfare Economics. *Quarterly Journal of Economics, 52*, 310-334. http://dx.doi.org/10.2307/1881737

Bergstrom, J. C., Stoll, J. R., & Randall, A. (1990). The impact of information on environmental commodity valuation decisions. *American Journal of Agricultural Economics, 72*, 614-21. http://dx.doi.org/10.2307/124 3031

Bergstrom, T., Blume, L., & Varian, H. (1986). On the private provision of public goods. *Journal of Public Economics, 29*, 25-49. http://dx.doi.org/10.1016/0047-2727(86)90024-1

Berrens, R. P., Bohara, A. K., Jenkins-Smith, H. C., Silva, C. L., & Weimer, D. L. (2004). Information and effort in contingent valuation surveys: application to global climate change using national Internet samples. *Journal of Environmental Economics and Management, 47*, 331–363. http://dx.doi.org/10.1016/S0095-0696(03)000 94-9

Bishop, R. C., & Heberlein, T. A. (1979). Measuring values of extra-market goods: are indirect methods biased?. *American Journal of Agricultural Economics, 61*, 926-930. http://dx.doi.org/10.2307/3180348

Bithas, K. (2011). Sustainability and externalities: Is the internalization of externalities a sufficient condition for sustainability?. *Ecological Economics, 70*(10), 1703-1706. http://dx.doi.org/10.1016/j.ecolecon.2011.05.014

Braden, J. B., & Kolstad, C. D. (1991). *Measuring the demand for environmental quality.* North-Holland, Amsterdam.

Brown, C. T., & Gregory, R. (1999). Why the WTA-WTP disparity matters. *Ecological Economics, 28*, 323-335. http://dx.doi.org/10.1016/S0921-8009(98)00050-0

Cameron, T. (2004). Individual option prices for climate change mitigation. *Journal of Public Economics, 89*, 283-301. http://dx.doi.org/10.1016/j.jpubeco.2004.01.005

Coursey, D. L., Hovis, J. L., & Schulze, W. D. (1987). The Disparity between Willingness to Accept and Willingness to Pay Measures of Value. *The Quarterly Journal of Economics, 102*(3), 679-690. http://dx.doi.org/10.2307/1884223

EEA. (2012). *EEA greenhouse gas - data viewer.* European Environmental Agency. Retrieved from http://www.eea.europa.eu/data-and-maps/data/data-viewers/greenhouse-gases-viewer

Fleischer, A., & Sternberg, M. (2005). The economic impact of global climate change on Mediterranean rangeland ecosystems: A Space-for-Time approach. *Ecological Economics, 59*, 287-295. http://dx.doi.org/10.1016/ j.ecolecon.2005.10.016

IPCC. (2013). *Fifth Assessment Report, WGI-Climate Change 2013: The Physical Science Basis.* Intergovernmental Panel on Climate Change.

Kahneman, D., Knetsch, J. L., & Thaler, R. H. (1990). Experimental tests of the endowment effect and the Coase theorem. *Journal of Political Economy, 98*, 1325-1348. http://dx.doi.org/10.1086/261737

Karr, J. R. (1991). Biological integrity: a long-neglected aspect of water resource management. *Ecological Applications, 1*(1), 66-84. http://dx.doi.org/10.2307/1941848

Knetsch, J. L. (1995). Asymmetric valuation of gains and losses and preference order assumptions. *Economic Inquiry, 33*, 134-14. http://dx.doi.org/10.1111/j.1465-7295.1995.tb01851.x

Kohn, R. E. (1993). Measuring the existence value of wildlife: Comment. *Land Economics, 69*, 304-308. http://dx.doi.org/10.2307/3146596

Kolstad, D. C., & Guzman, M. R. (1999). Information and the divergence between willingness to accept and willingness to pay. *Journal of Environmental Economics and Management, 38*, 66-80. http://dx.doi.org/10.1006/jeem.1999.1070

Layton, D. F., & Gardner, B. (2000). Heterogeneous preferences regarding global climate change. *Review of Economics and Statistics, 82*, 616–624. http://dx.doi.org/10.1162/003465300559091

Mac Donald, H. F., & Bowker, J. M. (1994). The endowment effect and WTA: a quasi-experimental test. *Agricultural and Applied Economics, 26*(2), 545-551.

Neumayer, E. (2007). A missed opportunity: The Stern Review on climate change fails to tackle the issue of non-substitutable loss of natural capital. *Global Environmental Change, 17*(3–4), 297–301. http://dx.doi.org/10.1016/j.gloenvcha.2007.04.001

Nordhaus, W. D. (2008). A review of the Stern Review on the economics of climate change. *Journal of Economic Literature, 45*(3), 686–702. http://dx.doi.org/10.1257/jel.45.3.686

Nyborg, K. (2000). Homo economics and homo politicus: interpretation and aggregation of environmental values. *J. Econ. Behav. Organ, 42*, 305–322. http://dx.doi.org/10.1016/S0167-2681(00)00091-3

Parks, S., & Gowdy, J. (2013). What have economists learned about valuing nature? A review essay. *Ecosystem Services, 3*, 1-10. http://dx.doi.org/10.1016/j.ecoser.2012.12.002

Peterson, S. (2011). *Socrates and philosophy in the dialogues of Plato*. Cambridge University Press. http://dx.doi.org/10.1017/CBO9780511921346

Riera, P., Peñuelas, J., Farreras, V., & Estiarte, M. (2007). Valuation of climate-change effects on mediterranean shrublands. *Ecological Applications, 17*, 91–100. http://dx.doi.org/10.1890/1051-0761(2007)017 [0091:VOCEOM]2.0.CO;2

Rockström, J., Steffen, W., Noone, K., Persson, Å., Chapin, F. S., Lambin, E., ... Foley, J. (2009). Planetary boundaries: exploring the safe operating space for humanity. *Ecology and Society, 14*(2), 32.

Samuelson, P. A. (1977). Reaffirming the existence of 'reasonable' Bergson-Samuelson social welfare functions. *Economica, 44*, 81-88. http://dx.doi.org/10.2307/2553553

Samuelson, P. A. (1981). Bergsonian welfare economics. In S. Rosefielde (Ed.), *Economic Welfare and the Economics of Soviet Socialism: Essays in Honor of Abram Bergson* (pp. 223-266). Cambridge University Press. http://dx.doi.org/10.1017/CBO9780511895821.011

Spash, C. L. (2000). *Multiple value expression in contingent valuation: economics and ethics*. Cambridge Research for the Environment-Department of Land Economy University of Cambridge.

Spash, C. L. (2007). The economics of climate change impacts à la Stern: Novel and nuanced or rhetorically restricted? *Ecological Economics, 63*(4), 706–713. http://dx.doi.org/10.1016/j.ecolecon.2007.05.017

Spash, C. L., & Hanley, N. (1995). Methodological and ideological options- preferences, information and biodiversity preservation. *Ecological Economics, 12*, 191-208. http://dx.doi.org/10.1016/0921-8009(94)00056-2

Stern, N. (2007). *The economics of climate change: The Stern Review*. Cambridge University Press. http://dx.doi.org/10.1017/CBO9780511817434

Sterner, T., & Persson, U. M. (2007). An even Sterner Review: introducing relative prices into the discounting debate. Discussion paper 07–37. *Resources for the Future*. Washington DC.

Stevens, T. H., Echeverria, J., Glass, R. J., Hager, T., & More, T. A. (1991). Measuring the existence of value of wildlife: What do CVM estimates really show? *Land Economics, 67*, 390-400. http://dx.doi.org/10.2307/3146546

Stevens, T. H., More, T. A., & Glass, R. J. (1993). Measuring the existence value of wildlife: Reply. *Land Economics, 69*, 309-312. http://dx.doi.org/10.2307/3146597

Sugden, R. (2005). Coping with preference anomalies in cost–benefit analysis: A market simulation approach. *Environ. Resour. Econ., 32*, 129–160. http://dx.doi.org/10.1007/s10640-005-6031-5

Tintner, G. (1946). A note on welfare economics. *Econometrica, 14*, 69-78. http://dx.doi.org/10.2307/1905704

Tol, R. S. J. (2005). The marginal damage costs of carbon dioxide emissions: an assessment of the uncertainties. *Energy Policy, 33*, 2064–2074. http://dx.doi.org/10.1016/j.enpol.2004.04.002

Tol, R. S. J. (2009). The Economic Effects of Climate Change. *Journal of Economic Perspectives, 23*(2), 29-51. http://dx.doi.org/10.1257/jep.23.2.29

Toman, M. (2006). Values in the economics of climate change. *Environmental Values, 15*(3), 365-379. http://dx.doi.org/10.3197/096327106778226310

Van Houtven, G., Mansfield, C., Phaneuf, D. J., von Haefen, R., Milstead, B., Kenney, M. A., & Reckhow, K. H. (2014). Combining expert elicitation and stated preference methods to value ecosystem services from improved lake water quality. *Ecological Economics, 99*, 40-52. http://dx.doi.org/10.1016/j.ecolecon.2013.12.018

Van Kooten, G. C., & Schmitz, A. (1992). Preserving Waterfowl Habitat on the Canadian Prairies: Economic Incentives vs. Moral Suasion. *American Journal of Agricultural Economics, 74*(1), 79-89. http://dx.doi.org/10.2307/1242992

Vatn, A., & Bromley, D. W. (1994). Choices without prices without apologies. *Environmental Economics and Management, 26*, 129-148. http://dx.doi.org/10.1006/jeem.1994.1008

Weitzman, M. L. (2008). A review of the Stern Review on the economics of climate change. *Journal of Economic Literature, 45*(3), 703–724. http://dx.doi.org/10.1257/jel.45.3.703

Whitehead, J. C., & Blomquist, G. C. (1991). Measuring Contingent Values for Wetlands: Effects of Information about Related Environmental Goods. *Water Resources Research, 27*(10), 2523-2531. http://dx.doi.org/10.1029/91WR01769

Woodruff, P. (2010). *Plato's shorter ethical works: Crito the Stanford encyclopedia of philosophy*. The Metaphysics Research Lab, Stanford University.

Preliminary Investigation of Stream Sediments Contaminations Caused by Mining Activities in Ibodi and Its Environs, S/W Nigeria Using Geological and Geochemical Assessment Approach

A. I. Akintola[1], P. R. Ikhane[1], S. I. Bankole[2] & O. A. Mosebolatan[1]

[1] Department of Earth Sciences, Olabisi Onabanjo University, Nigeria

[2] Geosciences Department University of Lagos, Akoka Lagos, Nigeria

Correspondence: A. I. Akintola, Department of Earth Sciences, Olabisi Onabanjo University, Nigeria. Email: busayoakins@yahoo.com, a.i.akintola@student.utwente.nl

Abstract

Mining and related activities are sources of heavy metal contamination in streams, such as copper, zinc, cadmium, arsenic and lead . The study is focused on Ibodi, southwestern Nigeria which is located in the basement complex of Nigeria and it is to assess the stream sediments of Ibodi in order to decipher the environmental impact assessment of mining activities on the environment. A total of ten stream sediment were collected from the study area and its environs along major tributaries, air dried at room temperature, sieved with 75 micron sized sieve and analyzed in the laboratory using ICP-MS (inductively couple plasma mass spectrometry technique). The analytical results of the major elements analyses show that Iron oxide [Fe_2O_3] has the highest major element composition, value ranging from 2.36% - 10.61% with an average value of 5.262%. This highest concentration of Iron oxide was found in location 1, with a value of 10.61% and this can be attributed to the underlying geology of amphibolites' in the Ibodi study area, which are known to be rich in Iron as well as magnesium i.e ferromagnesian minerals. Magnesium oxide [MgO] range in composition from 0.11% - 0.92% with an average value of 0.349%, the highest concentration was found in location 3. [Al_2O_3] ranges in composition from 0.83% - 4.158% with an average value of 2.109% the highest value was also found at location 3, it is the next in abundance to Iron oxide in the Ibodi study area. Potassium oxide [K_2O] range in composition from 0.04% - 0.65% with an average value of 0.183%, other major oxides such as [P_2O_5], [TiO_2], [Na_2O] and [CaO] have average values of 0.109%, 0.096%,0.008% and 0.162% respectively; These values are generally low within the Ibodi study area, The analytical results for trace element geochemistry of Ibodi study area show that Vanadium [V] has a high concentrations and it range from 35.00 ppm - 202.00 ppm with an average mean value of 92.50 ppm, the highest concentration of this element was found in location 1 of the Ibodi study area. Arsenic [As] range from 0.10 ppm - 1.6 ppm with an average value of 0.644 ppm, there is a significant enrichment of Cobalt [Co] and Chromium [Cr] with concentrations ranging from 4.0 ppm - 53.50 ppm and 35.20 ppm -150.70 ppm respectively, with average mean values of 17.73 ppm and 88.78 ppm respectively; Manganese [Mn] has the highest concentration in the study area, with concentration value ranging from 86.00 ppm - 2165.00 ppm having an average value of 768.20 ppm; the highest concentration of this element was found in location 1 of the study area. [Ga], [Ni] and [Pb] show considerable enrichments within the study area with concentration values ranging from 3.10 ppm - 9.50 ppm, 5.30 ppm - 37.70 ppm and 6.75 ppm - 18.44 ppm, with average values of 6.08 ppm,19.80 ppm, 12.209 ppm respectively, Rubidium [Rb] range in concentration from 6.50 ppm - 30.90 ppm with an average value of 13.41 ppm, Strontium [Sr] has concentration values that range from 4.70 ppm - 37.20 ppm with an average value of 15.06 ppm. [Y], [Zr] and [Zn] has concentration values that range from 7.99 ppm - 21.10 ppm, 1.20 ppm - 4.00 ppm and 26.20 ppm-83.60 ppm respectively with average mean values of 15.065 ppm, 2.21 ppm and 46.58 ppm, the value of zinc [Zn] is considerably high in the study area with the highest value found at location 4 of the study area indicating some level of enrichment of this metals within the study area, also [Rb], [Y] and [Sr] show some considerable enrichments within the study area. from the study of the environmental parameter such as box plot and Geo-accumulation indexes the values of the selected trace elements are all less than 1, meaning that all the selected trace metals in Ibodi study area have values less than zero and are in the negative zone. In order to determine the pollution status of the study area, the values of the elements when compared to the Muller classes of geo-accumulation suggests that the study area is practically uncontaminated with the selected trace metals, the elements fall into the class 0 i.e. Practically unpolluted.

Keywords: zinc, geochemical, basement complex, cancer, mining

1. Introduction

Mining and related operations are the major anthropogenic source of heavy metal contamination which increases harmful metal content in the streams and negatively influence the environment. In Ibodi area of southwestern Nigeria, there is an intense artisanal mining operation going on which involves excavation of the alluvial soil (host rock) and washing away the waste directly into the stream. Heavy metals such as copper, zinc, cadmium, iron, arsenic, lead are been released into stream as a result of these mining activities it should be noted that uncontrolled direct dumping of mine tailings, domestic waste and discharge of domestic and industrial sewage water into the urban drainage systems are critical components of trace and heavy metal contamination (Tijani et al., 2004; Singh et al., 1990) especially in areas with lack of strict land-use plan and environmental protection regulations. Though sediments are said to represent the ultimate sinks for heavy metals in the environment (Gibbs et al., 1977), changing physico-chemical and environ-mental conditions may lead to remobilization and release of sediment-bound metal pollutants into the water column and consequently into the food chain system within an aquatic environment with serious health and environmental consequences. The environmental impact of mining includes erosion, formation of sinkholes, loss of biodiversity, and contamination of soil, groundwater and surface water by chemicals from mining processes. In some cases, additional forest logging is done in the vicinity of mines and chemicals like mercury, cyanide, sulphuric acid, arsenic and methyl mercury are used in various stages of mining; within the last quarter of the last century there were much interest on environmental pollution and in particular about geochemical distribution and fate of heavy metals in both water and sediment phases of urban drainage system. Though significant advances had been made in the developed regions of the world, there are still increasing concerns about the impacts of urbanization, agricultural, mining and industrial activities on drainage networks in the developing regions of the world, especially in areas with inadequate land use planning and proper waste disposal and management systems (Ajayi & Mombeshora, 1990). This research work is to appraisal the various rock types within the study area and to investigate the geochemical assessment of major and trace element within the stream sediments of the study area with a view to elucidate any form or extent of pollution within the study area. The overall evaluation is expected to give an insight into vulnerability of urban drainage networks in a typical developing region in response to poor sanitation and waste disposal facilities and other anthropogenic activities within the populated urban catchment of a developing country. The study area covers Ibodi and its environment, Southwestern Nigeria. The area lies within latitude 7°33'N and 7°36'N and longitude 4°39'E and 4°42'E other neighboring villages located within the studied area include Afon, Olorombo, Ile-oko, Ijano, Safari, Onigbogi, Isireyun, Oloyin, Aye-ile, and Iyemogun. The study area is easily accessible by complex road networks of major and minor roads as well as footpath linking one sampling point to the other (Figure 1) The climate is sub-humid tropical with average annual rainfall 1348.4 mm. The area is well drained the common rivers in the study area include Isireyun and Ileki rivers the drainage pattern is dendritic.

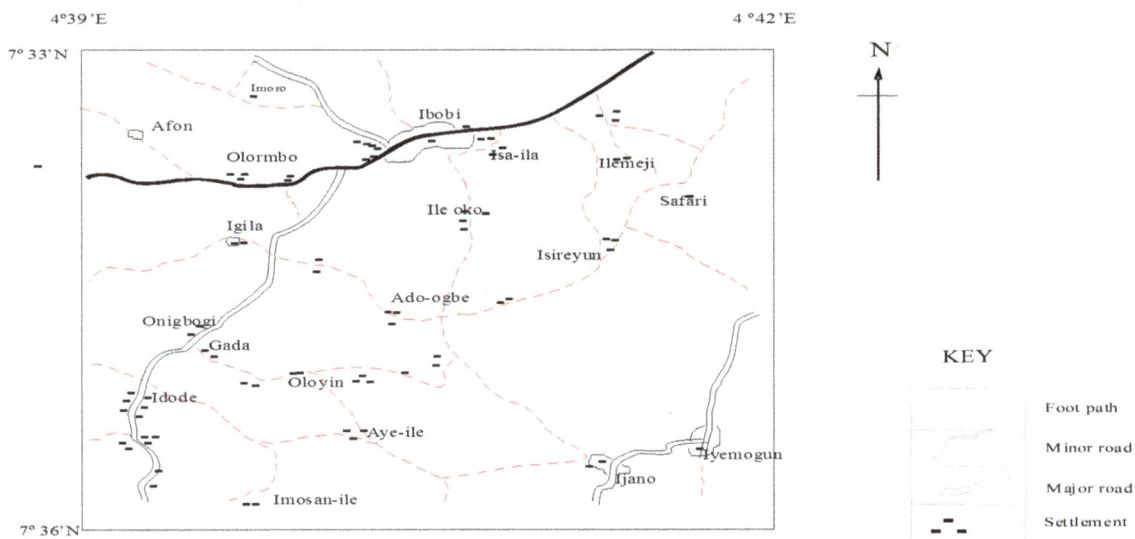

Figure 1. Map showing accessibility and location of the study area

2. Methodology of Study

Systematic geological mapping and stream sediment sampling of first order streams in other to represent weathered rocks in the drainage system was carried out, followed by thin section Petrographic studies of fresh whole rock samples was carried out. Ten stream sediments samples were then analyzed for major and trace elements using inductively-coupled plasma atomic emission spectrophotometry (ICP-AES), at ACME Laboratory Vancouver Canada. The geochemical analytical procedure involves addition of 5ml of Perchloric acid ($HClO_4$), Trioxonitrate (V) HNO_3 and 15 ml Hydrofluoric acid (Hf) to 0.5 gm of sample. The solution was stirred properly and allowed to evaporate to dryness after it was warmed at a low temperature for some hours. 4ml hydrochloric acid (HCl) was then added to the cooled solution and warmed to dissolve the salts. The solution was cooled; and then diluted to 50 ml with distilled water. The solution is then introduced into the ICP torch as aqueous - aerosol. The emitted light by the ions in the ICP was converted to an electrical signal by a photo multiplier in the spectrometer, the intensity of the electrical signal produced by emitted light from the ions were compared to a standard (a previously measured intensity of a known concentration of the elements) and the concentration then computed.

3. Geological Setting, Field Description and Petrography

Nigeria is underlain by Precambrian basement complex rocks, younger granites of Jurassic age and Cretaceous to Recent sediments. The basement rocks occupy about half of the land mass of the country, and is a part of the Pan-African mobile belt lying between the West African and Congo cratons (Black, 1980). There are however contrasting documentation of the evolution of the basement rocks. However loosely, the basement is grouped into three major groups lithostratigraphically viz: the Migmatite-Gneiss Quartzite Complex: comprising biotite and biotite hornblende gneisses, quartzites and quartz schist. Schist Belts, comprising paraschists and meta igneous rocks, which include schists, amphibolites, amphibole schists, talcose rocks, epidote rocks, marble and calc-silicate rocks. They are mainly N-S to NNE-SSW trending belts of low grade supracrustal (and minor volcanic) assemblages. Other secondary rocks used in delineating them are carbonates, calc gneiss and banded iron formation (BIF) and Older granites, which include granite, granodiorite, diorite charnockite, pegmatites and aplites. The study area is located within Ibodi, its geology consists of Precambrian rocks that are typical of Basement Complex of Nigeria and these rocks includes the following five lithologies: [i] Amphibolites [ii] Mica schist [iii] Granite gneiss [iv] Quartzite and [v] Talc (Figure 2).

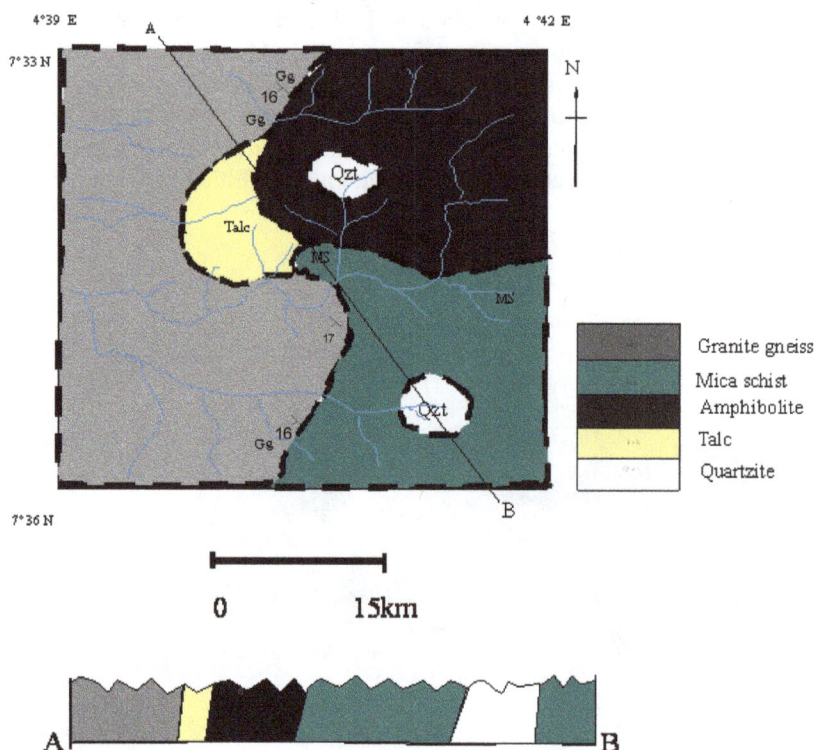

Figure 2. Geological map of Ibodi study area

3. Result and Discussion

The analytical results of the major elements are presented in (Tables 1 and 2). Table 1, shows the major elements oxides composition of the Ibodi study area in (Wt %) while Table 2, shows the statistical summary of major elements oxides with their range and average values respectively. From the analytical data and the various statistical plots figure 5, shows the line diagrams of graphical illustration for major elements oxide composition in stream sediments of Ibodi study area. The analyses show that Iron oxide [Fe_2O_3] has the highest major element composition with value ranging from 2.36% - 10.61% with an average value of 5.262%. This highest concentration of Iron oxide was found in location 1, with a value of 10.61%. and this can be attributed to the underlying geology of amphibolites' in the Ibodi study area, which are known to be rich in Iron as well as magnesium i.e ferromagnesian minerals. Magnesium oxide [MgO] range in composition from 0.11% - 0.92% with an average value of 0.349%, the highest concentration was found in location 3. [Al_2O_3] ranges in composition from 0.83% - 4.158% with an average value of 2.109% the highest value was also found at location 3, it is the next in abundance to Iron oxide in the Ibodi study area. Potassium oxide [K_2O] range in composition from 0.04% - 0.65% with an average value of 0.183%, other major oxides such as [P_2O_5], [TiO_2], [Na_2O] and [CaO] have average values of 0.109%, 0.096%, 0.008% and 0.162% respectively; These values are generally low within the Ibodi study area, (Figures 3 and 4) Show the 2D, 3D and geochemical maps of [Fe], [Ca], [Mn], [Ti] [Al], [K] and [Na] respectively within the study area. The correlation matrix (Table 3) shows a very strong and positive correlation of Ca-Fe, Mg-P, Ti-Mg, Na-Fe, S-Fe, S-Mg, S-Ti and S-Na with 'r' values of 0.854, 0.878, 0.893, 0.820, 0.942, 0.832, 0.906 and 0.842 respectively; indicating that they are governed by the same geochemical factors and are from the same source. Also, the correlation matrix showed that Ti-Fe, Ti-P, Na-Ca, Na-Ti, K-Ca, K-P, and S-P with 'r' values of 0.697, 0.682, 0.528, 0.637, 0.547, 0.601 and 0.646 respectively, has strong and positive correlation, indicating that they are also governed by the same geochemical factors and they are from the same source, (Figure 6) shows the scatter plots for correlation matrix of major elements within the study area. The analytical results for trace element geochemistry of Ibodi study area are presented in (Tables 4 and 5). Table 4, shows the trace element concentrations of Ibodi study area in (ppm) and Table 5, shows the statistical summary of trace elements with their range and average values respectively. From these tables, Vanadium [V] has a high concentrations and it range from 35.00 ppm - 202.00 ppm with an average mean value of 92.50 ppm, the highest concentration of this element was found in location 1 of the Ibodi study area. Arsenic [As] range from 0.10 ppm - 1.6 ppm with an average value of 0.644 ppm, there is a significant enrichment of Cobalt [Co] and Chromium [Cr] with concentrations ranging from 4.0 ppm - 53.50 ppm and 35.20 ppm -150.70 ppm respectively, with average mean values of 17.73 ppm and 88.78 ppm respectively; the 2D and 3D geochemical maps of [V], [As], [Co] and [Cr] are shown in (Figure 8a). Manganese [Mn] has the highest concentration in the study area, with concentration value ranging from 86.00 ppm - 2165.00 ppm having an average value of 768.20 ppm; the highest concentration of this element was found in location 1 of the study area. [Ga], [Ni] and [Pb] show considerable enrichments within the study area with concentration values ranging from 3.10 ppm - 9.50 ppm, 5.30 ppm - 37.70 ppm and 6.75 ppm - 18.44 ppm, with average values of 6.08 ppm, 19.80 ppm, 12.209 ppm respectively, Figure 8b shows the 2D and 3D geochemical maps of [Ga], [Mn], [Ni] and [Pb] within the Ibodi study area. Rubidium [Rb] range in concentration from 6.50 ppm - 30.90 ppm with an average value of 13.41 ppm, Strontium [Sr] has concentration values that range from 4.70 ppm - 37.20 ppm with an average value of 15.06 ppm. [Y], [Zr] and [Zn] has concentration values that range from 7.99 ppm - 21.10 ppm, 1.20 ppm - 4.00 ppm and 26.20 ppm-83.60 ppm respectively with average mean values of 15.065 ppm, 2.21 ppm and 46.58 ppm, the value of zinc [Zn] is considerably high in the study area with the highest value found at location 4 of the study area indicating some level of enrichment of this metals within the study area, also [Rb], [Y] and [Sr] show some considerable enrichments within the study area as reflected in Figure 8c, which show the 2D and 3D geochemical maps of [Rb], [Sr], [Y], [Zn] and [Zr] in Ibodi study area; while Figure 7 show Pie charts of various concentrations of trace Elements within the Ibodi stream sediments. The geo-accumulation index (Igeo) is an environmental parameter that enables the assessment of contamination by means of comparism. It is used in relation to bottom sediment (Muller, 1969). It is computed using the formula: Igeo=Log2(Cn/1.5*Bn); Where Cn is the measured concentration of the elements. Bn is the normal or average shale content 1.5 is the correcting or matrix factor for geo-accumulation. The Igeo consist of seven grades (Table 6) ranging from practically uncontaminated to extremely contaminated (Muller, 1969). The geo-accumlation values of the selected trace elements in the stream sediment samples of Ibodi study area is shown in (Table 7). From the box plot and Geo-accumulation Table, (Figure 9; Table 7) respectively, the values of the selected trace elements are all less than 1, meaning that all the selected trace metals in Ibodi study area have values less than zero and are in the negative zone. In order to determine the pollution status of the study area, the values of the elements when compared to the Muller classes of geo-accumulation (1969), suggests that the study area is practically uncontaminated with the selected trace metals, the elements fall into the class 0 i.e. Practically unpolluted (Table 6) of Muller (1969).

Table 1. Major Element Oxides Composition Of Ibodi Study Area (Wt %)

Location Nos	Fe_2O_3%	CaO%	P_2O_5%	MgO%	TiO_2%	Al_2O_3%	Na_2O%	K_2O%
Loc 1	10.61775	0.15389	0.178932	0.2145	0.088457	2.3058	0.00536	0.08435
Loc 2	4.15701	0.09793	0.066526	0.363	0.088457	2.457	0.0067	0.1205
Loc 3	6.64092	0.15389	0.094054	0.924	0.163562	4.158	0.01608	0.6507
Loc 4	2.59974	0.08394	0.032116	0.2475	0.085119	1.7955	0.00268	0.1446
Loc 5	5.36679	0.22384	0.130758	0.3135	0.105147	2.1546	0.00536	0.1205
Loc 6	2.36808	0.08394	0.04588	0.1155	0.086788	0.8316	0.00268	0.10845
Loc 7	6.22908	0.18187	0.110112	0.132	0.068429	1.6065	0.00402	0.0482
Loc 8	4.91634	0.26581	0.162874	0.495	0.113492	2.0223	0.00938	0.2169
Loc 9	3.28185	0.1399	0.089466	0.264	0.08457	1.2285	0.0004	0.1205
Loc 10	6.44787	0.23783	0.178932	0.429	0.076774	2.5326	0.0335	0.2169

Table 2. Summary of Major Element in the Stream Sediment

Elements	N	Minimum	Maximum	Mean	Std. Deviation
Fe_2O_3	10	2.36	10.61	5.2625	2.44181
CaO	10	0.08	0.26	0.1623	0.06468
P_2O_5	10	0.03	0.17	0.109	0.05319
MgO	10	0.11	0.92	0.3498	0.235
TiO_2	10	0.06	0.16	0.0961	0.02694
Al_2O_3	10	0.83	4.15	2.1092	0.90093
Na_2O	10	0.002	0.016	0.0086	0.00977
K_2O	10	0.04	0.65	0.1832	0.17246

Table 3. Correlation Co-Efficients for Major Element Oxides

	Fe	Ca	P	Mg	Ti	Al	Na	K	S
Fe	1								
Ca	0.854(**)	1							
P	0.106	-0.043	1						
Mg	0.360	0.069	0.878(**)	1					
Ti	0.697(*)	0.379	0.682(*)	0.893(**)	1				
Al	0.359	0.465	-0.651(*)	-0.540	-0.157	1			
Na	0.820(**)	0.528	0.077	0.341	0.637(*)	0.400	1		
K	0.486	0.547	0.601	0.574	0.437	-0.550	0.436	1	
S	0.942	0.646	0.420	0.832	0.906	0.176	0.842	0.293	1

Table 4. Trace Element Concentrations (ppm) in Stream Sediment of the Study Area

	V (ppm)	Cr (ppm)	Mn (ppm)	Co (ppm)	Ni (ppm)	Zn (ppm)	Ga (ppm)	As (ppm)	Se (ppm)	Rb (ppm)	Sr (ppm)	Y (ppm)	Zr (ppm)	Pb (ppm)
Loc 1	202.00	150.40	2165.00	53.50	37.70	65.10	9.10	1.60	0.20	8.80	37.20	11.74	2.90	17.31
loc 2	96.00	62.20	181.00	10.70	17.70	65.10	7.70	0.30	0.10	10.30	29.40	12.94	1.90	17.31
loc 3	120.00	100.30	1105.00	29.60	32.50	50.40	9.50	0.90	0.20	30.90	15.20	19.77	2.50	11.65
loc 4	115.00	150.70	86.00	4.00	11.90	83.60	7.00	<0.10	<0.10	6.70	8.20	20.27	4.00	13.4
loc 5	88.00	79.70	1550.00	24.60	24.20	26.20	6.30	1.10	<0.10	10.80	12.10	14.59	2.40	14.22
loc 6	35.00	35.20	381.00	8.40	5.30	42.60	3.10	0.30	<0.10	10.50	4.70	7.99	1.20	9.69
loc 7	95.00	92.40	561.00	15.30	14.20	28.00	5.80	1.10	0.10	6.50	13.70	10.83	2.20	6.75
loc 8	85.00	82.50	1292.00	21.80	22.40	41.70	5.30	0.70	0.10	16.70	18.90	11.05	2.30	18.44
loc 9	59.00	55.00	748.00	17.10	15.20	37.80	3.90	0.20	<0.10	10.90	12.40	11.01	1.40	9.72
loc 10	116.00	114.90	889.00	22.90	27.30	39.00	6.10	0.60	<0.10	15.40	18.00	21.10	2.10	10.63

Table 5. Summary of Selected Trace Elements Concentration of Stream Sediments

Elements	N	Minimum	Maximum	Mean	Std. Deviation
V	10	35.00	202.00	92.50	27.6697
Cr	10	35.20	150.70	88.78	33.6398
Mn	10	86.00	2165.00	768.20	475.121
Co	10	4.00	53.50	17.73	8.09253
Ni	10	5.30	37.70	19.80	8.36328
Zn	10	26.20	83.60	46.58	17.2757
Ga	10	3.10	9.50	6.08	1.81157
As	9	0.10	1.60	0.6444	0.33953
Se	4	0.10	0.20	0.125	0.05
Rb	10	6.50	30.90	13.41	7.0606
Sr	10	4.70	37.20	15.06	6.74754
Y	10	7.99	21.10	15.065	5.02623
Zr	10	1.20	4.00	2.21	0.75491
Pb	10	6.75	18.44	12.209	3.63893

Table 6. Geo-Accumulation Index Classes (Muller 1981)

Igeo class	Values	Sediment quality
0	Igeo<o	Practically uncontaminated
1	0<Igeo<1	Uncontaminated to moderately contaminate
2	1<Igeo<2	Moderately contaminated
3	2<Igeo<3	Moderately to heavily contaminated
4	3<Igeo<4	Heavily contaminated
5	4<Igeo<5	Heavily to extremely contaminated
6	5<Igeo<6	Extremely contaminated

Table 7. Geo-Accumulation Distribution of the Elements Against the Locations

	As	Co	Ce	Sb	Mn	Mo	Se	Zn	Pb	Ni
L1	-2.90689	-2.90689	-3.90689	-4.90689	-2.20645	-4.71425	-3.90689	-7.22882	-1.58496	-6.71425
L2	-3.90689	-2.90689	-3.90689	-3.90689	-1.44746	-5.71425	-3.90689	-7.81378	-1.58496	-6.71425
L3	-3.90689	-2.90689	-2.90689	-4.90689	-1.58496	-5.71425	-2.90689	-7.22882	-1.09954	0
L4	-3.90689	-2.90689	-3.90689	-3.32193	-2.20645	-5.12928	-2.90689	-7.22882	-1.58496	-6.71425
L5	-3.90689	-1.90689	-3.90689	-3.90689	-2	-5.12928	-3.90689	-7.81378	-1.09954	-6.71425
L6	-3.90689	-1.32193	-2.90689	-3.90689	-2	-4.71425	-3.90689	-6.81378	-1.09954	0
L7	-2.32193	-2.90689	-2.32193	-3.90689	-2.20645	-6.71425	-3.90689	-6.22882	-1.09954	-6.71425
L8	-3.90689	-1.90689	-3.90689	-4.90689	-1.58496	-5.71425	-2.90689	-7.22882	-0.90689	0
L9	-3.90689	-2.32193	-3.90689	-3.90689	-1.44746	-5.71425	-2.32193	-6.81378	-1.09954	-6.71425
L10	-3.90689	-2.90689	-3.90689	-3.90689	-1.44746	-5.12928	-3.90689	-7.81378	-1.58496	0

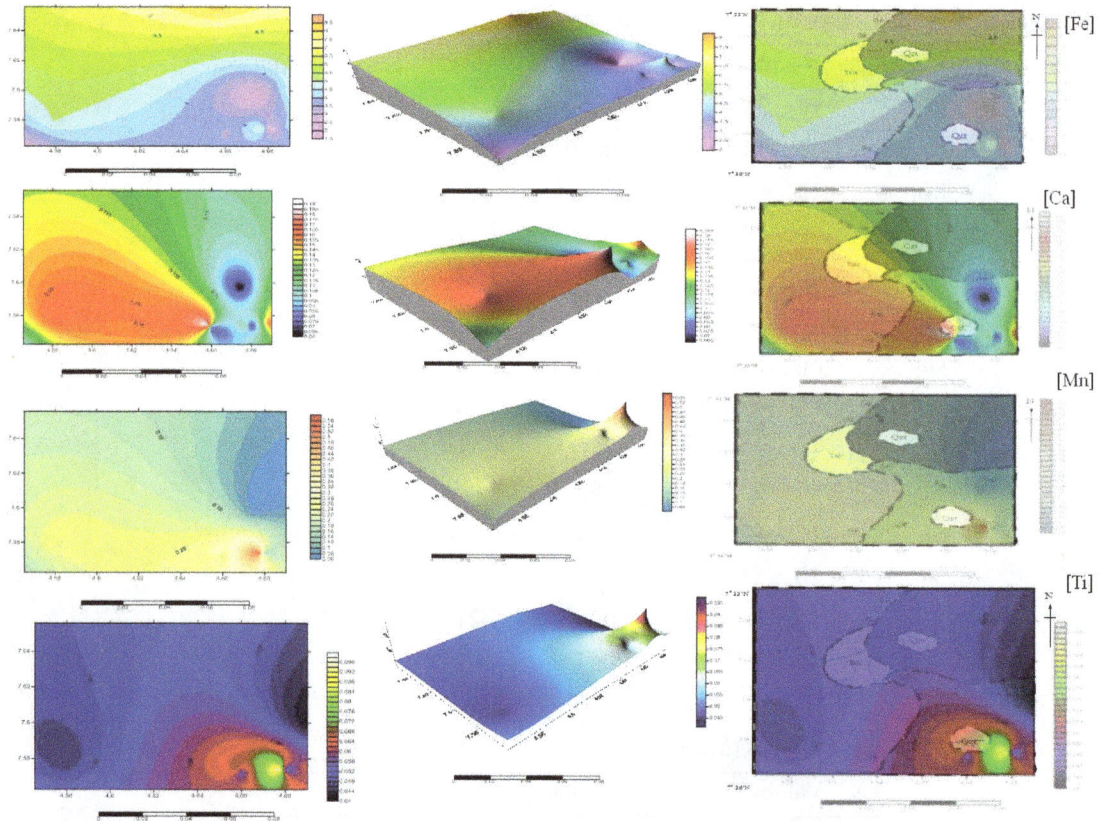

Figure 3. Showing 2D, 3D and geochemical maps of [Fe], [Ca], [Mn] and [Ti] respectively

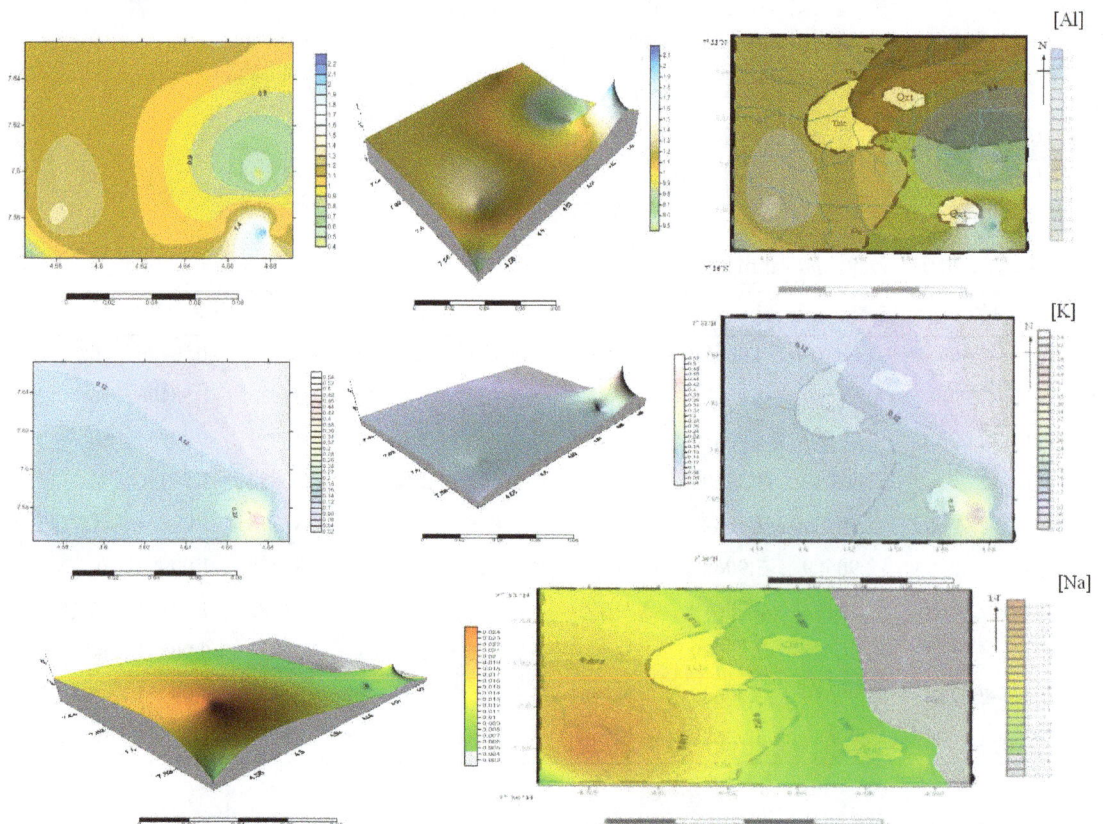

Figure 4. Showing 2D, 3D and geochemical maps of [Al], [K] and [Na]respectively

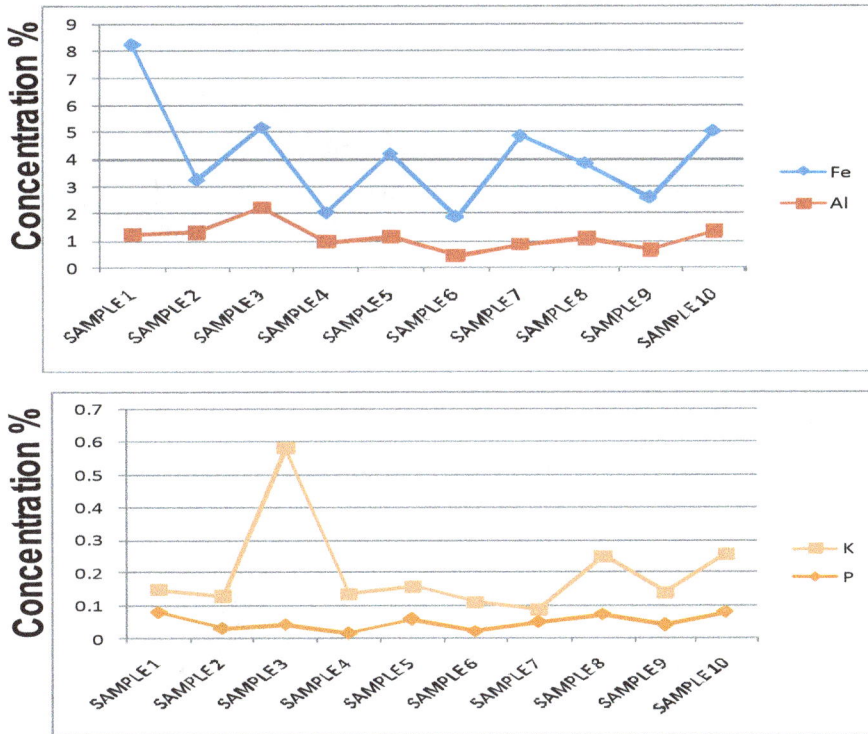

Figure 5. Line diagrams showing Descriptive Statistics of major Concentration in Stream Sediments around the Study Area

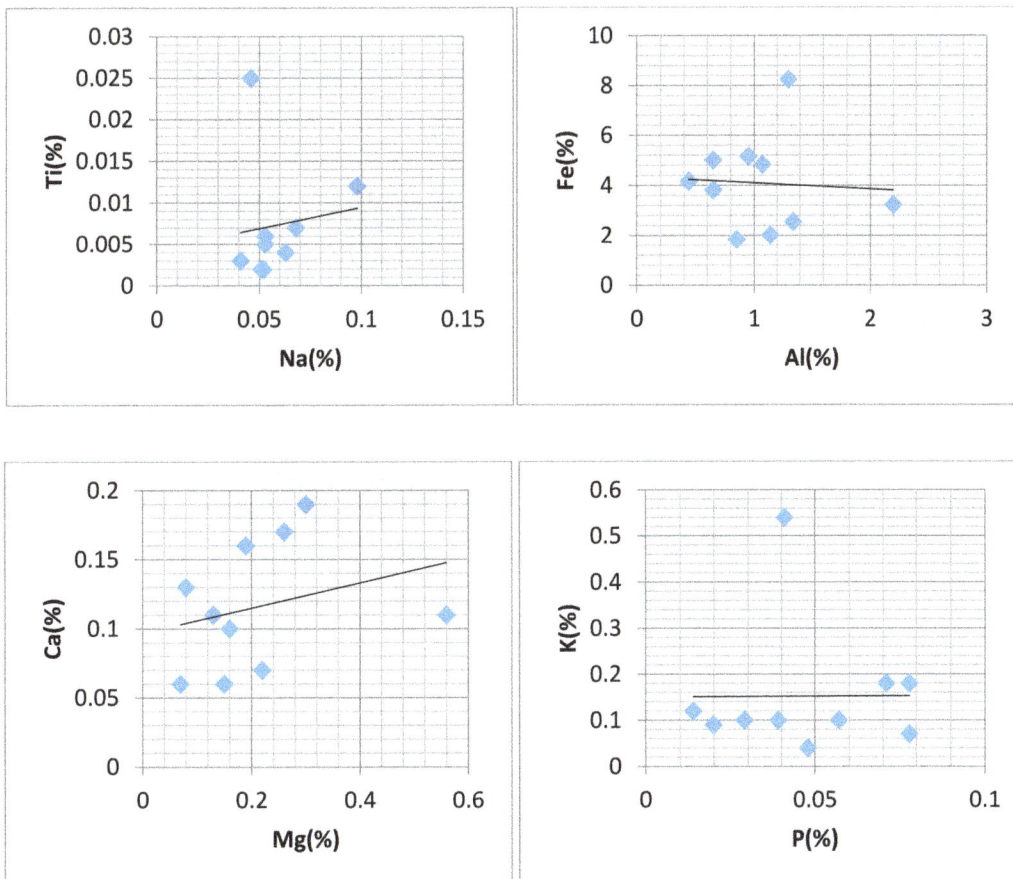

Figure 6. Scatter plots for correlation matrix of major element

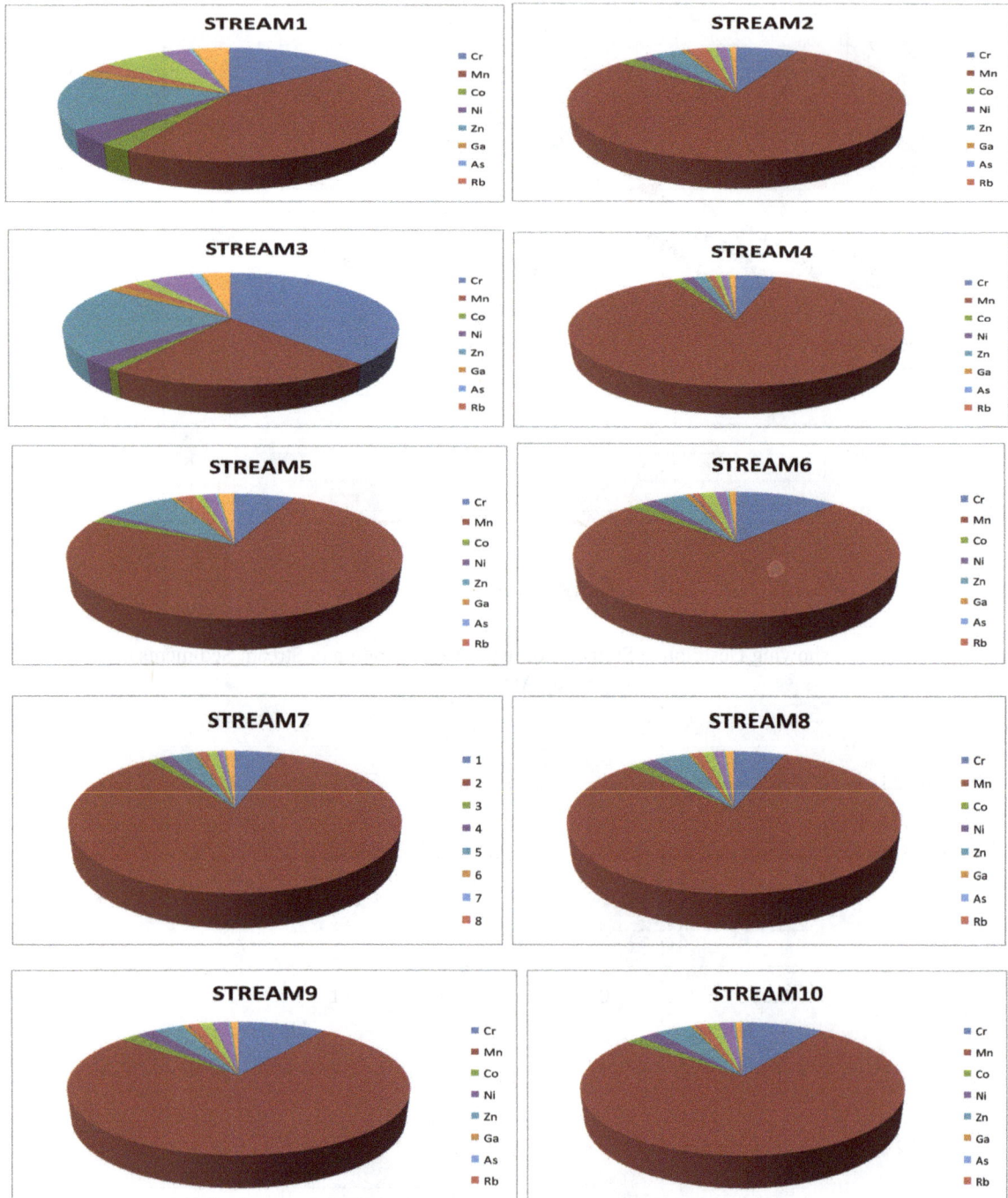

Figure 7. Pie charts showing various concentration of trace Elements within the Ibodi stream sediments

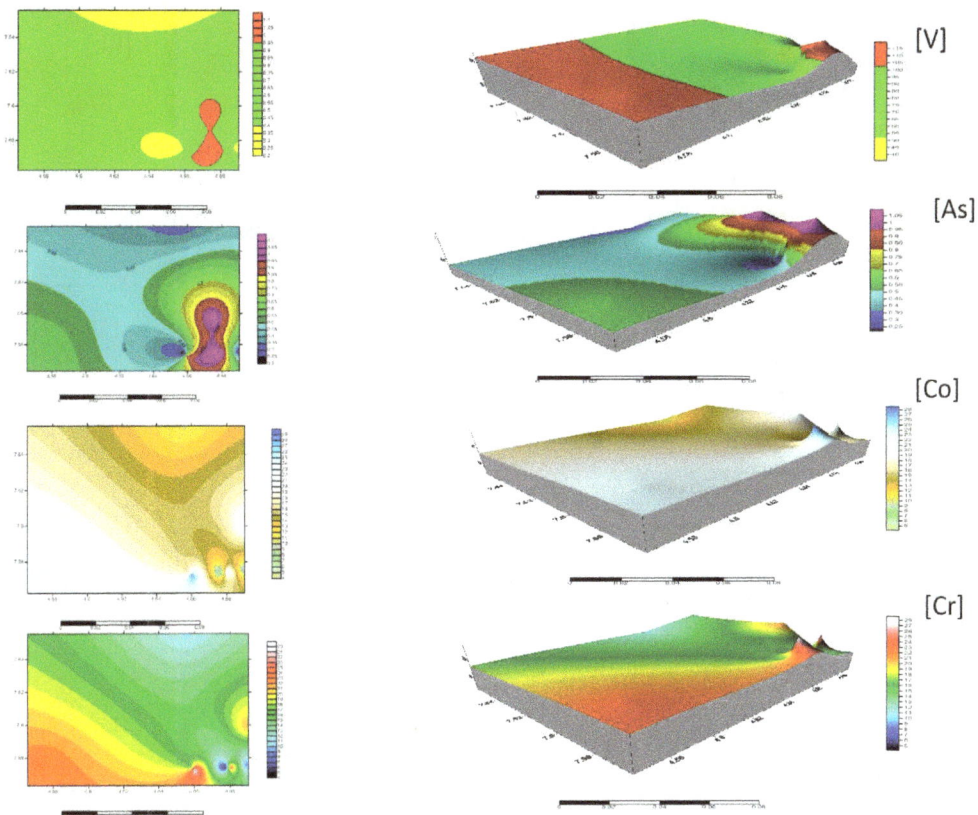

Figure 8a. showing 2D and 3D geochemical maps of [V], [As], [Co] and [Cr] respectively

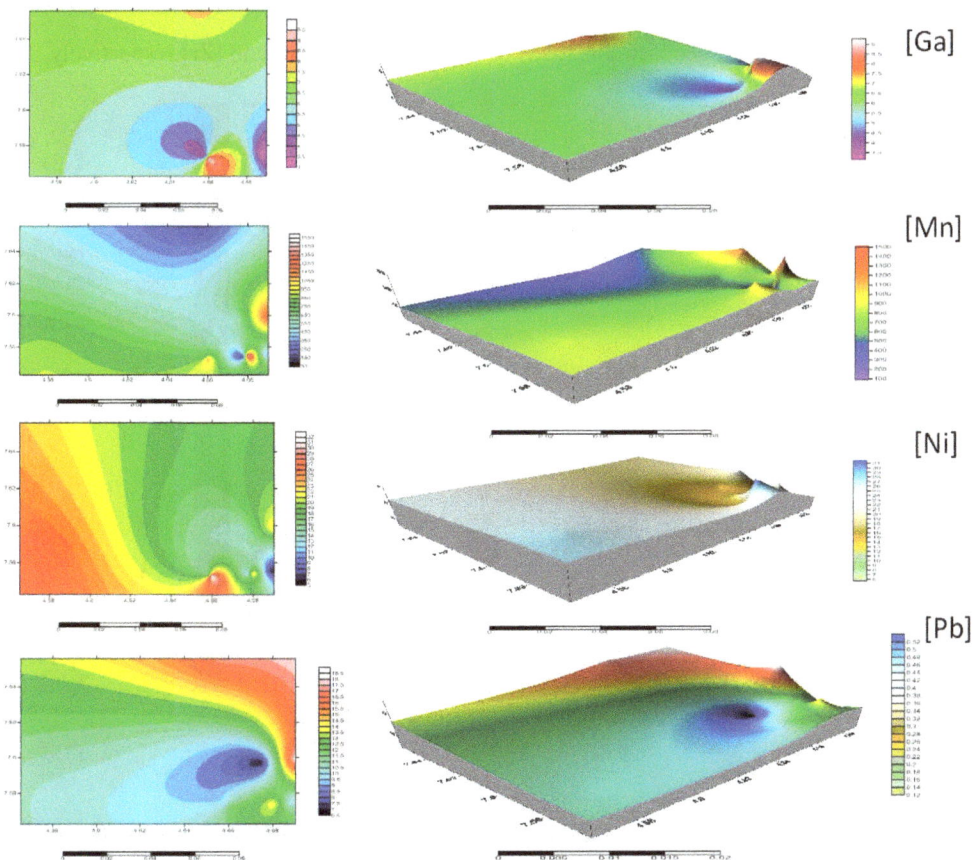

Figure 8b. showing 2D and 3D geochemical maps of [Ga], [Mn], [Ni] and [Pb] respectively

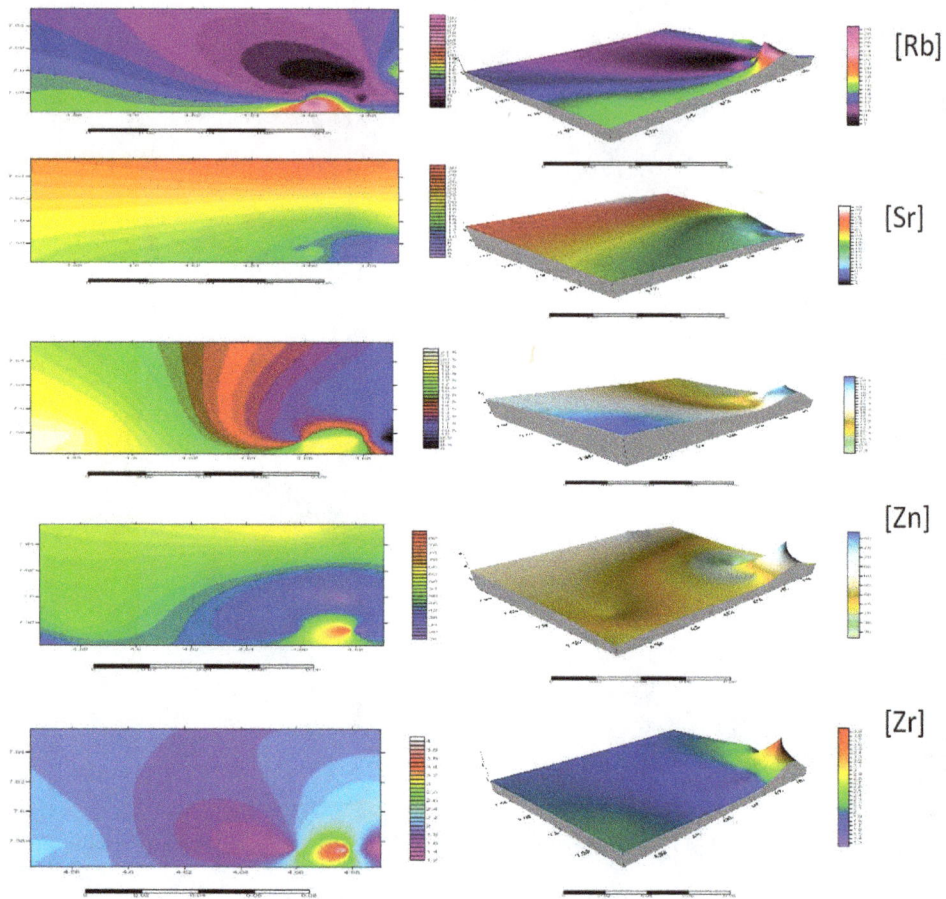

Figure 8c. showing 2D and 3D geochemical maps of [Rb], [Sr], [Y], [Zn] and [Zr] respectively

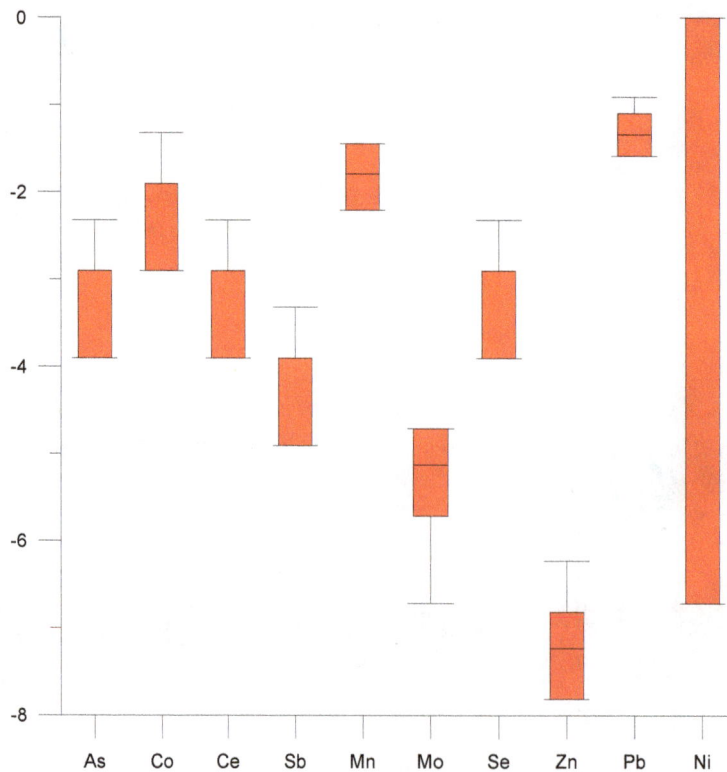

Figure 9. Box plot showing the geo-accumulation index of the selected trace elements

5. Conclusion

The study area is characterized by lithologies such as Amphibolites, Mica schist, Granite gneiss, Quartzite and Talc they are Precambrian rocks that are typical of the Basement Complex of Nigeria. The result of major elements concentration shows that there are natural concentrations of Major elements such as [Fe], [Mg], [Al], [K], [Ti], [Na] and [Ca] in some locations within the study area which indicates abundance of Ferro-magnesia and Al-rich minerals present in the Amphibolite and other rocks in the study area. Also the trace element geochemistry shows the various concentrations of trace elements [V], [As], [Co]. [Cr], [Ga], [Mn], [Ni], [Pb], [Rb], [Sr], [Y], [Zn], [Zr] in the study area. The result of the correlation co-efficient suggests a common source between these elements. It can be inferred from the geo-accumulation index which is an environmental parameter that enables the assessment of contamination by means of comparism that result from the Ibodi study area has proven that values of selected trace elements are all less than 1, meaning that all the trace metals in Ibodi study area have values less than zero and are in the negative zone suggesting that the study area is practically uncontaminated with the selected trace metals with the elements falling into the class of 0 'Practically uncontaminated' according to Muller, 1969. The result has clearly shown the level of uncontamination in the area under investigation. It is therefore suggested that regular geochemical research work should be carried out in this study area to determine future rise in contamination level as a result of mining activities going on in this environment in addition; Mining activities in the area should be controlled to minimize the amount of pollutants released into the streams and water bodies within the study area in other to guide against health hazards caused by high concentrations of some of these metals on both man and animals.

Acknowledgement

The authors acknowledge the assistance of Mr Mafoluku, chief technologist of the Department of Geology University of Ibadan for his cooperation during the production of the thin section slides for petro graphic studies, in addition Dr Okunlola Olugbenga of the Department of Geology University of Ibadan is also highly appreciated for his numerous support.

References

Ajayi, S. O, Mombeshora, C., & Osibanjo, O. (1990). Pollution studies on Nigerian rivers: The onset of lead pollution of surface waters in Ibadan. *Int. Journal on Environ. Science, 9*, 81-92

Black, R. (1980). *Precambrian of west Africa Episode, 4*, 3-8

Gibbs, A., Bruce, R., George, C., Bacuta jr, A., Robert, W., Kay, A., & Allan, K. (1977). Platinum – group element abundance distribution in chromite deposits of the Acoje block, Zambales Ophiolite complex, Philippines. *Journal of geochemical exploration, 37*(1), 113-145

Muller, G. (1969). Index of Geoaccumulation in sediments of Rhine River. *Geol. J., 2*, 108-118

Singh, M., Ansari, A. A., Muller, G., & Singh, I. B. (1990). Heavy metals in freshly deposited sediments of Gomatti River (A tributary of gange river), effect of human activities. *Environ. Geol., 29*(3), 246-252.

Tijani, M. N, Jinno, K., & Hiroshiro, Y. (2004). Environmental impact of Heavy metal distribution in water and sediments of Ogunpa River Ibadan Area, southwestern Nigeria. *Journal of Mining and Geology, 40*(1), 73-83. http://dx.doi.org/10.4314/jmg.v40i1.18811

3

A Fuzzy Linear Optimization Investigation for Cropping Patterns and Virtual Water Transfers under Water Scarcity in Gansu Province, Northern China

Chen Zhang[1] & Edward A. McBean[1]

[1] School of Engineering, University of Guelph, Canada

Correspondence: Chen Zhang, School of Engineering, University of Guelph, Canada. E-mail: czhang03@uoguelph.ca

Abstract

Water scarcity is occurring worldwide due to factors including climate change, population increases, and economic activity. The situation is most severe in arid areas. Based on a water supply stress index and virtual water for production, a fuzzy linear programming model is used to provide a set of solutions that contain a range of possible values rather than a specific optimal solution. The model is used to investigate cropping patterns that can obtain acceptable stress levels on local water resources for both current and future climate scenarios. Results in Gansu Province indicate that existing cropping patterns are causing extreme stress on local water resources, especially in eastern Gansu, and the stress will continue in the future if cropping patterns are not adjusted. The optimization model indicates that a considerable amount of water used is not necessarily needed, and deficits of virtual water demands can be covered by virtual water transfers. Optimal cropping patterns show that in the current scenario, total cropping areas in eastern Gansu should be reduced by up to 55% to attain acceptable water stress, and 24.2 and 22.5% for western Gansu and mid Gansu, respectively. The enormous cropping reduction in certain crops and the relatively larger decreases of cropping areas in the eastern region are indicated as a direction, to improve the balance between irrigation and municipal water needs. Virtual water transfer analysis shows that mid Gansu can be a major export region, while eastern Gansu will require large amounts of virtual water imports.

Keywords: fuzzy linear programming, virtual water, water scarcity, Gansu

1. Introduction

Increases in population, urbanization, and economic activities as well as climate change have led to increasing pressures on freshwater resources (Dalin et al., 2012). Water scarcity influences food security, human health and natural ecosystems, effects which are particularly severe in arid regions (Seckler et al., 1999). In response, the concept of virtual water is valuable, involving reducing water stress in a region by importing more food, especially food which requires significant water, instead of producing these foods (Allan, 1998). Since irrigation water represents a major use of freshwater, it is preferable in arid regions that crops using more water be imported and crops using less water can be exported, for purposes of saving water (Hoekstra & Hung, 2005).

Virtual water for a crop is the amount of water used to produce the crop and "embedded" in the crop. With transfer of the crop from one place to another, the embedded water is transferred in a virtual sense. Virtual water has been applied as a water-saving strategy in many areas around the world. For example, both Morocco and the Netherlands depend more on external water resources – 14 and 95% from outside of the two countries, respectively; and an amount of 640 million m^3 of water is saved per year for Morocco since the products are imported instead of being produced (Hoekstra & Chapagain, 2006). Another study assessing worldwide consumption of cotton production indicates that approximately 2.6% of the global water footprint is for the production of cotton products, and thus represents a large impact on water resources in nations such as China, USA, Mexico, and Germany (Chapagain et al., 2006). From the analysis of water economic productivity for the Mancha Occidental Region in Spain, it is noted that crops with high virtual water and low economic value are major crops in the region, and the region relies mostly on its domestic water resources, although import of virtual water would be beneficial to local water bodies (Aldaya et al., 2010). The findings of the water footprint analysis are indications of the current water policy, by indicating that water can be allocated and used more effectively.

Minimum water stress can be obtained through optimization of the cropping structure and virtual water transfers. One of the most effective techniques for optimization is linear programming (LP), which consists of a linear objective function and linear equality and/or inequality constraints. LP will return a crisp solution, meaning a number that is typically a single value. According to Klir et al. (1997), practical problems commonly contain uncertainties resulting from a variety of factors such as vagueness, which a crisp number is not able to capture. Compared with crisp numbers, a fuzzy number is a set of possible values that have different degrees between 0 and 1 showing the closeness of the value in the fuzzy set to the given crisp number, and hence reflects the sensitivities which exist in real problems. In other words, parameters in LP models are generally best estimates, regardless of the dynamic natural environments which, on the other hand, can be best handled by fuzzy numbers.

LP models require precise data which would add difficulties in the process of data collection. Additionally, the solution of an LP model commonly relies only on a limited number of constraints thus leading to a waste of information (Rommelfanger, 1996). However, in real world situations, parameters in objective functions and constraints are often not entirely precise and contain vague information (Tanaka & Asai, 1984; Inuiguchi & Ramk, 2000; Maleki et al., 2000). In this case, fuzzy linear programming (FLP) provides a more reasonable and appropriate solution. In FLP models, parameters of objective functions and constraints are fuzzy numbers which contain a set of possible values of each parameter. These fuzzy numbers can avoid the crisp situation of "Yes" and "No", thus making the linear programming model less crude (Zimmermann, 1978; Herrera & Verdegay, 1995). As well, satisfaction criteria such as "approximately" and "nearly" are more desirable than the precisely defined criteria such as "exactly" and "absolutely" for a realistic optimization problem (Tanaka & Asai, 1984).

FLP is now widely applied to decision problems. Sahoo et al. (2006) developed a fuzzy optimization model for managing and allocating land and water resources system in Mahanadi-Kathajodi Delta in eastern India with three objectives – maximization of production, maximization of net annual return, and minimization of labour cost for the study area. Compared with models such as maximizing production and maximizing net return, a multi-objective fuzzy optimization model can provide the optimal solution when all three objectives are at the same priority level (Sahoo et al., 2006). In the research of Biswas and Pal (2005), production of seasonal crops is optimized using a priority-based fuzzy goal programming model. Biswas and Pal indicate that a better cropping plan is developed through use of fuzzy goal programming.

Being one of the driest province in China, Gansu Province is facing the issue of water stress due to reasons such as climate change and inappropriate water use. It is useful to investigate strategies to reduce water stress. The impacts of cropping patterns on local water resources involve uncertainties due to factors such as climate conditions, locations and crop types. In such a case, parameters in the optimization model for cropping patterns should be fuzzy numbers instead of fixed numbers. Considering the merits of FLP indicated above, and that FLP for cropping patterns related to water stress have not been utilized for Gansu Province, the optimization of cropping patterns can be modified from "minimum water stress" to "acceptable water stress".

The objectives of this paper were to: (1) develop an FLP model for cropping patterns to achieve the goal of attaining acceptable stress on local water resources using virtual water constraints under two scenarios – current and future climate scenarios; and (2) given the attainment of acceptable water stress, use the optimized cropping patterns to determine virtual water transfers for the major crops.

2. Methods
2.1 Study Areas and Required Data
Gansu Province is one of the most arid regions in China and has high stress on water resources. Three regions from the west to the east portions of Gansu Province were selected as study areas (see Figure 1). The West region is the most arid region, the Mid region has a moderate arid climate, while the East region is relatively humid. Typical stations representing the three regions were used based on the quality of required data (see Table 1). Major crops in the three areas are: wheat, corn, cotton, potatoes, and soy.

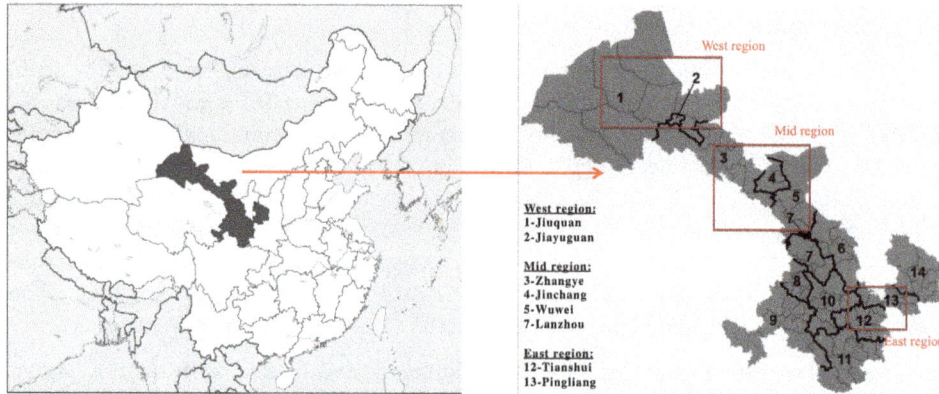

Figure 1. Location of Gansu Province and Selected Stations

Table 1. Selected stations

Portion	Station Name	LAT(°)	LON(°)	ELEV(m)
West	Jiuquan	39.77	98.48	1478
	Jiayuguan	39.81	98.30	1558
	Zhangye	39.08	100.28	1462
Mid	Wuwei	37.93	102.64	1535
	Lanzhou	36.05	103.88	1518
East	Tianshui	34.58	105.75	1143
	Pingliang	35.55	106.67	1135

Agricultural data including annual cropping areas and crop yield of major crops, as well as population data were obtained from Gansu Yearbooks on China Data Online (CDO, 2013), water resources data from Gansu Water Resources Bulletin, irrigation water demands of current and future scenarios as calculated in Zhang and McBean (2014) using data from National Climatic Data Centre (NCDC, 2013), and virtual water contents (VWC) generated in the study of Zhang et al. (2014).

2.2 Fuzzy Linear Programming Model for Cropping Patterns

The Water Supply Stress Index (WaSSI), representing water supply stress from environmental and anthropogenic sectors as proposed by McNulty et al. (2010), characterizes the scarcity of water resources caused by the growing of the major crops. When WaSSI is 0.2, there is a stress; and when WaSSI increases to 0.4, the stress becomes scarcity (Brown & Matlock, 2011). WaSSI for the selected areas is expressed in Equation 1 as:

$$WaSSI_j = \frac{\sum WD_{ij}}{WS_j} \tag{1}$$

Where:

$WaSSI_j$ = water supply stress index in region j;
WD_{ij} = irrigation water demand of crop i in region $j (m^3)$; and
WS_j = total water supply volume to the crop field in region j (m^3).

To obtain acceptable stress on local water resources while ensuring the basic crop needs are met, the FLP model was developed that maximizes crop yields of the five crops, subject to constraints including acceptable WaSSI and appropriate cropping areas based on blue virtual water for producing each crop (See Equation 2.2). Blue virtual water refers to the amount of irrigation water including both surface and ground water, used for crop growth. Two climate scenarios were analyzed – (1) current scenario considered the time period from 2000 to 2010; and (2) future scenario considered the time period from 2030 to 2050.

$$\text{Maximize} \quad \sum \widetilde{y_{ij}}\widetilde{A_{ij}}$$

$$\text{subject to} \quad \sum (\widetilde{d_{ij}}\widetilde{A_{ij}})/WS_j = \widetilde{0.22} \cdot \alpha_j,$$

$$\widetilde{d_{ij}}\widetilde{A_{ij}} \geq VWp_{ij}, \; or \tag{2}$$

$$\widetilde{d_{ij}}\widetilde{A_{ij}} \leq VWp_{ij} \; and \; \widetilde{d_{ij}}\widetilde{A_{ij}} \geq \widetilde{VWn_{ij}}$$

$$\widetilde{A_{ij}} > 0$$

Where:

$\widetilde{y_{ij}}$ = crop yield of crop i per cropping area in region j ($1000kg/1000Ha$);

$\widetilde{A_{ij}}$ = optimized cropping area of crop i in region j($1000Ha$);

$\widetilde{\alpha_j}$ = ratio of effective irrigation area and cropping area (%) in region j, specifically, $\alpha_{West} = \alpha_{Mid} = 1, and\alpha_{East} = 22\%$;

$\widetilde{d_{ij}}$ = irrigation need for crop i in region $j(mm)$;

$\widetilde{VWp_{ij}}$ = blue virtual water for producing crop i in region $j(m^3)$ before optimization;

$\widetilde{VWn_{ij}}$ = blue virtual water needs for crop i in region $j(m^3)$, which is determined from: $\widetilde{VWn_{ij}} = \widetilde{POP_j} \cdot p_{ij} \cdot \widetilde{VWC_{ij}}$

where:

$\widetilde{POP_j}$ = population in region j (10^4);

p_{ij} = need for crop i per person in region j ($kg/person$); and

$\widetilde{VWC_{ij}}$ = virtual water content of crop i in region j ($m^3/1000kg$).

i = crop type: wheat, corn, cotton, potatoes, soy; and

j = region of the Province: West, Mid, and East.

Table 2 describes the criterion of whether the constraint of $\widetilde{d_{ij}}\widetilde{A_{ij}}$ should be greater or less than the blue virtual water for producing crop i in region j before optimization. Table 2 is primarily based on the suggestions for cropping patterns from Zhang et al. (2014), combined with existing cropping structures and crop needs by local people. For example, as suggested by Zhang et al. (2014), cropping of potatoes should be increased due to the low virtual water content and high economic value for the virtual water used. However, in fact, the production of potato is exceeding local demands; this indicates reduction in potato cropping is desirable instead of increased cropping. Based on Table 2, if crop i is to be decreased, the blue virtual water used for production of crops should be less than the current use of blue virtual water amount (VWp_{ij} before optimization), otherwise, more blue virtual water is expected to be applied for production, in comparison with existing VWp_{ij}.

Table 2. Criterion of $\widetilde{d_{ij}}\widetilde{A_{ij}}$ in FLP models

	West region	Mid region	East region
Crops needing to be decreased	Wheat	Wheat	Wheat
	Corn	Soy	Corn
	Soy	Potatoes	Soy
	Potatoes		Potatoes
Crops can be increased	Cotton	Corn	-
Crops unchanged	-	Cotton	Cotton

Some parameters are quantified as fuzzy numbers instead of fixed numbers since these parameters can be closely related to factors such as climate conditions and time period which commonly contains uncertainties. Specifically, the type of fuzzy numbers in the FLP model are LR triangular fuzzy numbers. LR fuzzy numbers are expressed as , $\widetilde{M}=(m,\alpha,\beta)_{LR}$, where m is the mean value of $\widetilde{M}$, α is the left spread, and β is the right spread. In addition, $\widetilde{M}$

is positive and $m - \alpha > 0$ (Fan et al., 2009; Dehghan et al., 2006). The membership function of $\widetilde{M}$ is described in Equation 3, and $\widetilde{M}$ is a LR triangular fuzzy number if $L(x)$ and $R(x)$ are linear functions (Fan et al., 2009).

$$\begin{cases} L(\quad (m - x)/\alpha \quad) & x \leq m, \quad \alpha > 0 \\ R(\quad (x - m)/\beta \quad) & x \geq m, \quad \beta > 0 \end{cases} \qquad (3)$$

Each parameter varies over time due to climate variability. The mean values m herein are the mean values of each parameter during the selected time period, with the lower and upper values being the minimum and maximum values during the time period, respectively. For the future time period, mean values and extreme values were determined from projections for future scenarios. Taking the irrigation need (d) for example, an LR fuzzy number of $\widetilde{d}$ can be expressed as $\widetilde{d} = (d, \ d - d_{min}, \ d_{max} - d)_{LR}$. The objective is to set WaSSI to be approximately 0.22 for the three regions – i.e. $\overline{0.22} = (0.2, 0.22, 0.28)$, which, in the LR fuzzy number format, is $(0.22, 0.02, 0.06)_{LR}$.

For the future scenarios of A1B, B1, and A2, $\widetilde{d}$ was from the General Circulation Models (GCM models) predictions described in Zhang and McBean (2014). Parameters such as population, yield per cropping area, and cropping area before optimization were estimated using the moving average method. Since water supply delivered to the fields is very near the limit of the supply capacity of total water resources (as high as 95%), delivery capacities were assumed to be unchanged for the future scenario. The crop that a person comsumes is another parameter which was assumed as not changing.

The solving of the FLP followed the approach of converting the FLP model to a crisp linear model, as indicated by Fan et al. (2009) with the assistance of LINGO®.

2.3 Water savings and virtual water transfers

Optimization for cropping patterns for the three regions within Gansu Province will generate estimates of the potential saving in local water resources (Equation 4). Additionally, the difference (D_{ij}) of virtual water for production after optimization (VWp_{opt-ij}) and virtual water needs (VWn_{ij}) can be transferred amongst the regions (Equation 5). If D_{ij} is positive, the extra amount of virtual water for crop i can be exported to other regions; otherwise, if D_{ij} is negative, imports of virtual water for crop i are required from outside of a region j.

$$S_j = \sum VWp_{ij} - \sum VWp_{opt-ij} \qquad (4)$$

Where:

S_j = blue virtual water saving in region j;

VWp_{ij} = blue virtual water for production of crop i in region j before optimization(m^3); and

VWp_{opt-ij} = blue virtual water for production of crop i in region j after optimization(m^3).

$$D_{ij} = VWp_{opt-ij} - VWn_{ij} \qquad (5)$$

3. Results

3.1 Fuzzy linear optimization for cropping patterns

Table 3 and Figures 2 (a) through (e) are the optimization results for the existing cropping areas of the five crops in the three regions. Existing WaSSI for the West, Mid, and East regions are 0.29, 0.28, and 0.49, respectively. The results show that total cropping areas in all three regions should be reduced to obtain acceptable WaSSI magnitudes of 0.22. Specifically, the West and Mid region need to reduce the total cropping areas by 24.2 and 22.5%, respectively; and the East region by 55.4%. Further, the East region will also have a relatively higher reduction in every crop except for potatoes, compared with the West and Mid regions. Major reductions of cropping areas in the East region involve decreases in soy (approximately 90%), followed by wheat (approximately 45%). Potatoes and soy in the West region, as well as wheat and soy in the Mid region will need to be decreased as well (see Table 3).

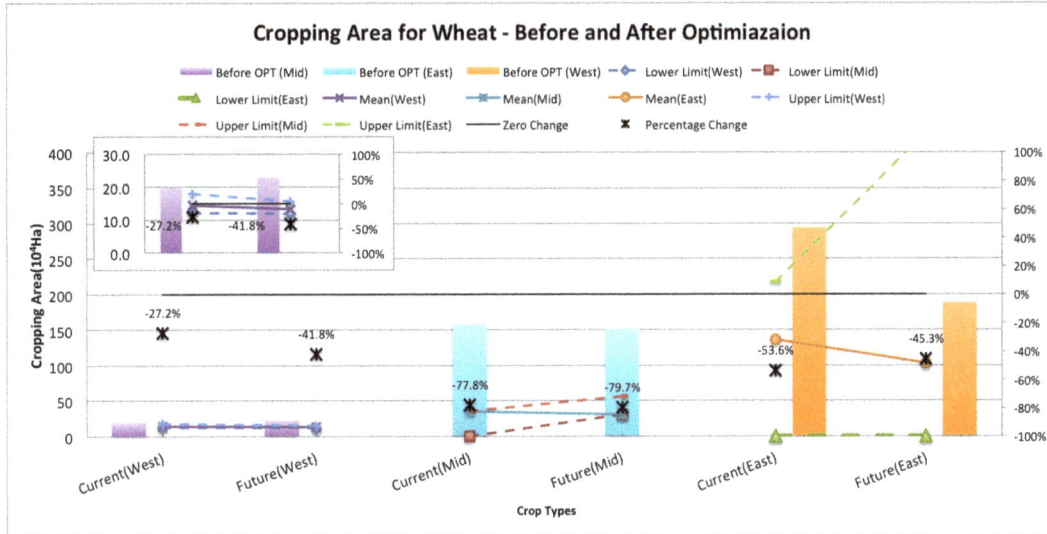

a1

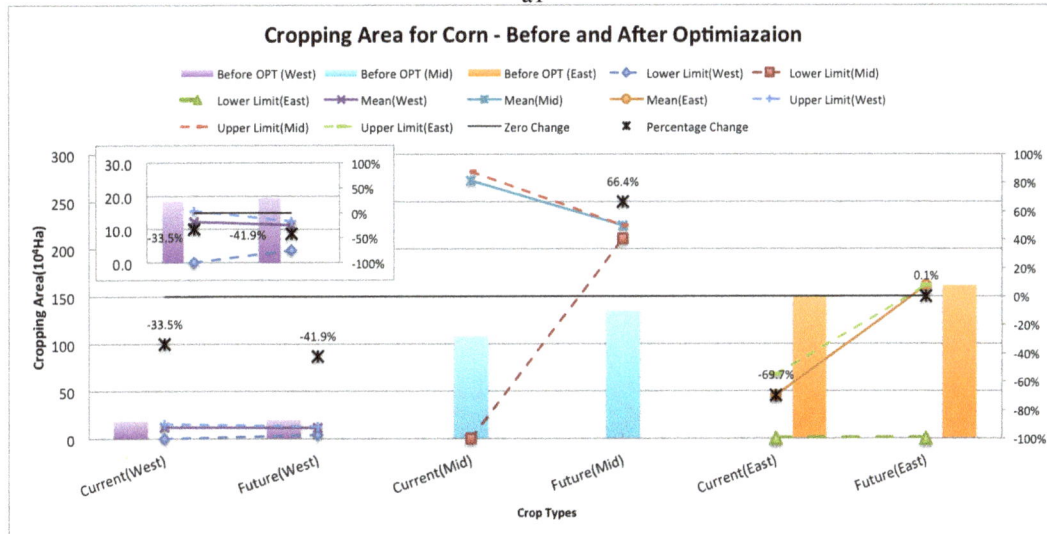

a2

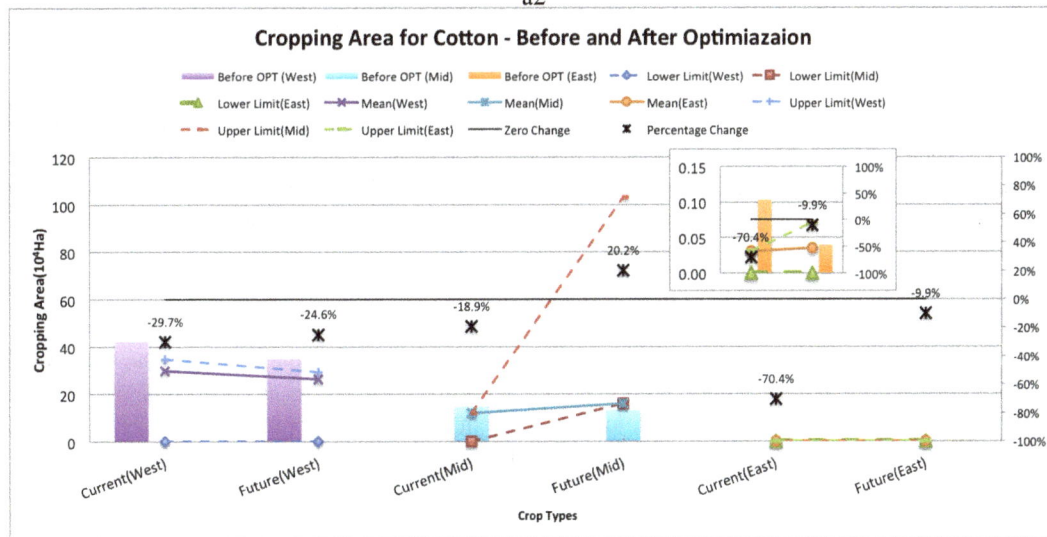

a3

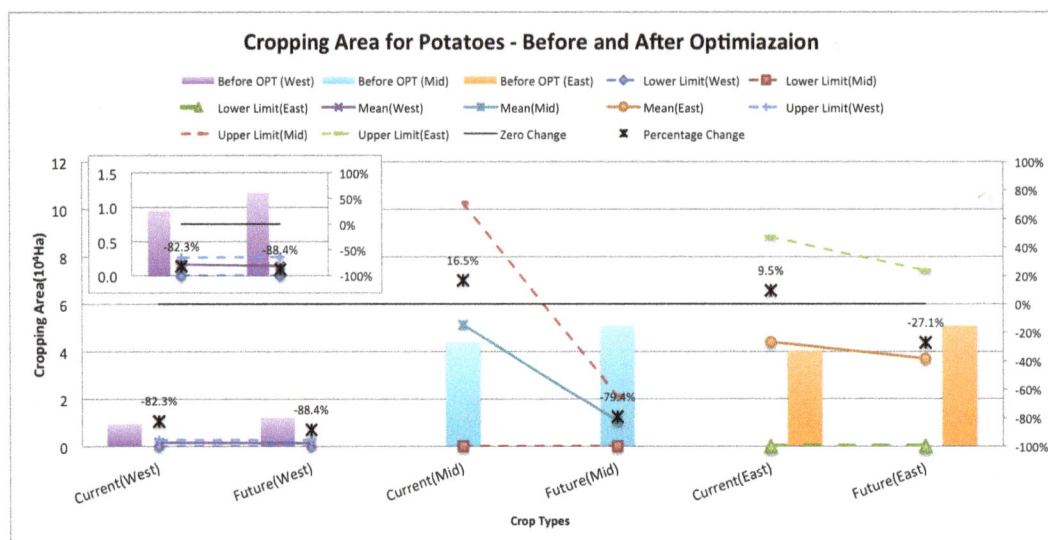

a4

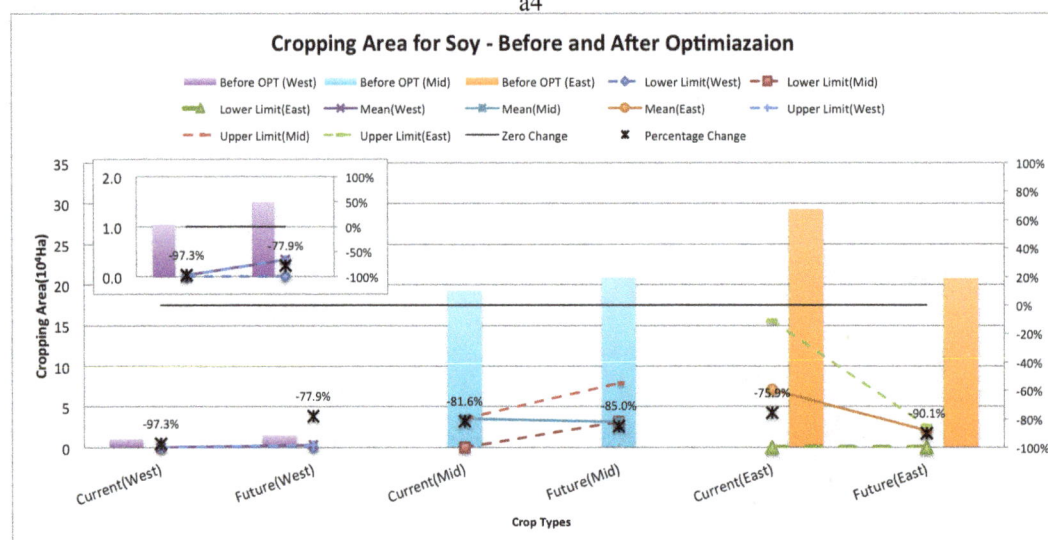

a5

Figure 2. Cropping Area of for the major crops in the three region(before and after optimization)

In the future scenario, the predicted WaSSI for the West region will remain at the same level as existing. There is going to be a slight increase of WaSSI from 0.28 to 0.33 in the Mid region, due to climate change. WaSSI in the East region will decrease from 0.49 to 0.37. Results in Table 4 show that optimization for cropping patterns is still necessary in the future as a result of climate change, with reduction of 27.7, 33.0, and 40.3% in total cropping areas for the West, Mid, and East regions, respectively. Individual crops which need major reductions in the current scenario will still need major reductions in the future.

Changes in optimized cropping areas from the current scenario to the future scenario for each crop tend to be larger for the Mid and East regions shown in Figures 2 (a) through (e), especially for corn, potatoes, and soy (see Figure 2 (b), (d), and (e)). Commonly, the changes in lower and upper bounds have the same trends as the mean values change from the current scenario to the future scenario (e.g. wheat and potatoes in the West region). However, for some crops such as the crops in the East region, lower bounds of optimized cropping areas are zeroes; and for some crops such as wheat in the East region and corn in the Mid region (see Figure 2 (a) and (b)), the trends are particularly steep.

3.2 Savings in blue water
Tables 5 and 6 show savings of virtual water (S) if cropping patterns are optimized. For both current and future scenarios, there will be significant savings in water use for all three regions. The highest savings of water use are for the East region. From Table 5, if the East region changes the cropping patterns, the total water used would be

decreased by $5495 \times 10^5 \mathrm{m}^3$ (55.4%) from existing cropping patterns; the major savings are from decreased soy cropping – i.e. 75.9%. The savings are apparent for other crops in the East region as well. For example, savings from decreased cropping in cotton and corn will be approximately 70%, and will be approximately 54% for wheat cropping. There will be 2098×10^5 and $3063 \times 10^5 \mathrm{m}^3$ of water saved for the West and Mid regions, respectively. Among the changes, soy and potatoes in the West region are relatively larger, accounting for 97.3 and 82.3% respectively, followed by soy and wheat in the Mid region (81.6 and 77.8%). Cropping for corn in the Mid region, on the other hand, will involve a large increase in water used ($4755 \times 10^5 \mathrm{m}^3$), adopting the optimized cropping patterns.

Table 3. Comparison of cropping areas before and after optimization - current scenario (1000Ha)

	Crop Types	Before OPT	Lower Limit	Mean	Upper Limit	%Change for mean value
West	Wheat	19.9	12.1	14.5	18.0	-27.2%
	Corn	18.5	0.00	12.3	15.7	-33.5%
	Cotton	42.3	0.00	29.7	34.6	-29.7%
	Potatoes	0.95	0.00	0.17	0.27	-82.3%
	Soy	1.05	0.00	0.03	0.03	-97.3%
	sum	82.7	12.1	56.7	68.5	-31.4%
	WSI	0.29	0.05	0.22	0.28	-24.2%
Mid	Wheat	159	0.00	35.3	35.3	-77.8%
	Corn	108	0.00	273	282	152%
	Cotton	14.7	0.00	11.9	11.9	-18.9%
	Potatoes	4.41	0.00	5.14	10.3	16.5%
	Soy	19.3	0.00	3.55	3.55	-81.6%
	sum	305	0.00	329	343	7.56%
	WSI	0.28	0.00	0.22	0.28	-22.5%
East	Wheat	294	0.00	136	217	-53.6%
	Corn	149	0.00	45.3	67.5	-69.7%
	Cotton	0.10	0.00	0.03	0.03	-70.4%
	Potatoes	4.03	0.00	4.42	8.83	9.47%
	Soy	29.3	0.00	7.06	15.6	-75.9%
	sum	477	0.00	193	309	-59.5%
	WSI	0.49	0.00	0.22	0.57	-55.4%

Table 4. Comparison of cropping areas before and after optimization - future scenario (1000Ha)

	Crop Types	Before OPT	Lower Limit	Mean	Upper Limit	%Change for mean value
West	Wheat	22.9	12.0	13.3	15.7	-41.8%
	Corn	19.5	3.71	11.3	12.5	-41.9%
	Cotton	34.8	0.00	26.2	29.1	-24.6%
	Potatoes	1.21	0.00	0.14	0.28	-88.4%
	Soy	1.50	0.00	0.33	0.33	-77.9%
	sum	80	15.7	51.4	57.9	-35.8%
	WSI	0.30	0.09	0.22	0.28	-27.7%
Mid	Wheat	151	30.5	30.5	56.1	-79.7%
	Corn	135	211	225	225	66.4%
	Cotton	13.2	15.8	15.8	103	20.2%
	Potatoes	5.10	0.00	1.05	2.10	-79.4%
	Soy	20.9	3.13	3.13	7.85	-85.0%
	sum	325	260	275	394	-15.3%
	WSI	0.33	0.13	0.22	0.42	-33.0%
East	Wheat	188	0.00	103	429	-45.3%
	Corn	161	0.00	162	162	0.1%
	Cotton	0.04	0.00	0.04	0.07	-9.9%
	Potatoes	5.08	0.00	3.70	7.41	-27.1%
	Soy	20.8	0.00	2.06	2.62	-90.1%
	sum	375	0.00	270	601	-28.0%
	WSI	0.37	0.00	0.22	1.32	-40.3%

From Table 6, in the future climate change scenario, the savings in water use will be larger for the West and Mid regions (34.5 and 33.0%). The major contributions will be the decreased cropping of potatoes and soy. On the contrary, savings will be smaller for the East region (40.3%) compared with the current scenario, and major contributions will be from reduced cropping in soy (90.1%) and wheat (45.3%).

Table 5. Blue virtual water savings for current scenario ($10^5 m^3$)

	West		Mid		East	
	S	% Change	S	% Change	S	% Change
Wheat	544	27.2%	7274	77.8%	4744	53.6%
Corn	370	33.5%	-4755	-152%	538	69.7%
Cotton	1103	29.7%	105	18.9%	0.69	70.4%
Potatoes	9.31	82.3%	-5.67	-16.5%	-1.73	-9.5%
Soy	72.3	97.3%	444	81.6%	215	75.9%
Sum	2098	30.4%	3063	22.5%	5495	55.4%

Table 6. Blue virtual water savings for future scenario ($10^5 m^3$)

	West		Mid		East	
	S	% Change	S	% Change	S	% Change
Wheat	1042	41.8%	7716	79.7%	2801	45.3%
Corn	541	41.9%	-3078	-66.4%	-1	-0.1%
Cotton	844	24.6%	-123	-20.2%	0.04	9.9%
Potatoes	15.3	88.4%	38.5	79.4%	7.43	27.1%
Soy	92.0	77.9%	624	85.0%	182	90.1%
Sum	2534	34.5%	5177	33.0%	2990	40.3%

3.3 Virtual water transfers

Table 7 is the result for virtual water transfers in the current scenario if optimization solutions are adopted. It is noted that the East region will require major virtual water imports ($3225 \times 10^5 m^3$), of which wheat and cotton take the larger reductions. On the contrary, the Mid region is a major virtual water export area – an amount of 8209×10^5 virtual water can be exported in total. Additionally, there will still be an extra, remaining virtual water, of $7808 \times 10^5 m^3$ consisting of virtual water for corn, potatoes, and soy that can be transferred to other areas outside Gansu in addition to exporting to the West and the East regions of Gansu. The majority of virtual water gaps can be filled through virtual water transfers amongst the three regions. However, there will still exist an import requirement of $107 \times 10^5 m^3$ of virtual water import for wheat in the West region, and $1483 \times 10^5 m^3$ ($186 \times 10^5 m^3$ of virtual water for wheat and $1297 \times 10^5 m^3$ of virtual water for cotton) in the East region, which can be transferred from outside the Province.

In the future scenario, the import demands are lower for the West and Mid regions, while there will be an increase for the East region ($4240 \times 10^5 m^3$). However, the capability of virtual water export will also decrease in the future. In other words, through virtual water transfers, the level of virtual water that still needs to be imported is higher for each region, and the extra remaining level of virtual water that can be exported to outside the Province is lower, in comparison with the current scenario.

Table 7. Blue virtual water transfers for current scenario ($10^5 m^3$)

	Crop Type	Import needs	From	Qim	Still in needs	Can be exported	To	Qex	Remaining	To	Qex	Extra Remaining
	Wheat	-413	Mid	307	-107							
	Corn					376			376			376
	Cotton					2049	Mid	386	1662	East	1662	0.00
West	Potatoes	-0.16	Mid	0.16	0.00							
	Soy	-15.1	Mid	15.1	0.00							
	Sum	-429		322	-107	2425						376
	Wheat					307	West	307	0.00			0.00
	Corn					7783	East	79.6	7703			7703
Mid	Cotton	-386	West	386	0.00							
	Potatoes					38.7	West	0.16	38.6			38.6
	Soy					80.6	West	15.1	65.5			65.5
	Sum	-386		386	0.00	8209						7808
	Wheat	-186			-186							
	Corn	-79.6	Mid	79.6	0.00							
East	Cotton	-2959	West	1662	-1297							
	Potatoes					16.0						16.0
	Soy					33.8						33.8
	Sum	-3225			-1483	49.8						49.8

Table 8. Blue virtual water transfers for future scenario ($10^5 m^3$)

Crop Type	Import needs	From	Qim	Still in needs	Can be exported	To	Qex	Remaining	To	Qex	Extra Remaining
Wheat	-394	Mid	98.4	-296							
Corn					631			631			631
Cotton					1706	Mid	148	1558	East	1558	0.00
Potatoes					0.71			0.71			0.71
Soy											6.28
Sum	-394		98.4	-296	2344						638
Wheat					98.4	West	98.4	0.00			0.00
Corn					7597			7597			7597
Cotton	-148	West	148	0.00							
Potatoes					8.71						8.71
Soy					89.6	East	16.0	73.6			73.6
Sum	-148		148	0.00	7794						7680
Wheat	-1117			-1117							
Corn					667						667
Cotton	-3107	West	1558	-1549							
Potatoes					15.8						15.8
Soy	-16.0	Mid	16.0	0.00							
Sum	-4240		1574	-2666	686						686

4. Discussion

From the analysis of optimization for cropping patterns through virtual water aspects, the results indicate that current cropping patterns are causing stress on the local water resources, and the objective of saving water and reducing the water stress to an acceptable level can be achieved by adjusting the cropping patterns and introducing virtual water transfers.

The selected five crops are basic crops that are necessarily needed to be grown in the three regions. However, existing cropping patterns are mainly based on market value and market demands, rather than environmental concerns. There are many studies showing that water is becoming scarce due to overexploitation of water resources (e.g. Jiang (2009) and Kang et al. (2008)), and strategies need to be considered such as applying water-saving irrigation techniques and deficit irrigation, desalination, and water reuse policies (e.g. Belder et al., 2004), Fereres & Soriano, 2007, and Toze, 2006). Although introducing virtual water to respond to water scarcity, and analyses of water footprint are also mentioned as one of the strategies (e.g. Zheng et al., 2012), there is limited knowledge on how cropping patterns are influencing the allocation of water resources. Through optimizing cropping patterns and transferring virtual water, in fact, a considerable amount of local water does not need to be used, making it available for other uses such as municipal supply.

Before optimization, the East region has the highest WaSSI (0.49) although the climate is the least arid amongst the three regions. It is possible for humid areas to have water scarcity, since major water resources can be used for agriculture, industry, and domestic purposes (Molden, 2007; Brown & Matlock, 2011). WaSSI for the West and the Mid regions are moderate (0.29 and 0.28). WaSSI is predicted to remain at the same level for the West region and to slightly increase to 0.33 for the Mid region. WaSSI for the East region will decrease to 0.37 in the future, due to the decreasing cropping areas in the future based on the moving average estimates. However, larger reductions in cropping areas still need to be obtained to attain acceptable water stress.

It is noticeable that relatively large reductions in total cropping acreages are required in the East region (55.4% for the current scenario and 40.3% for the future scenario) and for some crops such as soy and potatoes in the West region (approximately 97 and 82%), as well as soy and wheat in the Mid region (approximately 81 and 78%), which seems not to be a feasible solution. The East region has the advantage of a relatively humid climate, and is the major crop contributing area of Gansu Province. However, the crop yield per unit area for crops such as wheat in the East region is only one-third of that in the West region. The West region has more sunshine hours and larger differences in daily temperature, which are helpful for crop growth. Therefore, the volume of water used to produce one kilogram of crop (e.g. wheat) in the East is larger than the volume used in the West region. In this case, inadequate water supply and inappropriate water management lead to excessive WaSSI in the East region. Alternatives include changing the irrigation water supply volumes, and reduce WaSSI to a lesser degree. However, if water supply volumes are increased for irrigation use, there will be shortages for municipal water use; if existing cropping patterns are not changed, there will be failures to meet the expected crop yields, or there would be risks of no irrigation water that can be applied. Large decreases for individual crops result for two reasons – (i) the crop is produced at levels more than local demands (e.g. potatoes and soy in the West region); and (ii) the crop has low economic value for the water used, thus imports of virtual water instead of producing locally are recommended, (e.g. wheat and soy in the East region). Although the idea is to lower local cropping by virtual water imports, in order to avoid the huge sudden changes and to create a balance between irrigation water use and municipal water use, it is suggested that the large reduction can be a direction over time. For example, a solution between the mean values and the upper limit values of the FLP solution sets would make more sense than the solution which would change circumstances too rapidly.

Research has been showing that climate change will have impacts on irrigation water needs and water resources in the future (e.g. Zhang & McBean,2014; Daz et al., 2007; Arnell, 2004; Christensen et al., 2004). The impacts of increasing volumes of blue virtual water for production are evident in relation to future scenarios, ranging from 3.0% to 60% regarding crop types and regions, and decreases are displayed for several crops as well – e.g. wheat and cotton in the East region. This is consistent with similar studies on impacts of climate change on crop water requirements and irrigation water needs (e.g. Wang et al., 2011; Chung et al., 2011).

Water use for existing cropping patterns indicates that the water for major crop production mainly depends on local water resources although water supply stress already exists. These modelling results confirm that Gansu Province has high self-sufficiency in water in the current situation. However, the high self-sufficiency reflects high stress on water resources. Wang et al. (2005) note that self-sufficiency for water has reached 99.52% in 2003 under the situation of water scarcity. The water self-sufficiency of Gansu Province decreased to 97.2% in 2007 (Ge et al., 2011). From this analysis of virtual water transfers, water stress can be reduced and local crop demands can be satisfied by transferring virtual water amongst the three regions. In addition, the remaining virtual water after transfers can be used to produce crops that have a need of virtual water import, or be transported to other areas or to outside the Province for economic benefits.

5. Conclusion

The fuzzy linear programming models (FLP) provide a range of possible solutions for cropping patterns to achieve the goals of saving water and attaining acceptable water stress. Cropping patterns were optimized using FLP models for three regions in Gansu Province from the West to the East under two climate scenarios – the current scenario and a future scenario. The results indicate that water is most scarce in the East region, since the irrigation water needs have exceeded the water supply in this region as a result of far too much cropping, in combination with municipal water demands. In this case, the East region is facing the risk of failing to meet the expected crop yields and inadequate applicable irrigation water. Although the East region is the least arid region amongst the three regions, water scarcity is likely to appear due to over-committing of water resources. The results show that water stress is occurring in both current and future scenarios if cropping patterns are not adjusted. Furthermore, existing cropping patterns are resulting in high self-sufficiency in water resources, especially for the East region. The results also show that the optimized cropping patterns can save local water resources to a substantial extent - as high as 50% of current use. Additionally, the deficits after the optimization, can be covered by virtual water transfers amongst the three regions, instead of producing the crops locally. There will still be remaining virtual water for some crops in the three regions which can be exported to other areas, or can be used to produce crops that can be managed by virtual water imports. The East region is the major virtual water import region if the optimization for cropping patterns is adopted, while the Mid region is the major virtual water export region. However, transfers of virtual water for crops such as wheat cannot satisfy local demands among the three regions; virtual water will need to be imported from other areas either in the Province or outside the Province. Huge, sudden changes in cropping patterns are not recommended although the idea is to save local water resources for minimized uses through virtual water transfers for crops. The optimization results can be a direction over time rather than a required change to avoid rapid circumstance changes.

References

Aldaya, M. M., Mart??nez-Santos, P., & Llamas, M. R. (2010). Incorporating the water footprint and virtualwater into policy: Reflections from the mancha occidental region, spain. Water Resources Management, 24, 941-958. http://dx.doi.org/10.1007/s11269-009-9480-8

Allan, J. A. (1998). Virtual water: A strategic resource global solutions to regional deficits. Groundwater, 36, 545-546. http://dx.doi.org/10.1111/j.1745-6584.1998.tb02825.x

Arnell, N. W. (2004). Climate change and global water resources: Sres emissions and socio-economic scenarios. Global environmental change, 14, 31-52. http://dx.doi.org/10.1016/j.gloenvcha.2003.10.004.

Belder, P., Bouman, B., Cabangon, R., Guoan, L., Quilang, E., Yuanhua, L., Spiertz, J., & Tuong, T. (2004). Effect of water-saving irrigation on rice yield and water use in typical lowland conditions in asia. Agricultural Water

Management, 65, 193-210.

Biswas, A., & Pal, B. B. (2005). Application of fuzzy goal programming technique to land use planning in agricultural system. Omega, 33, 391-398. http://dx.doi.org/10.1016/j.omega.2004.07.003

Brown, A., & Matlock, M. D. (2011). A review of water scarcity indices and methodologies. The Sustainability Consortium, White paper (p. 19).

CDO. (2013). Retrieved from http://chinadataonline.org/

Chapagain, A., Hoekstra, A., Savenije, H., & Gautam, R. (2006). The water footprint of cotton consumption: An assessment of the impact of worldwide consumption of cotton products on the water resources in the cotton producing countries. Ecological economics, 60, 186-203. http://dx.doi.org/10.1016/j.ecolecon.2005.11.027.

Christensen, N. S., Wood, A. W., Voisin, N., Lettenmaier, D. P., & Palmer, R. N. (2004). The effects of climate change on the hydrology and water resources of the colorado river basin. Climatic change, 62, 337-363. http://dx.doi.org/10.1023/B:CLIM.0000013684.13621.1f.

Chung, S., Rodr??guez-D??az, J., Weatherhead, E., & Knox, J. (2011). Climate change impacts on water for irrigating paddy rice in south korea. Irrigation and Drainage, 60, 263-273. http://dx.doi.org/10.1002/ird.559.

Dalin, C., Konar, M., Hanasaki, N., Rinaldo, A., & Rodriguez-Iturbe, I. (2012). Evolution of the global virtual water trade network. Proceedings of the National Academy of Sciences, 109, 5989-5994.

Dehghan, M., Hashemi, B., & Ghatee, M. (2006). Computational methods for solving fully fuzzy linear systems. Applied Mathematics and Computation, 179, 328-343. http://dx.doi.org/10.1016/j.amc.2005.11.124.

D??az, J. R., Weatherhead, E., Knox, J., & Camacho, E. (2007). Climate change impacts on irrigation water requirements in the guadalquivir river basin in spain. Regional Environmental Change, 7, 149-159.

Fan, Y., Huang, G., Li, Y., Cao, M., & Cheng, G. (2009). A fuzzy linear programming approach for municipal solid-waste management under uncertainty. Engineering Optimization, 41, 1081-1101.

Fereres, E., & Soriano, M. A. (2007). Deficit irrigation for reducing agricultural water use. Journal of Experimental Botany, 58, 147-159.

Ge, L., Xie, G., Zhang, C., Li, S., Qi, Y., Cao, S., & He, T. (2011). An evaluation of china's water footprint. Water resources management, 25, 2633-2647. http://dx.doi.org/10.1007/s11269-011-9830-1.

Herrera, F., & Verdegay, J. (1995). Three models of fuzzy integer linear programming. European Journal of Operational Research, 83, 581-593. http://dx.doi.org/10.1016/0377-2217(93)E0338-X.

Hoekstra, A. Y., & Chapagain, A. K. (2006). The water footprints of morocco and the netherlands, .

Hoekstra, A. Y., & Hung, P. Q. (2005). Globalisation of water resources: international virtual waterflows in relation to crop trade. Global environmental change, 15, 45-56. http://dx.doi.org/10.1016/j.gloenvcha.2004.06.004.

Inuiguchi, M., & Ram?k, J. (2000). Possibilistic linear programming: a brief review of fuzzy mathematical programming and a comparison with stochastic programming in portfolio selection problem. Fuzzy sets and systems, 111, 3-28. http://dx.doi.org/10.1016/S0165-0114(98)00449-7.

Jiang, Y. (2009). China's water scarcity. Journal of Environmental Management, 90, 3185-3196.

Kang, S., Su, X., Tong, L., Zhang, J., & Zhang, L. (2008). A warning from an ancient oasis: intensive human activities are leading to potential ecological and social catastrophe. The International Journal of Sustainable Development & World Ecology, 15, 440-447. http://dx.doi.org/10.3843/SusDev.15.5:5.

Klir, G. J., St Clair, U., & Yuan, B. (1997). Fuzzy set theory: foundations and applications.

Maleki, H., Tata, M., & Mashinchi, M. (2000). Linear programming with fuzzy variables. Fuzzy sets and systems, 109, 21-33. http://dx.doi.org/10.1016/S0165-0114(98)00066-9.

McNulty, S. G., Sun, G., Myers, J. M., Cohen, E., & Caldwell, P. (2010). Robbing peter to pay paul: trade-offs between ecosystem carbon sequestration and water yield. In Watershed Management (pp. 23-27). http://dx.doi.org/10.1061/41143(394)10.

Molden, D. (2007). Water for food, water for life: a comprehensive assessment of water management in agriculture: summary. International Water Management Institute (IWMI).

NCDC. (2013). Retrieved from http://www7.ncdc.noaa.gov/cdo/cdoselect.cmd

Rommelfanger, H. (1996). Fuzzy linear programming and applications. European journal of operational research, 92, 512-527. http://dx.doi.org/10.1016/0377-2217(95)00008-9.

Sahoo, B., Lohani, A. K., & Sahu, R. K. (2006). Fuzzy multiobjective and linear programming based management models for optimal land-water-crop system planning. Water Resources Management, 20, 931-948. http://dx.doi.org/10.1007/s11269-005-9015-x.

Seckler, D., Barker, R., & Amarasinghe, U. (1999). Water scarcity in the twenty-first century. International Journal of Water Resources Development, 15, 29-42. http://dx.doi.org/10.1080/07900629948916.

Tanaka, H., & Asai, K. (1984). Fuzzy linear programming problems with fuzzy numbers. Fuzzy sets and systems, 13, 1-10. http://dx.doi.org/10.1016/0165-0114(84)90022-8.

Toze, S. (2006). Reuse of effuent water-benefits and risks. Agricultural water management, 80, 147-159.

Wang, H. L., Wang, R. Y., Zhang, Q., Niu, J.-Y., & Lu, X.-D. (2011). Impact of warming climate on crop water requirement in gansu province. Chinese Journal of Eco-Agriculture, 4, 028.

Wang, X. H., Xu, Z. M., & Li, Y. H. (2005). A rough estimate of water footprint of gansu province in 2003. Journal of Natural Resources, 6, 015.

Zhang, C., & McBean, E. A. (2014). Adaptation investigations to respond to climate change projections in gansu province, northern china. Water Resources Management, 28, 1531-1544. http://dx.doi.org/10.1007/s11269-014-0554-x.

Zhang, C., McBean, E. A., & Huang, J. (2014). A virtual water assessment methodology for cropping pattern investigation. Water Resources Management, 28, 2331-2349. http://dx.doi.org/10.1007/s11269-014-0618-y.

Zheng, Z., Liu, J., Koeneman, P., Zarate, E., & Hoekstra, A. (2012). Assessing water footprint at river basin level: a case study for the heihe river basin in northwest china. Hydrology and earth system sciences, 16, 2771-2781.

Zimmermann, H. J. (1978). Fuzzy programming and linear programming with several objective functions. Fuzzy sets and systems, 1, 45-55. http://dx.doi.org/10.1016/0165-0114(78)90031-3

Land Management for Sustainable Agriculture Under Climate Change in the Congo-Basin Countries of Central Africa

Ernest L. Molua[1]

[1] Department of Agricultural Economics and Agribusiness, Faculty of Agriculture & Veterinary Medicine, University of Buea, Cameroon

Correspondence: Ernest L. Molua, Department of Agricultural Economics and Agribusiness, Faculty of Agriculture & Veterinary Medicine, University of Buea, Cameroon. Tel: 237-9949-4393. E-mail: emolua@cidrcam.org

Abstract

Degradation of ecosystem services is evident in the Congo-Basin Countries of Central Africa. This paper examines land management for sustainable agriculture in the context of environmental change and climatic stress. The interaction between drivers of change, institutional arrangements and the actions of stakeholders are discussed to highlight local, regional and global benefits of sustainable agricultural land management. The paper notes that tenure security has to be at the heart of any agricultural development plans, and managing agricultural land in the context of environmental change and climate stress requires that land managers ensure the long-term productive potential. It is recommended that land-based entrepreneurs adopt land use systems which enable them to maximize the economic and social benefits from land while enhancing the ecological support functions of land resources. The paper concludes that land policy reforms which aim to promote sustainable land management should address issues related to land rights and institutions driven by inefficient and inequitable historical legacies, including access to land by women, indigenous groups; as well the inefficiencies and inequities that arise from poor legal and administrative systems.

Keywords: central African sub-region, agricultural land management, sustainable agriculture

1. Background

The Congo Basin, acknowledged as the second largest tropical rainforest on earth and the lungs of Africa, has a rich and diverse ecosystem which provides food, fresh water, shelter and medicine for tens of millions of people, as well as home to some critically endangered species. Its rivers, swamps, trees and savannahs significantly contribute to sustain life across the whole planet, playing a critical role in regulating the global climate for the benefit of the entire biosphere. Agriculture is the economic foundation of the Central African countries within the Congo Basin, employing about 60 percent of the workforce and contributing an average of 30 percent of gross domestic product. However, agricultural growth rates have stagnated and food insecurity remains a concern, with malnourishment significant in some countries (Bush, 2010). With the exception of few countries such as Gabon and Equatorial Guinea that rely on oil to drive their economies, the predominant livelihood activity in the subregion is smallholder semi-subsistence farming. Households rely on cash and subsistence incomes from a number of sources that include rainfed cultivation, livestock production and non-timber forest exploitation. The agricultural activities are affected not only by unfavourable climatic conditions, poor markets and infrastructure services, but also by unfavourable physical conditions (poor soils, land degradation caused by cultivation on sloping land, deforestation) (Aklilu & de Graaff, 2007).

Some thousands of hectares of land is used for crop production in Central Africa, millions hectares more of potentially arable land competes with other land uses for infrastructure and human settlement along with land set aside for, e.g. forest ecosystems in natural reserves. Figure 1 shows the dominant ecosystem classes and agricultural production potentials across Africa. We note for Central Africa sub-region the domination of forest, woodland and grassland particularly suitable for rainfed production of various crops (e.g. food, fibre, fodder crops and pasture grasses) under different input and management techniques. Compared to other regions in Africa, Central Africa's intensive cropland is largely across Cameroon, Central African Republic and Burundi, with unused potential in the other countries in the sub-region. Millions of hectares of land in the sub-region has cultivation potential (whether very suitable, suitable or moderately suitable) required for the cultivation of typical crops such as grain maize. However, this suitability coincides with land dominantly classified as forest ecosystem. For a country like Gabon, this is important where a significant proportion of prime arable land is under protected forests and conservation programmes, thus constraining farmers' access to land. Overall, however, the potential of Central Africa subregion demonstrates the feasibility of intensifying production, regenerating and preserving soils, and maintaining biodiversity through agroecological technologies and indigenous methods which enable farmers to take advantage of local resources and reduce dependency on external capital-intensive inputs.

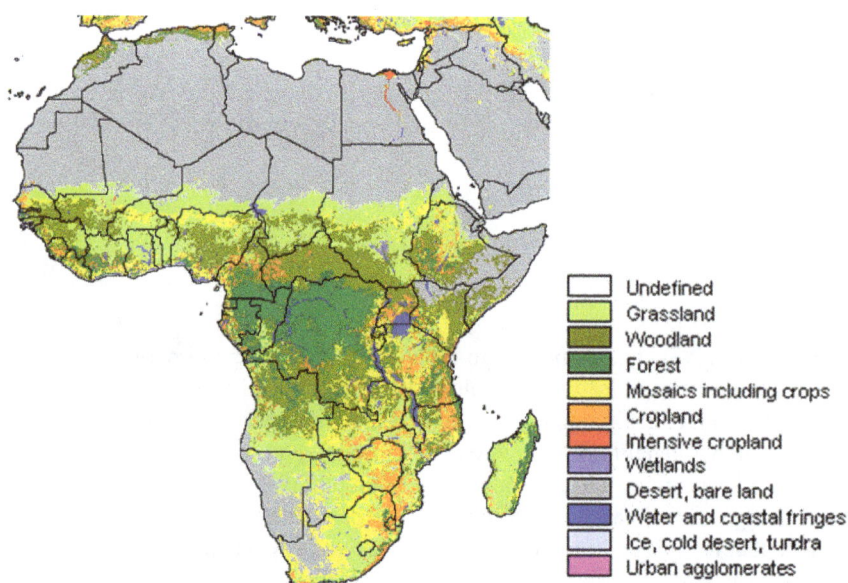

Figure 1. Dominant ecosystem classes across Africa (Fischer et al., 2000)

This paper examines land management for sustainable agriculture under climate change in the Congo-Basin Countries of Central Africa. This is deemed important because of the dominant view that land reforms could enhance security of tenure and promote investment in agriculture, thus lead to increased growth and development (Byres, 2004; Barrows & Roth, 1990). However, such land reform programmes failed to develop the smallholder agriculture sector. The coexistence of various forms of tenure in the countries - state, communal, customary, individual - suggests the need to review and develop complex policy and analytic models focusing on the pertinent relationship between land tenure and sustainable development (DFID, 1999). The rest of this paper examines land-use and tenure issues in the sub-region in section two, it then reviews the challenges of agricultural land management under climatic change in section three and makes recommendations, in section four, for some policy safeguards for sustainable agricultural land capital management for countries in the sub-region.

2. Land Use and Tenure for Agricultural Development

When considering major crop types, particularly cereals; about 40% of the Central African land surface can be regarded as suitable for crop cultivation. Despite this optimistic aggregate picture, there are hotspots such as in Northern Cameroon and Chad where the rain-fed cultivation potential is nearly fully exhausted or has already been exceeded, and there is need for irrigation. Table 1 shows the per capita land in use for cultivation and net rain-fed cultivation potential for cereals with projected population levels between 1995 and 2050. Contemporary Land use for cultivation in the middle belt of Africa is 3.8% compared to rainfed net potential land of 40.9%. The arable land per person is about 0.3 ha, the world average in 1995 being about 0.26 ha/person for a world population of almost 5.7 billion people. High potential land is estimated to cover 3.22 ha/person and 0.97 ha/person in 2050. Overall, land in use in Africa for cultivation averages 6.6% compared to global average of 11.2% of the total land. About 26.65% of African land is high potential rain-fed land, higher than the world average of 21.8%. The per capita land use for suitable land in Africa was 1.12 ha/person in 1995 and projected to decline to 0.45 ha/person in 2050. This compares favourably with the global figure of 0.51 ha/person in 1995 and expected to decline to 0.33 ha/person in 2050. This suggests a considerable availability of resources suitable for agricultural uses in Africa, including Central Africa. This, nonetheless, highlights the need for a land management policy which promotes efficient input use and technology leading to higher average per hectare output in lands currently under exploitation, and more effort to better manage competition with other non-agricultural uses with safeguards for bio-diversity and the carbon cycle.

Table 1. Per capita land in use for cultivation and net rain-fed cultivation potential for cereals, populations of 1995 and projected populations in 2050

Region	Population (million)		Total land	Land in use forcultivation (1994-96)		Rain-fed VS+S+MS net potential land (mixed input)		Per Capita land use (ha/pers.)	Per Capita VS+S+MS land (mixed input)	
	1995	2050	$(10^6\,ha)$	$(10^6\,ha)$	%	$(10^6\,ha)$	%	1994-96	1995	2050
Eastern Africa	219.5	593	639.5	46.0	7.2	237.4	37.1	0.21	1.08	0.40
Middle Africa	*83.3*	*274.6*	*657.1*	*24.8*	*3.8*	*268.5*	*40.9*	*0.3*	*3.22*	*0.97*
Northern Africa	157.8	303.2	794.1	44.1	5.6	92.9	11.7	0.28	0.59	0.30
Southern Africa	47.3	65.5	266.4	17.4	6.5	28.4	10.6	0.37	0.60	0.43
Western Africa	209.4	526.3	633.0	65.4	10.3	174.3	27.5	0.31	0.83	0.33
Africa	*717.3*	*1762.7*	*2990.1*	*197.7*	*6.6*	*801.4*	*26.8*	*0.28*	*1.12*	*0.45*
Developing countries	4510.8	7746.6	8171.5	909.6	11.1	1971.5	24.1	0.2	0.44	0.25
Developed countries	1171.7	1155.0	5228.0	595.5	11.4	952.5	18.2	0.51	0.81	0.82
World	*5682.4*	*8901.6*	*13399.5*	*1505.2*	*11.2*	*2924.0*	*21.8*	*0.26*	*0.51*	*0.33*

Notes: VS = very suitable, S = suitable, MS = moderately suitable. Calculation of net rain-fed land with cultivation potential makes an allowance for infrastructure and settlement, taken to be 10% in 1995 and 15% in 2050 of the gross suitable area. Source: Fisher et al. (2000). http://www.iiasa.ac.at/Research/LUC/GAEZ/index.html

Figure 2 shows the agricultural area relative to the total national land area for selected countries in the region from 1965 to 2009. For Chad more than 35% of country land area is utilized as agricultural land, and almost 30% for the Republic of Congo, 20% for Gabon and more than 15% for Cameroon. This agricultural land is likely to expand in order to satisfy the needs of the rapidly growing population. Such expansion for food and raw material needs may transform the natural ecosystem of the region, particularly the Congo Basin which is of global significance in its climate change mitigation potentials. Combining the information emanating from figures 1 and 2 and table 2, we may conclude that the way land resource is used in the subregion could be decisive for future social and economic well-being as well as for the sustained quality of land resources in Central Africa. Therefore, if more production is required from the existing land resource while cultivable areas are being shifted towards non-agricultural uses, land policy and land administration must play important roles in land use dynamics through suitable institutional mechanisms, scientific management, conservation and development of land resources.

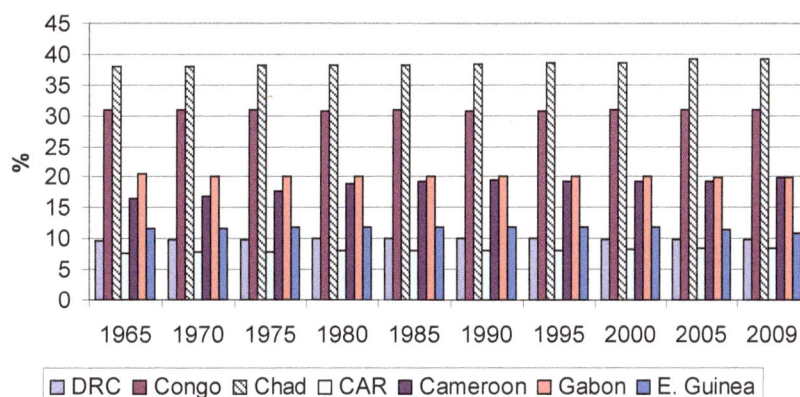

Figure 2. Agriculture area per unit of national land area (%)

Conservation and development of land resources are important processes that determine the value and total economic importance of African land. Whether it is the scramble for African land or the intensification of agriculture, the over-exploitation of soil resources and the expansion into marginal lands has led to soil fertility depletion (Pasquini & Alexander, 2005). Fertility depletion is common in North Cameroon and Chad, whilst the 'land grab' is significant in Southern Cameroon and the Democratic Republic of Congo. The limited adoption of fertilizer replenishment strategies and soil and water conservation measures and decline in the use and length of fallow periods is impacting negatively on their agricultural production (Aklilu & de Graaff, 2007; Scoones et al., 1996). Land grab means that prime agricultural land carved out deprives the community land resources (Deininger, 2011). In addition to this challenge, for smallholder farmers, the prevailing global financial crisis and low farm returns has meant a decline in farm input investment, including fertilizers, seeds, and technology adoption against soil degradation. This has significant impact on agriculture and poverty reduction (Griffin et al., 2002). For the poorest people, GDP growth originating in agriculture is more effective in raising incomes of extremely poor people than GDP growth originating outside the sector. Therefore, the land question in relation to access and management is an important problem to resolve if these countries and their rural peasantry are to emerge from economic and social crisis.

Tenure security which has to be at the heart of any agricultural development plans, is achieved when property rights are clarified and widely acknowledged (Andersson, 2007). In most cases, progress will consist of (a) the reconciliation of diverse and conflicting claims, (b) the clarification of latent or overlapping rights in resources, and (c) the reconciliation of statutory and customary regimes. The question of considering customary rights in sustainable land management has been identified in almost all the countries in Central Africa. Currently, customary land tenure is not adequately recognized in the majority of Central African countries. However, in reality most people in the region occupy their land under a customary system. This means absence of formal tenure rights and consequently insecurity of land tenure. Concerns about population growth and pressure on land in urban areas and coastal zones have been raised in countries like Cameroon, Congo, Gabon and Equatorial Guinea (Gilland, 2002; Pender, 1998). Forced evictions, expropriations and related land issues are also critical issues in Central Africa.

Pender et al. (2006) and Andersson (2007) have shown that property rights and secure access to and control over land and natural resources can generate critical incentives for conservation and sustainable use, management, and governance of natural resources. Insecure, unclear, limited or short-term property rights can inhibit sustainable land and natural resource management and discourage stakeholders from acting as long-term stewards of land and natural resources (Adesina & Baidu-Forson, 1995). Property rights affect outcomes such as agricultural productivity, household income, and land degradation (Mbaga-Semgalawe & Folmer, 2000). This is captured in Figure 3 which shows how agricultural production and land conditions are affected by land management practices, including both private decisions made by farm households and collective decisions made by groups of farmers and communities. Pender et al. (2006) notes that farm households may make decisions about land use (e.g. grazing land), the crop types to plant, the amount of labor to use, and the types and amounts of inputs, investments, and agronomic practices to use to conserve soil and water, improve soil fertility or reduce pest losses. Communities also can influence land management through their collective decisions. They may make investments on communal land areas (e.g. erosion controls on degraded lands, tree planting) or private lands (e.g. drainage investments as part of watershed conservation and development efforts) or regulate use of communal land (e.g. restrictions on use of grazing areas) or private lands (e.g. bylaws limiting burning or cutting of trees). Agricultural income strategies and land management decisions are affected by these different factors operating at different scales (Pender, 1998; Pagiola, 1996). Local institutions also have important influences on income strategies and land management. In much of Central Africa, customary land tenure institutions determine what land use rights and land management obligations farmers have, how secure those rights are, whether those rights may be transferred or used as collateral, how conflicts are resolved, and other questions. Such institutions can have substantial effects on land management by regulating land use.

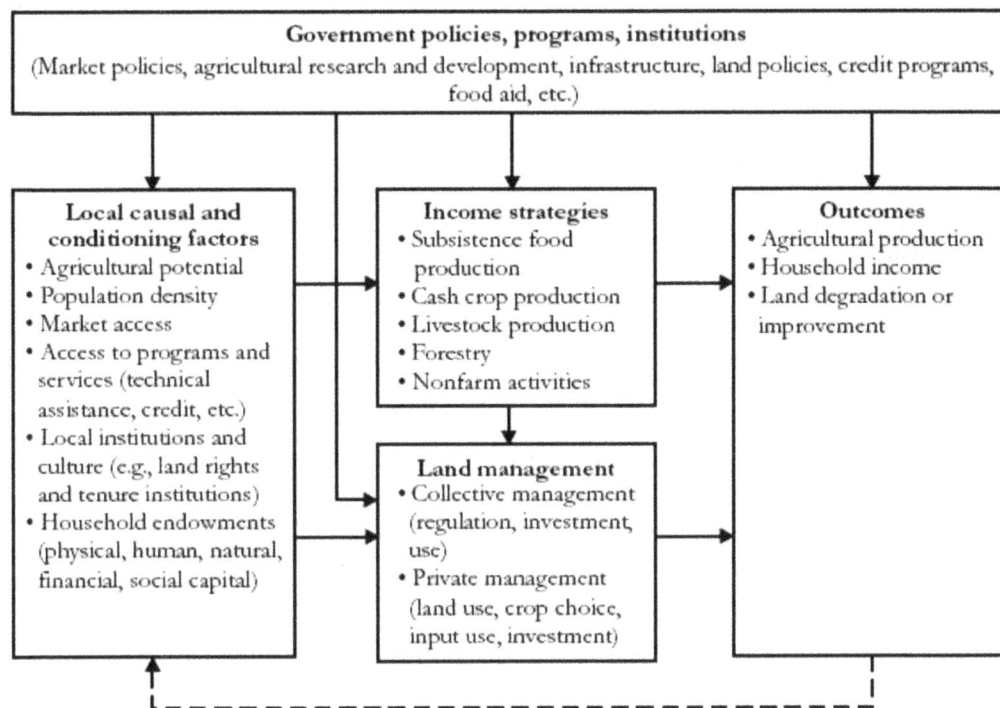

Figure 3. Income strategies and agricultural land management (Source: Pender et al., 2006)

3. Agricultural Land and Environmental Stress Under Climate change

Studies of anthropogenic biogeophysical climate change in the Congo basin point to a regional surface warming and a decrease in precipitation, largely as a consequence of tropical deforestation (Polcher & Laval, 1994; McGuffie et al., 1995; Semazzi & Song, 2001). Recently, Nogherotto et al. (2013) and Akkermans et al. (2014) show a strong spatial correlation between deforestation and global warming. Akkermans et al. (2014), for instance, reveal that countries in the Central Africa sub-region by 2050 will be an average of 1.4 °C hotter than today as a result of global greenhouse gas emissions. Deforestation will add an extra 0.7 °C to that figure. Their findings also show a strong spatial correlation between deforestation and global warming. In certain deforestation 'hot spots', increases caused by deforestation could rise to 1.25 °C, in addition to the warming caused by greenhouse gases.

Managing agricultural land in the context of environmental stress and climate change is therefore an imperative. This therefore requires that land managers use of land resources for the production of goods to meet changing human needs, while simultaneously ensuring the long-term productive potential of these resources and the maintenance of their environmental functions. This implies that land-based entrepreneurs should adopt land use systems that, through appropriate management practices, enable them to maximize the economic and social benefits from the land while maintaining or enhancing the ecological support functions of the land resources. In countries within Central Africa where the economy is heavily based on agriculture, development of the agricultural sector through sustainable land management could be the most efficient poverty reduction measure. Yet, agricultural expansion for food production and economic development which comes at the expense of soil, water, biodiversity or forests is highly vulnerable to prevailing environmental stress and climate change (Shiva, 2008; Scoones et al., 1996). In the face of ensuing climate change, the impacts may not be uniform in the sub-region since Central Africa is characterized by extreme ecological diversity, stretching from the north of Chad that is characterized by permanent drought and the humid forest in the equatorial region to the south of Gabon. Being a typical equatorial region on both sides of the equator, Central Africa hosts one of the world's richest forest biodiversity. Given its latitudinal position, the region therefore experiences the main ecosystems types of the continent. Three other ecological regions are found in the sub-region: the wooded savannah found mostly in Central African Republic, Cameroon and Angola; the arid region stretches from the 9° north of the equator to the Chad-Libyan border; and the Sahelian region of Central Africa and Northern Cameroon that suffers from permanent drought with its consequences on livelihoods.

The fragility of its ecology means that the region faces significant challenges from climate variability and change. Rainfall variability experienced in sub-Saharan Africa already has detrimental impacts on crop production and agriculture as a whole (Schlenker & Lobell, 2010). Indeed, too much or too little water due to erratic rainfall and insufficient storage capacity wields adverse impacts on food security (Schmidhuber & Tubiello 2007; Jones, 2003). Already, some farmers are experiencing more frequent and intense storms that cause erosion, rainwater run-off, and crop damage, while others experience more frequent droughts. At the same time rainfall patterns are becoming more variable with delayed onset and length of the rainy season, and in some cases, drought. These unpredictable patterns make it difficult for farmers to plan and manage their crops (Schmidhuber & Tubiello, 2007). All these impact the total economic value of land.

Future climate change is widely predicted to impact on rainfall variability in sub-Saharan Africa, with the effect of increasing droughts and floods (Müller et al., 2011). Even using optimistic lower-end projections of temperature rise, climate change may reduce crop yields by 8–20 percent by the 2050s, with more severe losses in some regions (Schlenker & Lobell 2010; Schmidhuber & Tubiello, 2007). Warmer temperatures, more variable rainfall, and higher incidence of extreme events all magnify stresses on the farming systems. Increasing frequencies of heat stress, drought and flood events, will result in yet further deleterious effects on productivity and wellbeing (Patt et al., 2010). It is likely that price and yield volatility will continue to rise as extreme weather continues. In fact, world food prices for some of the main grain crops are likely to rise sharply half of the 21st century, unlike the price declines witnessed in the 20th century (Rosegrant et al., 2001). This will have serious consequences on food security. Climate change will also impact on agriculture through effects on pests and disease. These interactions are complex and as yet the full implications in terms of productivity are uncertain. It is therefore essential for policy planners to find ways to cope with existing climate stressors, as well as other effects of future climate change. African producers have already developed a number of indigenous coping mechanisms to support survival in the face of climate variability. However, global climate change increases the risks that African farmers must efficiently manage, and policy makers must respond with the appropriate measures to facilitate adaptation required to uphold the value of land.

Fortunately, there are a range of land-based management practices and technologies that can be applied on-farm to increase agricultural resilience to climate stress. However, property rights and secure access to and control over land and natural resources can generate critical incentives for conservation and sustainable use, management, and governance of natural resources (Lapar et al., 1999; DFID, 1999). Insecure, unclear, limited or short-term property rights can thus inhibit sustainable land and natural resource management and discourage stakeholders from acting as long-term stewards of land and natural resources. Land tenure challenges are age-old and climate change simply exacerbates the situation. The uncertainty of future climate variability and change requires greater flexibility in all land-based production systems. Given the central role of user rights in those systems, land and resource tenure will likewise require greater flexibility, thus raising a critical policy matter for many countries to maintain flexibility in customary and statutory tenure systems. Tenure security will be a critical factor in providing the incentives for mitigating greenhouse gas emissions and adapting to climate change (Howden et al., 2007). Because climate-induced migrations (as already seen in northern Chad and northern Cameroon) could lead to social tensions, climate change will challenge institutions responsible for the governance of natural resources such as land to establish inclusive processes to negotiate claims, regulate disputes, and establish new tenure systems.

Effective response to the challenges of climate variability and change will require an ecosystem approach, working at the landscape level to increase productivity, enhance resilience to changing temperatures and rainfall patterns, and reduce greenhouse gas emissions that contribute to climate change. This means significant changes in agricultural practice in coming years, promoting and adopting ecological practices that can provide "no regrets" insurance against climate change, e.g. mulching of crops to allow more rain to soak into the soil, slow down how quickly that soil moisture evaporates, and reduce the erosion of soil into streams. Some other measures include intercropping to take advantage of different plants having different patterns of root growth and different needs for nutrients, less tillage of soil to boost fertility and health of soil over time while also reducing wind and water erosion of soil because it is less exposed, agroforestry of mixed food crops and trees for provisioning and environmental services including managing micro-climates and water management both in streams and on fields. These measures require secure unfettered access to land.

In semi arid zones such as in Northern Chad and other already degraded ecosystems such as through slash-and-burn agriculture in Southern Cameroon and Burundi, a range of well-established restoration and management options can improve human livelihoods, repair ecosystems, and increase the resilience of both people and landscapes to climate change. This will include the following at the farm-scale: soil and nutrient management (e.g. through conservation agriculture, improved application of fertilizers, and increasing fertility by integrating legumes into farming systems); water harvesting and use (e.g. capturing rainwater, retention of soil moisture and increasing water productivity through irrigation); integrated pest and disease management; resilient ecosystems (e.g. through farm management practices that reduce erosion and rainwater run-off, increase on-farm habitat for beneficial insects, pollinators and wildlife, sequester carbon, and reduce conversion of natural habitat to agriculture support ecosystem resilience across the landscape); genetic resources for climate resilient varieties; Harvesting, processing and supply chains that reduce post-harvest losses and preserve quantity, quality, and nutritional value of food products; and diversification to increase the efficiency of farming systems and build their resilience to climate change. A proactive land policy making process and efficient land administration service create incentive for adoption of better land management measures.

Therefore, an important challenge for governments and policy makers with respect to appropriate land management is to promote the adoption of these environmentally friendly practices, at the micro and macro-scales (World Bank, 2006). Figure 4 below contextualizes the role of government in a framework for sustainable land management and agricultural transformation. Considerations in the schema include: adopting an ecosystem approach with cross-sectoral coordination and collaboration at the landscape scale as being essential to adapt to climate stresses; scaling up effective climate-smart practices; ensuring institutional, technical, and financial support for small-holder; reducing data, knowledge and technology gaps which exist and should be addressed to support improved technologies, methodologies, and climate resilient varieties; and harmonization of climate change, agriculture and food security policies is required at the national, regional and international levels (Schmidhuber & Tubiello, 2007; Holden et al., 2006).

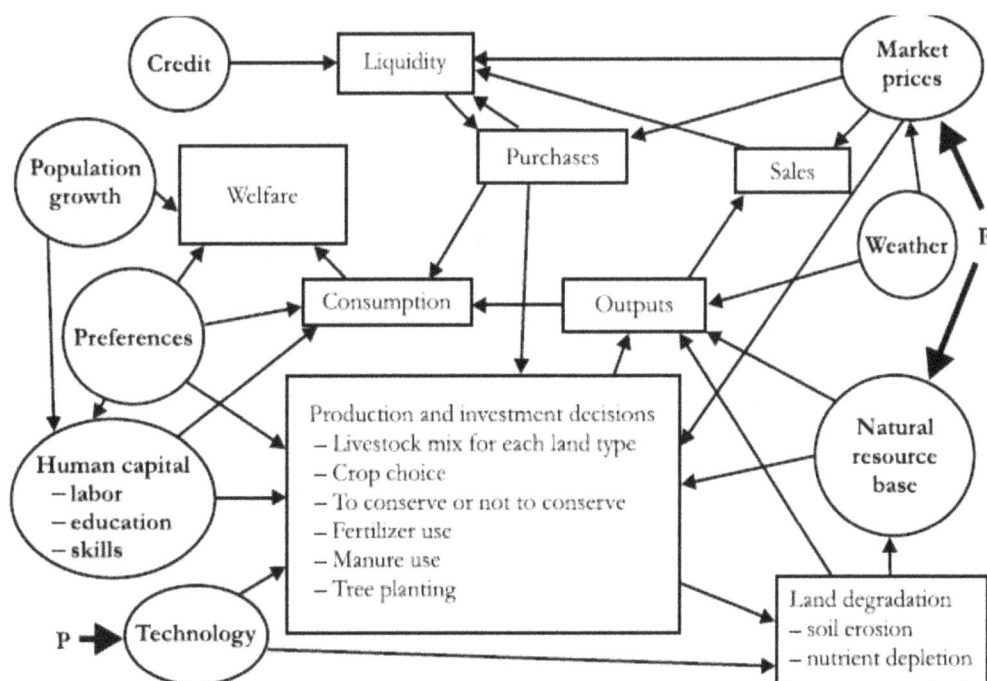

Figure 4.Household Dynamic Interaction for Sustainable Land Management,and Food Security
(Source: Holden et al., 2006)

In essence, the farm-level micro-scale challenges highlight the need to get land policy right and ensure better land administration for comprehensive access and user rights to land. As shown in Figure 3 households maximize their welfare subject to many constraints such as land degradation in the form of soil erosion and soil nutrient depletion. This may also include weather risk which affects production as well as prices (Molua, 2011; Pagiola, 1996). Any compromise through non-existent land policy or ineffective land administration jeopardizes important household production and consumption decisions. These decisions are based on expectations about prices and output and the risk involved, and thus need flexible access to agricultural land to be able to respond and minimize risks and maximize opportunities (Mbaga-Semgalawe & Folmer, 2000). Given the current situation where agricultural practitioners largely identify access to productive land as a constrain, the rising population growth which already strain land uses may also affect household welfare as more people have to share the outcome of a constant land area which is also affected by land degradation (Gilland, 2002; Pender, 1998). Holden et al. (2006) posit that this may lead to a Malthusian development path, and the ensuing poverty-environment trap can only be broken through availability of new technologies, improved access to markets, and better investment opportunities which all require unfettered access to prime productive land.

4. Policy Safeguards for Sustainable of Agricultural Land Capital Management

While the specific challenges to land capital management across countries in the central African sub-region are diverse, however, common among them is the concern about environmental sustainability and the need for successful land management interventions being increasingly recognized in national development plans and poverty reduction strategies. Translating this recognition into effective policies or programmes in the face of climate change is the Achilles heel to sustainable land management in the sub-region. Any policy to address land capital management for sustainable agriculture in the context of environmental change must capture three key principles, which include: (1) integrated use of natural resources at ecosystem and farming systems levels; (2) multilevel and multi-stakeholder involvement; and (3) targeted policy and institutional support including development of incentive mechanisms for land-user-driven and participatory approaches in sustainable land management and adoption at the local level.

Land policy reforms which aim to promote sustainable land management should address issues related to land rights and institutions driven by inefficient and inequitable historical legacies, including access to land by women, poor people, indigenous groups; as well the inefficiencies and inequities that arise from poor legal and administrative systems. This must also take cognizance of old and new challenges that arise in the context of increasing scarcity of land; and the increasing competition for their use from urbanization, environmental services and biofuels. Addressing the inequities, particularly related to gender is important for sustainable land management, not only because women are significant crop producers in the sub-region, but also that there are both economic and social benefits when women have secure access to valuable land. Hence, institutions for sustainable land management must take the gender question serious. Property rights in land – whether customary, formal, or religious – provide economic access to key markets and social access to non-market institutions such as household and community-level governance structures (Adesina & Baidu-Forson, 1995). Secure land rights confer direct economic benefits because land: (1) is a key input into agricultural production and enterprise development; (2) can be used as a source of income from rental or sale; (3) can provide collateral for credit where strong, well-regulated land markets exist; and (4) can increase the capacity to invest in coping mechanisms and adaptation practices to secure livelihoods. Without secure land rights, women are less able to take advantage of changes that may result from stresses such as climate change. Women may not fully reap these benefits if they do not have legal and socially-recognized rights to individually- or jointly-held land (Stockbridge, 2007). This therefore calls for proactive land policy and better functioning land administration. An effective land administration will correct imbalances by safeguarding active land markets, speeding the issuance of land titles, and accelerating plot surveys, implement land resettlement for displaced communities e.g. from large-scale land grabbing. These may not only lead to tenure security but also improve the efficiency of land management (Basu, 1996). In other words, reliable institutions are needed to promote timely adjudication and land registration, strengthening land administration institutions and organizing the national land register (Atwood, 1990).

Land policy should not only be limited to addressing inaccessibility to land or landlessness. Attention should be given to the need to increase productivity per unit of land through better land utilization (Migot-Adholla et al., 1991). Better land utilization will require increased investment in agriculture, from both public and private sectors (Shiva, 2008). These investments are sustainable if they land-user driven at the local level. Farmers will therefore require incentives to invest in land improvements, drainage and irrigation, which are provided by increasingly secure ownership rights. To provide these rights and incentives, policy must encourage communal right systems which allocate inheritable usufruct rights, first to homesteads and nearby fields, then to the right to resume fallowed land, and any areas into which farmers have invested by planting trees, or investing in soil conservation, drainage and irrigation. The strength of communal rights systems is that they restrict sales to outsiders of the community, or subject them to community approval. Studies in Africa have shown that customary systems can provide sufficient tenure security for investment (Bruce & Migot-Adholla, 1994).

The vast majority of investment in agriculture in the sub-region is from private domestic sources. Public investment in individual countries is historically small. National and local level governments can enact and implement policies that encourage access of smallholders to additional sources of funding, including microcredit, revolving grants, and crop insurance. Remittances from abroad and international donor investment are also important funding sources. Local civil society organizations, such as farmer cooperatives and water user associations, must have access to financing not only to cover capital costs for irrigation infrastructure, per se, but also for operation and maintenance of these systems.

Capacity building is thus imperative, and this should be promoted by governmental and non-governmental institutions, particularly the international NGOs and UN-based system organizations with their enormous financial and intellectual capital. Support to countries in the sub-region from these organizations should be on a wide range of complementary approaches, through training, information, communications, tools and equipment, advisory services for institutional strengthening, policy reforms and national programming. Common among countries in the sub-region is ineffective underperforming agricultural research institutions. Significant achievements could be made by building the capacities of local agricultural research institutes with national research mandate to promote the sustainable use and management of agricultural land and natural resources. Such centres with objectives underpinned in knowledge generation on agricultural natural resources and agro-ecosystems as well as research application; innovative technology development; and technology transfer, must conduct research, drive technology development and transfer in order to promote agriculture and related industries; and ensure natural resource conservation for poverty alleviation and a better quality of life.

5. Conclusion

Land is a primary asset at the centre of Central Africa's development challenge, as three-quarters of its people depend on it directly for their subsistence. Land degradation thus increases poverty and vulnerability, impedes agricultural growth, and contributes to social tensions as well as threatening biodiversity. This dominance of land as a source of wealth for economies in the Central African sub-region makes land policy especially politically sensitive and effective policies need to be based on an understanding of the political economy context. For the sub-region, environmental change, climate and population pressure are important in explaining land use, particularly the area allocated for agriculture to promote production of higher-value enterprises. Land policies, therefore, need to effectively accommodate not only customary land tenure into national legislation while improving the ability of disadvantaged groups such as women to own land within formal and customary systems where they often lack rights, but also adaptation to climate change. It is acknowledged that for Central Africa to achieve its development goals, climate change adaptation is a priority. An effective land policy and land administration will provide support to the implementation of adaptation measures that enhance agriculture and peoples' resilience for increased food security. This entails clarifying not only existing tenure of multiple users of the land, but also helping stakeholders negotiate new rules of resource access and use in the face of climate-induced disturbance to the status quo.

References

Adesina, A. A., & Baidu-Forson, J. (1995). Farmers' perceptions and adoption of new agricultural technology: Evidence from analysis in Burkina Faso and Guinea, West Africa. *Agricultural Economics, 13*, 1-9.

Akkermans, T., Wim, T., & Van Lipzig, N. P. M. (2014). The Regional Climate Impact of a Realistic Future Deforestation Scenario in the Congo Basin. *Journal of Climate, 27*, 2714-2734. http://dx.doi.org/10.1175/JCLI-D-13-00361.1

Aklilu, A., & de Graaff, J. (2007). Determinants of adoption and continued use of stone terraces for soil and water conservation in an Ethiopian highland watershed. *Ecological Economics, 61*, 294-302.

Andersson, J. A. (2007). How much did property rights matter? Understanding food insecurity in Zimbabwe: a critique of Richardson. *African Affairs, 106*(425), 681-690.

Atwood, D. A. (1990). Land Registration in Africa: The Impact in Agricultural Production. *World Development, 18*(5): 659-71.

Barrows, R., & Roth, M. (1990). Land Tenure and Investment in African Agriculture: Theory and Evidence. *Journal of Modern African Studies, 28*(2), 265-297.

Basu. (1996). The Market for Land: An Analysis of Interim Transactions. *Journal of Development Economics, 20*(1), 163-177.

Bush, R. (2010). Food Riots: Poverty, Power and Protest. *Journal of Agrarian Change, 10*(1), 119-29.

Byres, T. J. (Ed.). (2004). Redistributive Land Reform Today. *Journal of Agrarian Change, 4*(1-2).

DFID. (1999). Land Rights and Sustainable Development in Sub-Sahara Africa: Lessons and Ways Forward in Land Tenure Policy. In *African Nations,* 16-19 February 1999, UK. Department for International Development, London.

Fischer, G., van Velthuizen, H., Nachtergaele, F., & Medow, S. (2000). *Global Agro-Ecological Zones (Global AEZ) 2000*. Food and Agriculture Organization of the United Nations (FAO) and the International Institute for Applied Systems Analysis (IIASA), Rome, Italy / Laxenburg, Austria. Retrieved from http://www.iiasa.ac.at/Research/LUC/GAEZ/index.html

Gilland, B. (2002). World Population and Food Supply: Can Food Production Keep Pace with Population Growth in the Next Half-century? *Food Policy, 27*(1), 47-63.

Griffin, K., Khan, A., & Ickowitz, A. (2002). Poverty and the Distribution of Land. *Journal of Agrarian Change, 2*(3), 279-330.

Holden (2006). Policies for Poverty Reduction, Sustainable Land Management, and Food Security: A Bioeconomic Model with Market Imperfections. In J. Pender, F. Place, & S. Ehui (Eds.), *Strategies for Sustainable Land Management in the East African Highlands* (Washington, DC: International Food Policy Research Institute, pp. 333-356).

Howden, S. M., Soussana, J. F., Tubiello, F. N., Chhetri, N., Dunlop, M., & Meinke, H. (2007). Adapting agriculture to climate change. *Proceedings of the National Academy of Sciences, 104*(50), 19691-19696.

Lapar, M., Lucila, A., & Pandey, S. (1999). Adoption of soil conservation: The case of the Philippine uplands. *Agricultural Economics, 21*, 241-256.

Mbaga-Semgalawe, Z. M., & Folmer, H. (2000). Household adoption behavior of improved soil conservation: The case of the North Pare and West Usambara Mountains of Tanzania. *Land Use Policy, 17*, 321-336.

McGuffie, K., Henderson-Sellers, A., Zhang, H., Durbidge, T., & Pitman, A. (1995). Global climate sensitivity to tropical deforestation. *Global Planet. Change, 10*, 97-128. http://dx.doi.org/10.1016/0921-8181(94)00022-6

Migot-Adholla, S. E., Hazel, P., Beroit, B., & Place, F. (1991). Indigenous Land Rights Systems in sub-Saharan Africa: A Constraint on Productivity. *World Bank Economic Review, S*(1), 155-173.

Molua, E. L. (2011). Farm income, gender differentials and climate risk in Cameroon: typology of male and female adaptation options across agroecologies. *Sustainability Science, 6*(1), 21-31. http://dx.doi.org/10.1007/s11625-010-0123-z

Müller, C., Cramer, W., Hare, W. L., & Lotze-Campen, H. (2011). Climate change risks for African agriculture. *Proceedings of National Academy of Sciences of the United States, 108*(11), 4313-4315.

Nogherotto, R., Coppola, E., Giorgi, F., & Mariotti, L. (2013). Impact of Congo Basin deforestation on the African monsoon. *Atmospheric Science Letters, 14*(1), 45-51. http://dx.doi.org/10.1002/asl2.416.

Pagiola, S. (1996). Price policy and returns to soil conservation in semi-arid Kenya. *Environmental and Resource Economics, 8*(3), 225-271.

Pasquini, M. W., & Alexander, M. J. (2005). Soil fertility management strategies on the Jos Plateau: the need for integrating 'empirical'and 'scientific'knowledge in agricultural development. *The Geographical Journal, 171*(2), 112-124.

Patt, A. G., Tadross, M., Nussbaumer, P., Asante, K., Metzger, M., Rafael, J., ... & Brundrit, G. (2010). Estimating least-developed countries' vulnerability to climate-related extreme events over the next 50 years. *Proceedings of the National Academy of Sciences, 107*(4), 1333-1337.

Pender, J. L. (1998). Population growth, agricultural intensification, induced innovation and natural resource sustainability: An application of neoclassical growth theory. *Agricultural Economics, 19*(1), 99-112.

Pender, J., Ehui, S., & Place, F. (2006). Conceptual framework and hypotheses. In J. Pender, F. Place, & S. Ehui (Eds.), *Strategies for sustainable land management in the East African highlands* (pp. 31-58).

Pender, J., S. Ehui, and F. Place, 2006. 'Conceptual Framework and Hypotheses.' In: Pender, J., F. Place and S. Ehui (eds.), *Strategies for Sustainable Land Management in the East African Highlands* (Washington, DC: International Food Policy Research Institute, pp. 31-58).

Polcher, J., & Laval, K. (1994). The impact of African and Amazonian deforestation on tropical climate. *Journal of Hydrology, 155*(3), 389-405. http://dx.doi.org/10.1016/0022-1694(94)90179-1

Rosegrant, M. W., Paisner, M. S., Meijer, S., & Witcover, J. (2001). *2020 Global food outlook: Trends, alternatives, and choices* (Vol. 11). Intl Food Policy Res Inst.

Schlenker, W., & Lobell, D. B (2010). Robust negative impacts of climate change on African agriculture. *Environmental Research Letters, 5*, 1-8. http://dx.doi.org/10.1088/1748e9326

Schmidhuber, J., & Tubiello, F. N. (2007). Global food security under climate change. *Proceeding of National Academy of Sciences of the United States, 104*, 19703-08.

Scoones, I., Reij, C., & Toulmin, C. (1996). Sustaining the Soil: Indigenous Soil and Water Conservation in Africa. London: *IIED Issue Paper* 57.

Semazzi, F., & Song, Y. (2001). A GCM study of climate change induced by deforestation in Africa. *Climate Res., 17*, 169-182. http://dx.doi.org/10.3354/cr017169

Shiva, V. (2008). *Soil Not Oil: Climate Change, Peak Oil and Food Insecurity*. London: Zed Books.

The World Bank. (2006). *Sustainable Land Management*. Challenges, Opportunities, and Trade-offs. Washington, DC.

5

Slow Pyrolysis as a Method for the Destruction of Japanese Wireweed, *Sargassum muticum*

John J Milledge[1], Alan Staple[1] & Patricia J Harvey[1]

[1] Faculty of Engineering & Science, University of Greenwich, UK

Correspondence: John J Milledge, Algae Biotechnology Research Group, School of Science, University of Greenwich at Medway, Central Avenue, Chatham Maritime, Kent, ME4 4TB, UK.
E-mail: j.j.milledge@gre.ac.uk

Abstract

Japanese wireweed, *Sargassum muticum* is an invasive species to Great Britain, which might be controlled by harvesting it for energy and chemicals. Pyrolysis is the thermal decomposition of the organic components of dry biomass by heating in the absence of air. The distribution of matter between solid, liquid and syngas depends on the biomass and the pyrolysis temperature and time. Slow pyrolysis with lower temperatures ($\sim 400^0$ C) tends to produce more solid char. Pyrolysis char can be an effective soil ameliorant, a sequestration agent due to its stability or burned as a fuel.

The research attempts to answer the question: Could slow pyrolysis be an energy efficient means for the destruction of Japanese wireweed and produce a potential product, biochar? A simple test rig was developed to establish the yield of biochar, biocrude and syngas from the slow pyrolysis of *Sargassum muticum*. An energy balance was calculated using compositional data from the analysis of the seaweed feedstock, higher heating values (HHV) from bomb-calorimetry and literature values.

The energy required to heat 1 kg of dry seaweed by 400^0 C for slow pyrolysis was estimated at 0.5 MJ. The HHV of syngas and biocrude produced from the pyrolysis totalled 2.9 MJ. There is, therefore, sufficient energy in the biocrude and syngas fractions produced by the pyrolysis of seaweed to power the process and produce useful biochar, but insufficient energy for drying.

Keywords: *Sargassum muticum*, biochar, energy balance, pyrolysis, seaweed

1. Introduction

The environmental and economic impacts of biological invasions of non-native species were estimated to be in early part of the last decade ~ US$ 1.4 trillion per year, equivalent 5 % of the world economy (Engelen & Santos, 2009). The estimated cost of non-native species in the Great Britain for 2010 was £ 1.7 billion per year with the cost of invasive marine species to shipping and aquaculture estimated to be in excess of £ 40 million per year (Cook et al., 2013).

Sargassum muticum, Japanese wireweed, is native to the northwest Pacific region (Edwards et al., 2014). It first was 'introduced' outside of it natural range to British Columbia and has become the dominant species at the low-tide level in many areas on the west coast of North America (Fletcher & Fletcher, 1975). It first appeared in Europe in the early 1970s, and is now found on shorelines from Norway to Portugal (Engelen & Santos, 2009). Since its first recorded find in the UK, on the coast of the Isle of Wight, it has spread along the south-coast and around the British Isles (Davison, 2009; Gibson, 2011). It has been described as very invasive and perhaps the most 'successful' invasive species in the UK in terms of its rate of spread (Davison, 2009). Under the EU's Water Framework Directive the UK has identified it as a species of high priority (Davison, 2009).

It is causing considerable problems in certain areas of the Kent coast, especially on chalk ledges, and is spreading, possibly displacing native algae (Kent Wildlife Trust, 2006; Medway Swale Estuary Partnership, ND; The River Stour (Kent) Internal Drainage Board, 2012). The destruction of this seaweed is an issue in southern England and could have considerable financial and energy costs (CABI, 2011; Williams et al., 2010). The eradication of *Sargassum muticum* has been attempted and, although not successful, (Lodeiro, Cordero, Grille, Herrero, & de Vicente, 2004) results in the need to dispose of large quantities of seaweed biomass (Davison, 2009). Although

Sargassum muticum has been exploited for aquaculture in China (Liu, Pang, Gao, & Shan, 2013) and as a traditional food in Korea (E. J. Yang, Ham, Lee, Lee, & Hyun, 2013), there is currently no commercial exploitation of this biomass in Europe (Lodeiro et al., 2004). The valorisation of *Sargassum muticum* biomass for fuel and other products could encourage its harvesting and control.

One potential product could be biochar, defined as a solid material obtained from the carbonisation of biomass used to improve soil properties (Meyer, Glaser, & Quicker, 2011). The difference between charcoal and biochar is primarily in the end use with charcoal being used as a fuel and biochar as a nonfuel (Tenenbaum, 2009). Biochar is produced by pyrolysis, the thermal decomposition of the organic component of dry biomass by heating in the absence of air (McKendry, 2002; Saidur, Abdelaziz, Demirbas, Hossain, & Mekhilef, 2011). Pyrolysis can produce high volumes of fuel relative to the biomass feed and the process can be modified to favour the production of bio-oil, syngas or solid char (Miao, Wu, & Yang, 2004). The distribution between solid, liquid and syngas depends on the biomass and the pyrolysis temperature and time. Lower temperatures (around 400 °C) tend to produce more solid char (slow pyrolysis). Higher temperatures produce a higher proportion of liquid (biocrude) and gas. Pyrolysis processes can be classified by temperature and processing time. While there are no formal definitions, slow pyrolysis is characterised by long residence times (from minutes to days for solids) at low reactor temperatures (< 400 °C) with very low rates of heating (0.01 - 2 °C s^{-1}) (Milledge & Heaven, 2014; Peacocke & Joseph, ND), with slow pyrolysis resulting in higher yields of char rather than the liquid or gaseous products from higher temperature process (Brennan & Owende, 2010; Ghasemi et al., 2012). Charcoal was traditional produced in earth kilns with pyrolysis, gasification, and combustion processes occurring in kiln. However, biochar is now produced by slow pyrolysis in metal retorts, such as the Exeter retort (Rawle, 2014), where pyrolysis and combustion processes are physically separated (Meyer et al., 2011; Peterson & Jackson, 2014), that could co-produce solid char, liquid bio-oil and pyrolysis gas.

Char from pyrolysis may be utilised directly as a sequestration agent due to its stability, burned as an industrial fuel, used in a variety of applications such as carbon nanotubes and agrochemicals or further upgraded to a hydrogen-rich fuel (Rowbotham, Dyer, Greenwell, & Theodorou, 2012). During the last decade, biochar has received a lot of attention, both from scientists and policy makers as a soil enhancer to potentially increase agricultural yields and simultaneously sequester carbon to help mitigate climate change (Shackley & Sohi, 2010; van der Kolk & Zwart, 2013). In bioenergy–biochar systems (BEBCS) the evolved volatile and gaseous compounds from the pyrolysis of biomass, are utilised for biofuel or bioenergy production while the carbon-rich solid biochar is used as a soil improver (Woolf, Lehmann, Fisher, & Angenent, 2014). Although, increasing biochar production entails a reduction in bioenergy obtainable per unit of biomass feedstock, a model of BEBCS for dry terrestrial biomass (< 15 % moisture) suggests that it could produce net energy and biochar (Woolf et al., 2014).

Research on the pyrolysis of microalgae is 'quite extensive' and has achieved reliable and promising outcomes (Marcilla, Catalá, García-Quesada, Valdés, & Hernández, 2013), but there appears to be less work on seaweed and much of the work carried out has used fast rather than slow pyrolysis for the production of oil rather than char (Milledge & Heaven, 2014; Milledge, Smith, Dyer, & Harvey, 2014; Yanik, Stahl, Troeger, & Sinag, 2013; Zhou, Zhang, Zhang, Fu, & Chen, 2010). The char from the pyrolysis of algae, however, has been found to be an effective soil ameliorant and fertiliser and could be an additional revenue stream (Bird & Benson, 1987), but the thermal behaviour of seaweeds is complex with a myriad of diverse reactions and thermolysis pathways (Rowbotham, Dyer, Greenwell, Selby, & Theodorou, 2013). It is suggested considerable further study is required to discover if and how this could be exploited to commercialise seaweed for fuel and chemicals (Ross, Jones, Kubacki, & Bridgeman, 2008). Could the destruction of the invasive species to the UK, *Sargassum muticum,* by slow pyrolysis be energy efficient and produce a valuable product biochar?

2. Method

2.1 Species and Sample Collection

Sargassum muticum (Japanese wireweed) (Figure 1), was collected as part of a project to remove it from the Kent coast during March 2014. It was harvested with the holdfast to prevent regrowth and had natural contaminants such as mud, sand, chalk, small animals, other seaweeds etc. Excess chalk attached to *Sargassum muticum* was removed from the holdfast dried and weighed.

Figure 1. Dried *Sargassum muticum* showing natural contamination

Samples after collection were immediately placed in sealed bags and frozen and stored at ≤ 20 °C on return to the laboratory within 24 hours. Moisture and ash determinations were carried out. Seaweed fronds showing minimum contamination were removed from oven dried samples (105 °C for 24 hours) and determination carried out to establish calorific value; and C, H, S, N and ash content

2.2 Dry Weight Determination

The moisture content was established using the British Standards simplified oven drying method for the determination of moisture content in solid biofuels (BSI, 2009b). All measurements were repeated in triplicate and a mean value is reported. Samples after drying were stored in sealed container at 4 °C for further experimentation.

2.3 Ash Determination

The ash content of dried seaweed sample was measured using the British Standards method for determination of ash content in solid biofuels (BSI, 2009a). Again all measurements were carried out in triplicate and a mean value is reported. Ash was also examined by X-ray diffraction (XRD) analysis after grinding in a pestle and mortar to a fine powder <10μm

2.4 Calorific or High Heating Value Determination

Calorific values (CV) or Higher Heating Values HHV were measured using a Parr Model 1341 Bomb Calorimeter using the UKAS method for Determination of calorific value (BSI, 2010). The samples were oxidised by combustion in an adiabatic bomb containing oxygen under pressure and the HHV determined by measuring the temperature rise of a known mass of water. The dissolved sulphate and nitrate were calculated from titration to adjust for their contribution. A minimum of two determinations were carried out for each sample and a mean is reported.

2.5 Elemental Analysis

The carbon, hydrogen, nitrogen and sulphur content of the dried seaweed biomass were measured using Flash Dynamic Combustion (Flash EA1112 CHNS Elemental Analyser). The oxygen content was established by difference.

2.5 Slow Pyrolysis

Approximately 50 g of seaweed was subject to slow pyrolysis in a test-rig made from standard laboratory glassware, shown in Figure 2, which attempts to simulate the conditions in a commercial biochar retort, where biomass is heated externally in the absence of a flow of air. The temperature of the centre of seaweed biomass was taken at regular intervals during heating. The weight of initial dry seaweed, biochar and biocrude were measured.

Gas was collected in pre-weighed 'Tedlar' bags and weighed. Total gas weight yield was calculated using relative density of the sample gas and laboratory air, established by the Royal Society of Chemistry method in which a syringe of known volume is weighed 'empty' and full with gas or air (Nuffield Foundation, 2014).

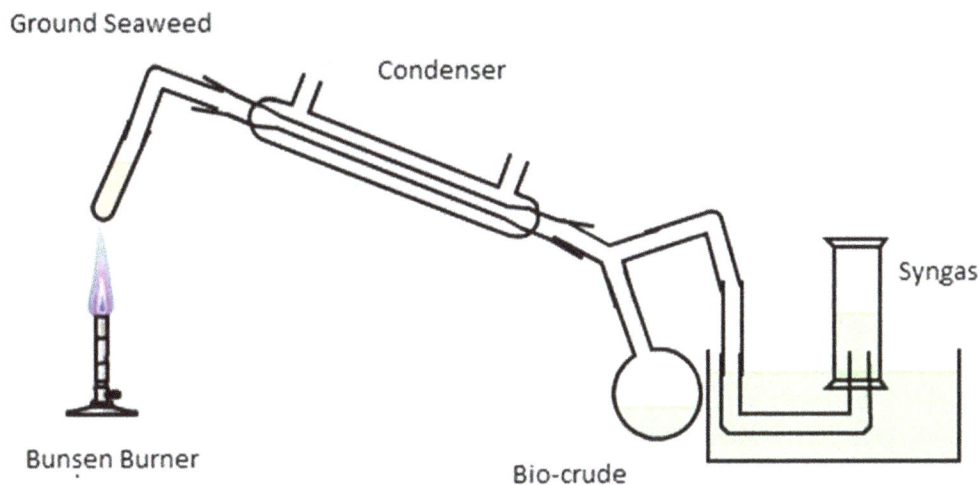

Figure 2. Slow Pyrolysis rig

3. Results and Discussion

3.1 Seaweed Composition

The moisture content of seaweed was found to be 79.9 % (Standard Deviation (SD) 1.7 %). The average proximate analysis, ultimate analysis and higher heating value (HHV) of the dried *Sargassum muticum* is shown in Table 1.

Table 1. Compositional Analysis of *Sargassum muticum*

Ash	Carbon	Hydrogen	Oxygen	Nitrogen	Sulphur	Higher Heating Value
% dw	% dw	% dw	% dw	% dw	% dw	MJ kg^{-1} dw
29.45	30.66	3.95	29.56	4.89	1.49	16.4

The compositional results are within the range reported in the literature for species of seaweed being considered as potential biofuels shown in Table 2.

Table 2. Compositional data for species of seaweed being considered as potential biofuels

	Ash	Carbon	Hydrogen	Oxygen	Nitrogen	Sulphur	HHV
	% dw	% dw	% dw	% dw	% dw	% dw	MJ kg^{-1} dw
Fucus vesiculosus[1]	22.82	32.88	4.77	35.63	2.53	2.44	15.0
Chorda filum[1]	11.61	39.14	4.69	37.23	1.42	1.62	15.6
Laminaria digitata[1]	25.75	31.59	4.85	34.16	0.9	2.44	17.6
Fucus serratus[1]	23.36	33.5	4.78	34.44	2.39	1.31	16.7
Laminaria hyperborea[1]	17.97	34.97	5.31	35.09	1.12	2.06	16.5
Macrocyctis pyrifera[1]	38.35	27.3	4.08	34.8	2.03	1.89	16.0
Enteromorpha prolifera[2]	30.1	28.75	5.22	32.28	3.65	0	
Laminaria saccharina[3]	24.2	31.3	3.7	36.3	2.4	0.7	

[1] (Ross et al., 2008) [2] (Zhou et al., 2010) [3] (Anastasakis and Ross, 2011)

Sargassum muticum has a much higher ash content than typical potential terrestrial biomass crops (2-7 %) (Ross et al., 2008) or wood (0.5-2 %) (Misra, Ragland, & Baker, 1993; Saidur et al., 2011). The high ash content of seaweed can provide both minerals and trace elements that are beneficial in both fertiliser and animal feed (McHugh, 2003; Philippsen, 2013), but ash content can be a considerable problem in direct combustion of biomass due to fouling of the boilers restricting the use of high ash content biomass (Demirbas, 2001). The metals in the ash of seaweed, however, have been shown to produce a significant catalytic effect on pyrolysis, and ash content (especially the potassium) strongly influences the product yields and bio-oil properties (Ross et al., 2008; Yanik et al., 2013). High potassium content caused high char yields and reduced bio-oil yields (Trinh et al., 2013; Yanik et al., 2013), but copper ions have been shown to promote the onset of pyrolysis in the alginate polymer and *Laminaria digitata* (Rowbotham et al., 2013). The result from XRD analysis of ash from *Sargassum muticum* is shown in Figure 3. Although the high temperature used to ash the seaweed (550^0 C) may have converted some of the mineralogy to anhydrous crystal structures the data shows that the ash contains a large proportion of potassium compounds which may influence pyrolysis yields.

Sargassum muticum contains significantly higher levels of both sulphur and nitrogen typical found in terrestrial plants used for biofuels (<0.2 % sulphur and 0.5-1% nitrogen) (Ross et al., 2008). High nitrogen and sulphur content in both macroalgae and microalgae has resulted in pyrolysis bio-oils containing nitrogen and sulphur compounds that require additional refining compared to those produced from terrestrial crops (Maddi, Viamajala, & Varanasi, 2011; Wang, Wang, Jiang, Han, & Ji, 2013)

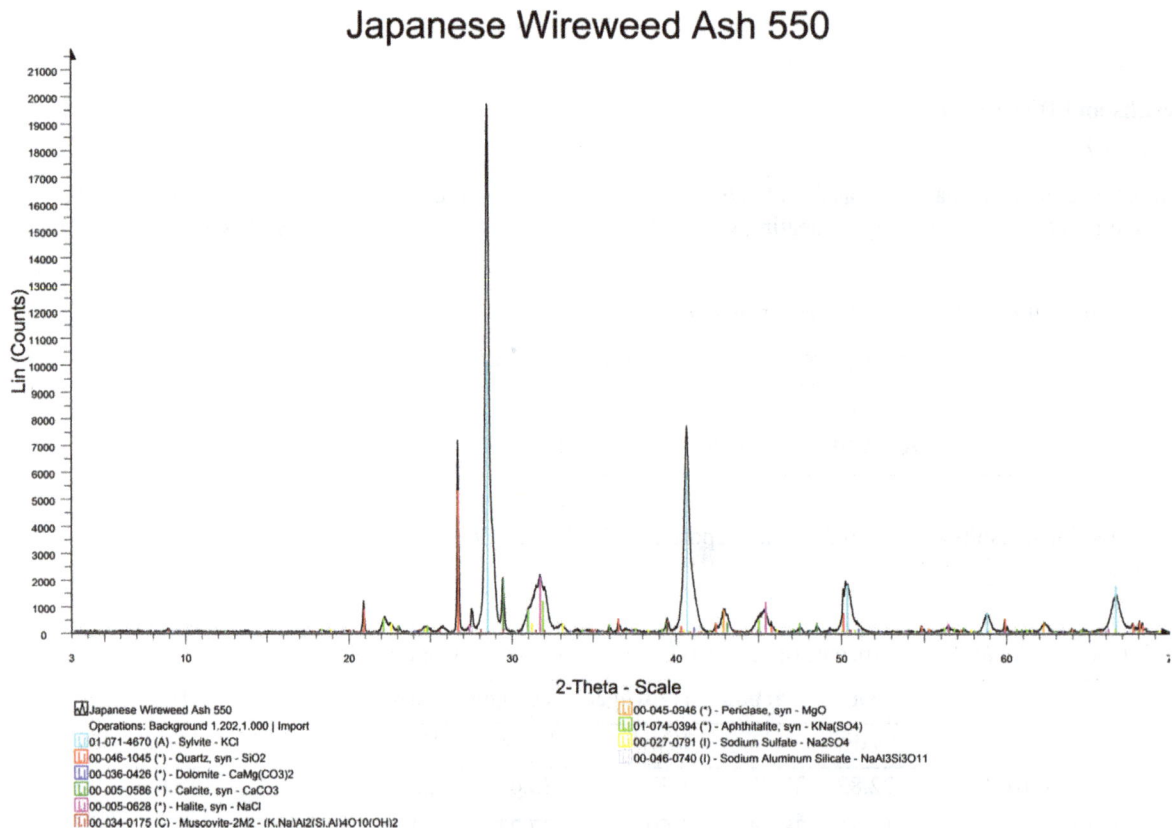

Figure 3. XRD analysis of ash produced from *Sargassum muticum*

3.2 Pyrolysis

The average temperature profile during pyrolysis of *Sargassum muticum* is shown in Figure 4.

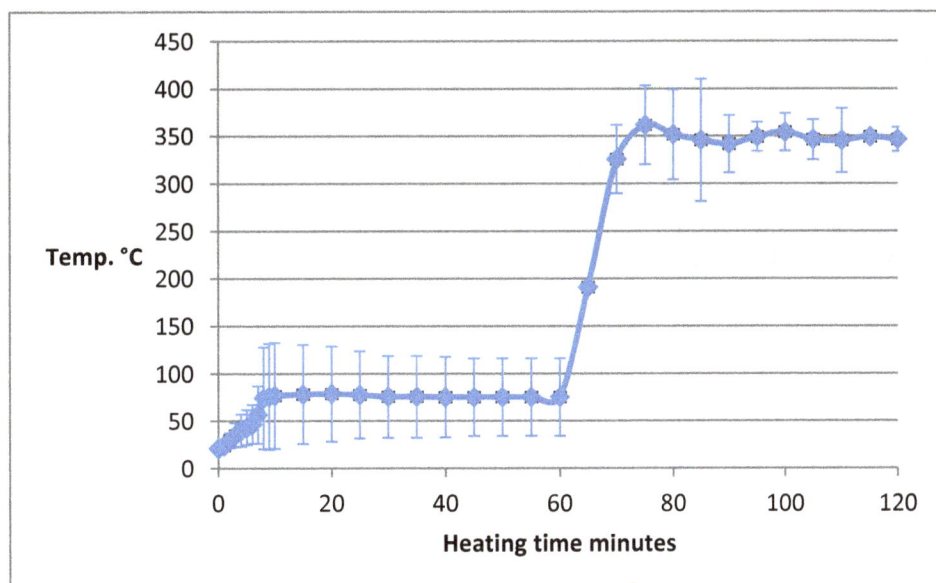

Figure 4. Average temperature profile during slow pyrolysis of oven-dried *Sargassum muticum*

The ash content of the dry seaweed used as feedstock for pyrolysis was higher 34.7 % (SD 1.18 %) than that from seaweed fronds analysed (Table 1) due to natural contaminants such as mud, sand and chalk (Figure 1). The percentage weight yield and HHV of the pyrolysis products are shown in Table 3

Table 3. Percentage yield and HHV of the pyrolysis products

	% of original DW	HHV (MJ kg^{-1} dw)
Biochar	67.6%	15.7
Syngas	16.5%	
Biocrude	11.3%	15.6
'Tar Hold-up'	4.6%	

Some viscous brown residues remained in the test rig, due to the use of standard laboratory glassware, including expansion and reduction connections, in construction of the test-rig. It has been termed 'tar hold-up' in Table 3. 79.9 % of the tar hold-up was removed by rinsing in hexane leaving dark brown tar like deposits that were believed to be long molecular weight organic compounds (high molecular weight hydrocarbon are known to be insoluble in Hexane (EPA, 1998)).

The ash content of the biochar was found to be 52.4 % (SD 1.8 %). Assuming that all the ash content remains within the biochar during pyrolysis this in close agreement with the calculated ash content of biochar of 51.5 %. HHV of the non-ash solids within the biochar, estimated from the measured HHV of the biochar and its ash content, is 33 MJ kg^{-1}, which is in close agreement with HHV for graphite of 33 MJ kg^{-1} (Plummer, 1930) and typical wood charcoal of 32 MJ kg^{-1} (Misginna & Rajabu, 2012) indicating that biochar is probably mainly comprised of carbon.

3.3 Energy Balance

The energy in 1 kg of *Sargassum muticum*, adjusting for the higher ash content of the samples pyrolysed compared to the sample used to establish HHV, was calculated at 15.2 MJ. Using the yield and HHV data in Table 3 the energy yield of biochar and biocrude from 1 kg of *Sargassum muticum* was calculated to be 10.6 MJ and 2.6 MJ. Using the yield data in Table 3 for syngas and 'tar-hold-up' and typical HHV data for syngas 2.9 MJ kg^{-1} (McKendry, 2002) and bio-oil derived from the pyrolysis of algae 30.1 MJ kg^{-1} (C. Yang, Jia, Chen, Liu, & Fang, 2011) the energy yield as syngas and 'tar-hold-up' from 1 kg of *Sargassum muticum* was estimated at 0.3 MJ and 1.4 MJ. The total energy yield of the 4 pyrolysis products, biochar, biocrude, syngas and 'tar hold-up was 14.9 MJ or 98 % or the energy in the original seaweed prior to pyrolysis. A recent EU-Interreg funded project found energy yields of up to 96 % for a range of 6 terrestrial biomass feedstocks and suggested that pyrolysis energy yields from biomass will be typically above 90 % and always below 100 % (Brownsort & Dickinson, 2010).

Pyrolysis of biomass comprises both exothermic and endothermic processes, with the net enthalpy change of the pyrolysis chemical reactions being positive or negative, depending both on feedstock characteristics and process conditions with factors encouraging char formation such as those in slow pyrolysis tending to produce an overall process that is exothermic (Woolf et al., 2014). The specific heat of seaweed is 0.8- 2.2 J $^0K^{-1}$ g^{-1} (Wang et al., 2014) with the specific heat of low moisture or dry seaweed being reported as 1.3 J $^0K^{-1}$ g^{-1} (Balingasa & Elepaño, 2009; Sarbatly, Wong, Bono, & Krishnaiah, 2010). Assuming no heat loss or gain as a result of reactions in pyrolysis (isoenthalpic reaction) and a specific heat of 1.3 J $^0K^{-1}$ g^{-1} the energy required to increase the temperature of 1 kg of dry seaweed by 400^0 C in slow pyrolysis was estimated at 0.5 MJ. The HHV of syngas and biocrude produced from the pyrolysis totalled 2.9 MJ. There is, therefore, sufficient energy in the biocrude and syngas fractions produced by the pyrolysis of seaweed to power the process and produce useful biochar.

Drying is required prior to pyrolysis. The removal of water from the algal biomass by evaporation often requires considerable energy. To heat and evaporate water at atmospheric pressure from a temperature of 20 °C, requires an energy input of approximately 2.6 MJ kg^{-1} or over 700 kWh m^{-3} (Mayhew & Rogers, 1972; Weast, 1985). To produce one kilogram of dry seaweed from seaweed containing 79.9 % moisture will require the evaporation of 3.97 kg water and 10.3 MJ of energy. Dryer typical efficiencies are 30- 80 % and vary with equipment, operating conditions and feed (Earle & Earle, 1983). Tunnel dryers have an efficiency ~50 % (Mujumdar, 2007). The energy within the seaweed (15.2 MJ) would therefore be insufficient to remove the water from the wet seaweed to produce dry biomass for pyrolysis. This finding is in agreement with the results of a recent model of pyrolysis which suggest that high water content biomass (>35 %) will not be energy positive for the production of biochar and bioenergy due to the high energy requirement for drying the feedstock (Woolf et al., 2014).

Solar drying does not require fossil fuel energy, but is weather dependent and can cause considerable denaturation of organic compounds. It is the least expensive drying option (Brennan & Owende, 2010), but large areas are required as only around 100 g of dry matter can be produced from each square metre of sun-dryer surface (Oswald, 1988). *Sargassum muticum* is unable to survive drying (Edwards et al., 2014) and therefore if sufficient areas of land adjacent to the areas from where it is to be removed could be found for solar drying there may be a low risk of the potential 're-infestation'.

4. Conclusion

There is sufficient energy in the biocrude and syngas fractions produced by the pyrolysis of seaweed to power the process and produce useful biochar. The high ash content of the biochar produced from the pyrolysis of seaweed, however, could restrict its use as a soil conditioning agent or as a charcoal fuel. There is insufficient energy within the seaweed for drying and solar drying or a similar low input-energy method prior to pyrolysis will be required

Acknowledgements

This work was supported by the EPSRC project number EP/K014900/1 (MacroBioCrude: Developing an Integrated Supply and Processing Pipeline for the Sustained Production of Ensiled Macroalgae-derived Hydrocarbon Fuels). Assistance of colleagues at the University of Greenwich: Dr Debbie Bartlett, Dr Ian Slipper, Mrs. Devyani Amin and Mr. Dudley Farman

References

Anastasakis, K., & Ross, A. B. (2011). Hydrothermal liquefaction of the brown macro-alga *Laminaria Saccharina*: Effect of reaction conditions on product distribution and composition. *Bioresource Technology, 102*(7), 4876-4883. http://dx.doi.org/10.1016/j.biortech.2011.01.031

Balingasa, C. R., & Elepaño, A. R. (2009). Studies on Engineering Properties of Red Seaweed (Kappaphycus spp.). *Philippines Journal of Agriculture and Biosystems Engineering, 7*, 59.

Bird, K. T., & Benson, P. H. (1987). *Seaweed cultivation for renewable resources.* Bergen, Norway: Elsevier.

Brennan, L., & Owende, P. (2010). Biofuels from microalgae--A review of technologies for production, processing, and extractions of biofuels and co-products. *Renewable and Sustainable Energy Reviews, 14*(2), 557-577. http://dx.doi.org/10.1016/j.rser.2009.10.009

Brownsort, P., & Dickinson, D. (2010). Mass and Energy Balances for Continuous Slow Pyrolysis of Six Feeds and Production of Biochar for Characterisation *Biochar: climate saving soils - Pyrolysis Studies Report*: UKBRC, University of Edinburgh and LTCB, Ghent University.

BSI. (2009a). Solid biofuels -determination of ash content *BS EN 14775:2009*.

BSI. (2009b). Solid biofuels. Determination of moisture content. Oven dry method. Total moisture. Simplified method *BS EN 14774-2:2009*.

BSI. (2010). Reaction to fire tests for products. Determination of the gross heat of combustion (calorific value) *BS EN ISO 1716:2010*.

CABI. (2011). *Sargassum muticum in   Invasive Species Compendium*. Retrieved from: http://www.cabi.org/isc/datasheet/108973

Cook, E. J., Jenkins, S., Maggs, C., Minchin, D., Mineur, F., Nall, C., & Sewell, J. (2013). Impacts of climate change on non-native species. *MCCIP Science Review*, 155-166. http://dx.doi.org/10.14465/2013.arc17.155-166

Davison, D. M. (2009). *Sargassum muticum* in Scotland 2008: a review of information, issues and implications. (Vol. Commissioned Report No.324 (ROAME No. R07AC707)): Scottish Natural Heritage.

Demirbas, A. (2001). Biomass resource facilities and biomass conversion processing for fuels and chemicals. *Energy Conversion and Management, 42*(11), 1357-1378. http://dx.doi.org/10.1016/S0196-8904(00)00137-0

Earle, R. L., & Earle, M. D. (1983). *Unit operations in food processing*: NZIFST (Inc.).

Edwards, M., Hanniffy, D., Heesch, S., Hernández-Kantun, J., Queguineur, B., Ratcliff, J., Soler-Vila, A., & Wan, A. (2014). *Microalgae fact-sheets*. Galway: NUI.

Engelen, A., & Santos, R. (2009). Which demographic traits determine population growth in the invasive brown seaweed *Sargassum muticum*? *Journal of Ecology, 97*(4), 675-684. http://dx.doi.org/10.1111/j.1365-2745.2009.01501.x

EPA. (1998). n-Hexane extractable material (hem) for sludge, sediment, and solid samples.

Fletcher, R. L., & Fletcher, S. M. (1975). Studies on the Recently Introduced Brown Alga *Sargassum muticum* (Yendo) Fensholt I. Ecology and Reproduction. *Botanica Marina, 18*(3), 149-156. http://dx.doi.org/10.1515/botm.1975.18.3.149

Ghasemi, Y., Rasoul-Amini, S., Naseri, A. T., Montazeri-Najafabady, N., Mobasher, M. A., & Dabbagh, F. (2012). Microalgae biofuel potentials (Review). *Applied Biochemistry and Microbiology, 48*(2), 126-144. http://dx.doi.org/10.1134/S0003683812020068

Gibson, C. E. (2011). Northern Ireland State of the Seas Report: Agri-Food and Biosciences Institute.

Kent Wildlife Trust. (2006). Have you seen you seen these species on the shores around Kent or Sussex? http://dx.doi.org/10.1007/s00343-013-2314-9

Liu, F., Pang, S. J., Gao, S. Q., & Shan, T. F. (2013). Intraspecific genetic analysis, gamete release performance, and growth of *Sargassum muticum* (Fucales, Phaeophyta) from China. *Chinese Journal of Oceanology and Limnology, 31*(6), 1268-1275. http://dx.doi.org/10.1007/s00343-013-2314-9

Lodeiro, P., Cordero, B., Grille, Z., Herrero, R., & de Vicente, M. E. S. (2004). Physicochemical studies of cadmium(II) biosorption by the invasive alga in europe, Sargassum muticum. *Biotechnology and Bioengineering, 88*(2), 237-247. http://dx.doi.org/10.1002/bit.20229

Maddi, B., Viamajala, S., & Varanasi, S. (2011). Comparative study of pyrolysis of algal biomass from natural lake blooms with lignocellulosic biomass. *Bioresource Technology, 102*(23), 11018-11026. http://dx.doi.org/10.1016/j.biortech.2011.09.055

Marcilla, A., Catalá, L., García-Quesada, J. C., Valdés, F. J., & Hernández, M. R. (2013). A review of thermochemical conversion of microalgae. *Renewable and Sustainable Energy Reviews, 27*(0), 11-19. http://dx.doi.org/10.1016/j.rser.2013.06.032

Mayhew, Y. R., & Rogers, G. F. C. (1972). *Thermodynamic and Transport Properties of Fluids*. Oxford: Blackwell.

McHugh, D. J. (2003). A guide to the seaweed industry *FAO Fisheries Technical Paper* FAO.

McKendry, P. (2002). Energy production from biomass (part 2): conversion technologies. *Bioresource Technology, 83*(1), 47-54. http://dx.doi.org/10.1016/S0960-8524(01)00119-5

Medway Swale Estuary Partnership. (ND). *In The Water*. Retrieved 22/04, 2014 from http://www.msep.org.uk/invasive-species-in-the-water.php

Meyer, S., Glaser, B., & Quicker, P. (2011). Technical, Economical, and Climate-Related Aspects of Biochar Production Technologies: A Literature Review. *Environmental Science & Technology, 45*(22), 9473-9483. http://dx.doi.org/10.1021/es201792c

Miao, X. L., Wu, Q. Y., & Yang, C. Y. (2004). Fast pyrolysis of microalgae to produce renewable fuels. *Journal of Analytical and Applied Pyrolysis, 71*(2), 855-863. http://dx.doi.org/10.1016/j.jaap.2003.11.004

Milledge, J. J., & Heaven, S. (2014). Methods of energy extraction from microalgal biomass: a review. *Reviews in Environmental Science and Bio/Technology, 13*(3), 301-320. http://dx.doi.org/10.1007/s11157-014-9339-1

Milledge, J. J., Smith, B., Dyer, P., & Harvey, P. (2014). Macroalgae-Derived Biofuel: A Review of Methods of Energy Extraction from Seaweed Biomass. *Energies, 7*(11), 7194-7222. http://dx.doi.org/10.3390/en7117194

Misginna, M. T., & Rajabu, H. M. (2012). *Yield and Chemical Characteristics of Charcoal Produced by TLUDND Gasifier Cookstove Using Eucalyptus Wood as Feedstock.* Paper presented at the Second International Conference on Advances in Engineering and Technology, Nagapattinam, India.

Misra, M. K., Ragland, K. W., & Baker, A. J. (1993). Wood ash composition as a function of furnace temperature. *Biomass & Bioenergy, 4*(2), 103-116. http://dx.doi.org/10.1016/0961-9534(93)90032-Y

Mujumdar, A. S. (2007). *Handbook of Industrial Drying,* (3 Ed.): CRC Press.

Nuffield Foundation. (2014). Determining relative molecular masses by weighing gases. Retrieved 2014, from http://www.nuffieldfoundation.org/practical-chemistry/determining-relative-molecular-masses-weighing-gases

Oswald, W. J. (1988). Large-scale algal culture systems (engineering aspects). In M. A. Borowitzka & L. J. Borowitzka (Eds.), *Micro-algal Biotechnology.* Cambridge: Cambridge University Press.

Peacocke, C., & Joseph, S. (ND). Notes on Terminology and Technology in Thermal Conversion. *IBI Information papers.* Retrieved April 15, 2014, from http://www.biochar-international.org/publications/IBI#Pyrolysis _guidelines.

Peterson, S. C., & Jackson, M. A. (2014). Simplifying pyrolysis: Using gasification to produce corn stover and wheat straw biochar for sorptive and horticultural media. *Industrial Crops and Products, 53*, 228-235. http://dx.doi.org/10.1016/j.indcrop.2013.12.028

Philippsen, A. (2013). *Energy Input, Carbon Intensity, and Cost for Ethanol Produced from Brown Seaweed.* (MASc), University of Victoria,, Victoria, BC, Canada.

Plummer, W. B. (1930). Heat of Combustion of Carbon1. *Industrial & Engineering Chemistry, 22*(6), 630-632. http://dx.doi.org/10.1021/ie50246a021

Rawle, R. (2014). *Economics of Biochar.* Paper presented at the Biochar Canterbury Christ Church, University.

Ross, A. B., Jones, J. M., Kubacki, M. L., & Bridgeman, T. (2008). Classification of macroalgae as fuel and its thermochemical behaviour. *Bioresource Technology, 99*(14), 6494-6504. http://dx.doi.org/10.1016/j.biortech. 2007.11.036

Rowbotham, J. S., Dyer, P. W., Greenwell, H. C., Selby, D., & Theodorou, M. K. (2013). Copper(II)-mediated thermolysis of alginates: a model kinetic study on the influence of metal ions in the thermochemical processing of macroalgae. *Interface Focus, 3*(1), 20120046. http://dx.doi.org/10.1098/rsfs.2012.0046

Rowbotham, J. S., Dyer, P. W., Greenwell, H. C., & Theodorou, M. K. (2012). Thermochemical processing of macroalgae: a late bloomer in the development of third-generation biofuels? *Biofuels, 3*(4), 441-461. http://dx.doi.org/10.4155/bfs.12.29

Saidur, R., Abdelaziz, E. A., Demirbas, A., Hossain, M. S., & Mekhilef, S. (2011). A review on biomass as a fuel for boilers. *Renewable & Sustainable Energy Reviews, 15*(5), 2262-2289. http://dx.doi.org/10.1016/j.rser. 2011.02.015

Sarbatly, R., Wong, T., Bono, A., & Krishnaiah, D. (2010). Kinetic and Thermodynamic Characteristics of Seaweed Dried in the Convective Air Drier. *International Journal of Food Engineering, 6*(5). http://dx.doi.org/10.2202/1556-3758.1600

Shackley, S., & Sohi, S. (2010). An assessment of the benefits and issues associated with the application of biochar to soil: A report commissioned by the United Kingdom Department for Environment, Food and Rural Affairs, and Department of Energy and Climate Change.

Tenenbaum, D. J. (2009). Biochar: Carbon Mitigation from the Ground Up. *Environmental Health Perspectives, 117*(2), A70-A73. http://dx.doi.org/10.1289/ehp.117-a70

The River Stour (Kent) Internal Drainage Board. (2012). Minutes of Board Meeting.

Trinh, T. N., Jensen, P. A., Dam-Johansen, K., Knudsen, N. O., Sorensen, H. R., & Hvilsted, S. (2013). Comparison of Lignin, Macroalgae, Wood, and Straw Fast Pyrolysis. *Energy & Fuels, 27*(3), 1399-1409. http://dx.doi.org/10.1021/ef301927y

van der Kolk, J., & Zwart, K. (2013). Pyrolysis in the Countries of the North Sea Region.Potentially available quantities of biomass waste for biochar production: A publication of the Interreg IVB project Biochar: climate saving soils. http://www.biochar-interreg4b.eu/images/file/WP44%20-%20Pyrolysis%20in%20 the%20Countries%20of%20the%20North%20Sea%20Region.pdf

Wang, S., Jiang, X. M., Wang, Q., Ji, H. S., Wu, L. F., Wang, J. F., & Xu, S. N. (2014). Research of specific heat capacities of three large seaweed biomass. *Journal of Thermal Analysis and Calorimetry, 115*(3), 2071-2077. http://dx.doi.org/10.1007/s10973-013-3141-0

Wang, S., Wang, Q., Jiang, X. M., Han, X. X., & Ji, H. S. (2013). Compositional analysis of bio-oil derived from pyrolysis of seaweed. *Energy Conversion and Management, 68*, 273-280. http://dx.doi.org/10.1016/j. enconman.2013.01.014

Weast, R. C. (Ed.). (1985). *Handbook of Chemistry and Physics.* Boca Raton Fl: CRC.

Williams, F. E., Eschen, R., Harris, A., Djeddour, D. H., Pratt, C. F., Shaw, R. S., ... Murphy, S. T. (2010). The Economic Cost of Invasive Non-Native Species on Great Britain: CABI.

Woolf, D., Lehmann, J., Fisher, E. M., & Angenent, L. T. (2014). Biofuels from Pyrolysis in Perspective: Trade-offs between Energy Yields and Soil-Carbon Additions. *Environmental Science & Technology, 48*(11), 6492-6499. http://dx.doi.org/10.1021/es500474q

Yang, C., Jia, L. S., Chen, C. P., Liu, G. F., & Fang, W. P. (2011). Bio-oil from hydro-liquefaction of *Dunaliella salina* over Ni/REHY catalyst. *Bioresource Technology, 102*(6), 4580-4584. http://dx.doi.org/10.1016/ j.biortech.2010.12.111

Yang, E. J., Ham, Y. M., Lee, W. J., Lee, N. H., & Hyun, C. G. (2013). Anti-inflammatory effects of apo-9 '-fucoxanthinone from the brown alga, *Sargassum muticum. Daru-Journal of Pharmaceutical Sciences, 21.* http://dx.doi.org/10.1186/2008-2231-21-62http://dx.doi.org/10.1186/2008-2231-21-62

Yanik, J., Stahl, R., Troeger, N., & Sinag, A. (2013). Pyrolysis of algal biomass. *Journal of Analytical and Applied Pyrolysis, 103*, 134-141. http://dx.doi.org/10.1016/j.jaap.2012.08.016

Zhou, D., Zhang, L., Zhang, S., Fu, H., & Chen, J. (2010). Hydrothermal Liquefaction of Macroalgae *Enteromorpha prolifera* to Bio-oil. *Energy & Fuels, 24*(7), 4054-4061. http://dx.doi.org/10.1021/ef100151h

6

The Effects of Polluted River Water to the Riverside Groundwater, Case in Niger River in Koulikoro

Drissa Traore[1] & Qian Hui[1,2]

[1] Chang An University, Xi'an, China

[2] Department Environmental Sciences and Engineering, Chang An University, Xi'an, China

Correspondence: Drissa Traore, Chang An University, Xi'an, China. E-mail: ibabosst@gmail.com

Abstract

Ground water demand is increasing in many African nations due to a number of factors. The growth of population, climate change, increase pollution of rivers, and insufficient number of purifying stations and waste water treatment (or almost nonexistent) have pushed to the water authorities for exploitation of underground water. These underground /groundwater have a relationship with surface water. Then what can be the effects of polluted River to its riverside groundwater? To explore the answer of this question and for the prevention sustainable and a better integrated management of water resources, we will do in-depth study on "the relationship between river water and riverside ground".

In Koulikoro region the results of this research show that Surface waters have poor bacteriological quality, the amount of total coliforms is very high, and accordingly Niger River's waters are not allowed for consumption without treatment. However the river water can be safely used for laundry, bath, sports and recreation. Generally the Groundwater quality is good despite increased salinity has been observed sporadically. We found also that for the entire region of Koulikoro the average infiltration rate is less than 19.8% of the gross rainfall.

Keywords: river water, groundwater, aquifer, infiltration, Mali, Koulikoro, Niger River

1. Introduction

1.1 The Problem Statement

Water is a precious natural resource in the world and without it life cannot exist. According to the latest assessment report of new age international publishers, there is about 97.2% to global water resources in the world is salt mainly in oceans and 2.8% is only available as fresh water. Out of this 2.8% of fresh water, about 2.2% is available as surface water (glaciers, icecaps, lakes, and an others forms) and 0.6% is groundwater. Groundwater is a huge resource but limited.

Groundwater demand in many African nations is greatly increasing due to a number of factors: the growth of population, climate change, increased pollution of rivers, and insufficient number of purifying station and wastewater treatment. Inadequate sanitation has led waterborne diarrheal and other enteric illnesses to become the leading cause of death on the continent, accounting for over 40% of all child deaths in the Sub-Saharan region (McMurray, 2007). Groundwater is also one of our most important sources of water for irrigation.

1.2 Exploration of Importance Problem

At one time, groundwater purity and availability were taken for granted, and it was a hidden resource; but over the past few years, it is undeniably true that polluted water is becoming the biggest threat to human live. For better management of fresh water resources, it is necessary for water scientists to understanding the effects of polluted river water to the riverside groundwater. This will raise the question of understanding the interaction between river water (surface water) and groundwater. In other words, understanding what can be the consequences of polluted River to riverside groundwater. However, it's very difficult to observe and measure the interaction. Thus we have to understand three phenomena to know: Infiltration, percolation and interflow.

Infiltration is movement of water into and through soil. The percolation is described as being the properties related to the connectivity of large numbers of objects which individually have some spatial extent, and for which their spatial relationships are relevant and statistically prescribed. The three basic assortments that percolation theory came are: bond, site, and continuum. In hydrology, interflow is the lateral movement of water in the unsaturated zone that returns to the surface or enters a stream prior to becoming groundwater.

1.3 Objectives

The major scope of this research will cover to the following.

- To characterize the aquifer system in Koulikoro region.

- To investigate and advice to riparian populations on groundwater related issues. And also regarding potential water problems.

- To create an analytical studies of aquifers in order to lay the foundations of sustainable and rational management of water resources in Koulikoro region of Mali.

1.4 Description of the Study Area

The region of Koulikoro is bordered by the Islamic Republic of Mauritania to the north, the region of Kayes to the west, by Guinea and to the region of Sikasso to the south, and by the region of Segou to the east. With an area of 90,210 km2, or 7.2% of the Mali's territory, Koulikoro Region is divided into seven (7) largest cities "cercles" (Table 1). The region is situated entirely within the tropical zone. In the region, there are four types of climates: the north Guinean climate, the south Sudanese climate, the North Sudanese climate and the South Sahelian climate, with alternating during the year a rainy season (May to October) and a dry season (November to April).The average annual rainfall is between 700 to 1200 mm in a normal year. The average temperatures are between 26 and 33 °C.

Koulikoro has a large river system which includes the Niger River with a long of 250 km; The Baoule, the Bagoe, the Baninfing, Sankarani and Bani with 120, 90, 70, 40 and 20 km long respectively. The Niger, Baoule and Sankarani are permanent streams. The 2012/2013 hydrological year has been marked by the maximum flow of about 199 occurred on 21st May 2013 and the minimum flow of 149 which occurred on 29th May 2013 with a year mean volume of 179. The total flow volume at Koulikoro, from 1st June 2012 to 30th April 2013 was about 35.68 Km^3 (http://nigerhycos.abn.ne January 13, 2014 07:05PM). The region also has ponds and lakes. Koulikoro water resources come mainly from the rains that are only meteoric waters which the region enjoys.

Table 1. Areas and populations

Cercles	Koulikoro	Banamba	Dioïla	Kangaba	Kati	Kolokani	Nara	Total
Area(Km^2)	7 260	7 500	12 000	5 150	16 300	12 000	30 000	90 210
Population	178 650	165 468	387 565	88 931	598 038	215 221	184 128	1 828 001

According to PIRT (**P**rojet **I**nventaire des **R**essources **T**errestres) the different types of soils in Koulikoro are characterized in their majority by their poverty, their acidity and their structural instability. Table 2 below gives the soils and their characteristics.

Koulikoro is a major trade and industrial town on the Niger River and provides opportunities for navigation.

Table 2. Soils and their characteristics

Soils characteristics	Lateritic and alluvial	Sandy lateritic and clayey	ferralitic and ferruginous	very sandy
Location	Kangaba, Kati, Kourouba, Baguinéba, Néguela, Sanankoraba, Siby, Kati	Koulikoro, Banamba, Toukoroba, kolokani	Dioïla	Nara, Banamba (Sébété, Boron)
Pluviometer (mm)	1200 to 900	950 to 750	1000 to 950	Around 450

2. Method

2.1 Description

To achieve our goals, we have developed a survey based on the documentation and on the collection of data. Some of our data were obtained from the Sigma 2 database. The Sigma 2 (Geographic Information System of Mali) database is an archive compiled by the Malian government containing information for over 20,000 wells drilled throughout the country since the 1970s, including their completion details, yield estimates, and water chemistry. We performed the analysis and synthesis of data by comparing the quality of groundwater with that of the world health organization and similarly for River water. In this context of our research, some groundwater data were taken from Bamako. To estimate the water quantity infiltrates, we have determined the average infiltration rate for the entire region of Koulikoro by using empirical equations established by the authors of the Hydrogeological Synthesis of Mali.

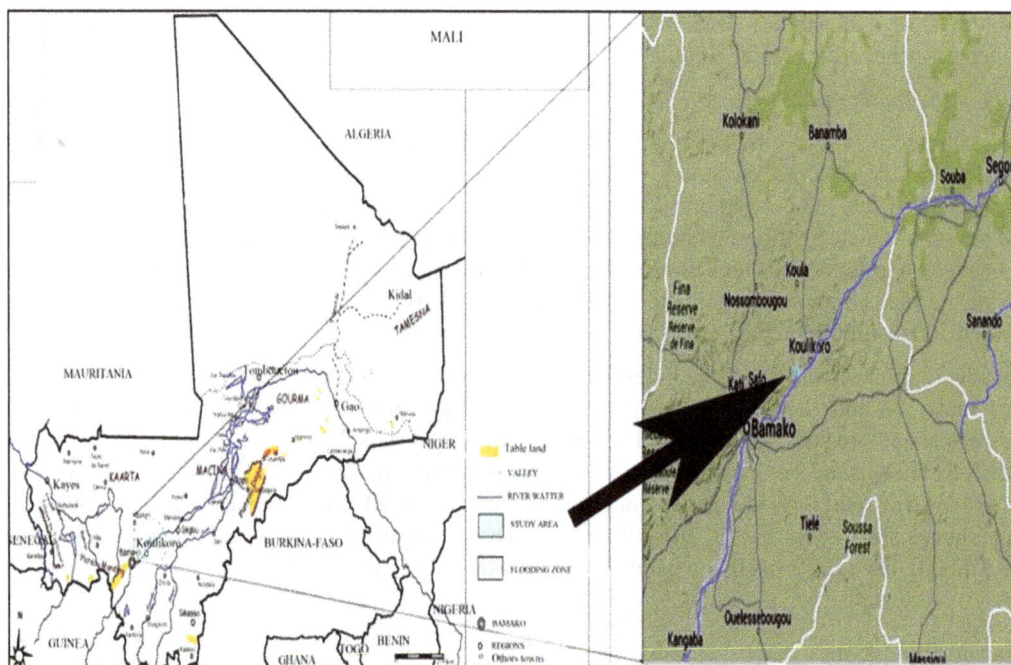

Figure 1. Study area

2.2 Data Base

2.2.1 Pollutions Sources

The pure water is clear colourless, no discernible taste or smell. A good knowledge of the physicochemical and bacteriological characteristics of the water can help us to determine the degree of pollution, in order to prevent contamination and public health. The water pollution is an alteration which makes water unsafe and disrupts the aquatic ecosystem. It can affect river water and groundwater. In the midst of research, we found that the worst environmental pollutants are: industrial pollution, agriculture pollution, outdoor garbage, interior living garbage and accidental pollution.

The River Niger is the fourth longest river in Africa (4 200 km).It takes its source in the Fouta Djalon mountains of Guinea and flows northeast through the Mali. The Niger River plays an important role in the life of vast population: its floodplains are used for cultivation cotton, rice and contributed to production of energy (Electricity power).Its important fish resources are benefited for the populations. We consider the Niger as a vital artery. Despite the function performed by the Niger River, it is more and more exposed to pollution.

Human and animal organic waste (domestic wastewater, dyeing, water from craft activities, and water from slaughterhouses) Figures 2 and 3.Industrial waste waters and effluents very often uncontrolled. Organic and chemical fertilizers and pesticides from agricultural areas. Pesticides include two class chemicals: insecticides and herbicides. There are also risks of an exceptional pollution: like the massive oil spill following incidents handling storage tanks or as a result of accident vehicles of transportation fuel.

2.2.2 River Water Quality Data

In the context of establishing a system of hydro-ecological monitoring of the Niger River, some measurements and samplings were carried out from 1995 to 1999.To these measurements are added to those made by the IRD (Institute of Research for the Development). These data allowed us to obtain the following Table 4.

Table 3. surface water quality (source UN-WATER/WWAP/2006/10/ Mali)

years	Average value/WHO average value (mg/l)								
	Dissolved Oxygen		NO_3^-		NH_4^+		PO_4^{3-}		Total coliforms (without adding system) colonies/100	
1980	12	-	0.01	3	0.0	0.5	0.00	-	-	0
1996	8	-	0.05	3	0.1	0.5	-	-	>100	0
1999	6.7	-	1.2	3	0.11	0.5	0.11	-	-	0

(-) No value Significant; WHO: World Health Organization.

2.2.3 Rainwater Quality Data

Hydrological year in Niger basin is always starting from June till May. The quality of rainwater reflects that of air. It can evolve significantly in space and time. The air pollution is caused by transportation, solid waste combustion, combustion in households (wood), and, soap factories (see Figure 4 below).

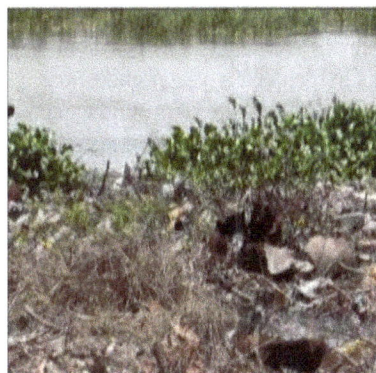

Figure 2. Pollution by garbages and plants

Figure 3. Environment pollution by dyeing

Figure 4. Local soap factory

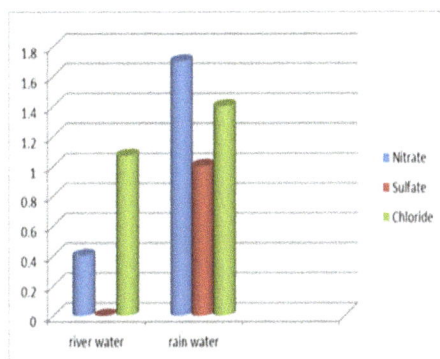

Figure 5. Average concentrations of Nitrate, Sulphate (sulfate) and Chloride of rainwater and of Niger River to Banankoro

Existing data on the quality of rainwater show that annual average mineralization of the waters of River Niger is superior to that of rain water. In contrast the average concentrations of chloride (Cl^-), nitrate (NO_3^-) and sulphate (SO_4^{-2}) in rainwater, are superior to those of river waters (Figure 5).

2.2.4 Groundwater Quality Data

An aquifer is a geologic formation with sufficient interconnected porosity and permeability to store and transmit significant quantities of water under natural hydraulic gradients[a]. The aquifers found in the Niger River Basin can generally be classified into fractured aquifers and inter-granular aquifers. In the fractured aquifers, the groundwater does not move through pores inside the rock body but through fractures and fissures. This is generally due to the absence of passable pores. In contrast to the fractured aquifers in the inter-granular aquifers, the groundwater flows mainly through passable pores.

Koulikoro region is characterized by the aquifer fractured semi-continuous type at the centre and south; the aquifer fractured discontinuous type at the northwest; and the aquifer continuous- generalize at the northeast. The quality of groundwater is related to lithology. In the region groundwater are few mineralized with dry residues less than 0.4mg/l and with conductivity less than 500 µS/cm. The necessary information on groundwater is given in Table 4.

Table 4. Groundwater quality

Element/substance	Nitrate(NO_3^-)	Fluoride(F)	Iron(Fe)	Uranium(U)	Iodine(I)	Manganese(Mn)
Groundwater	<0.1-7.3	<0.2-1.7	0.01-1.7	0.05-106	1-440	0.002-3.8
WHO's guideline	50 total nitrogen	1.5	-	1.410^3	-	0.5

All units in mg/l except for (U) and (I) in µg/l.

2.2.5 Groundwater Chemistry

Surface water and groundwater systems interact then groundwater chemistry and surface-water chemistry can not be dealt with separately. They interact throughout all landscapes from the mountains to the oceans. Water chemistry provides undoubtedly a lot of information in the knowledge of groundwater. Generally, the nature of groundwater depends on land which cross them. Its chemical composition is the result of the physical action of several agents.

2.2.6 The Water Supply

According to the Statistical Yearbook 2003, in Koulikoro region there are two types of aquifers: the alluvial aquifers (5 to 15 meters approximately in depth) and ground fracturing aquifers (20 to 40 meters in depth approximately). Aquifers are being exploited and generally of good quality. In Koulikoro region, only the cities of Koulikoro and Kati are supplied with water by the Energy Company of Mali (EDM: Energie Du Mali), and others towns and villages are supplied by boreholes and wells. It has 4,843 modern water points. The water needs of the population in the region of Koulikoro by city (2005-2008) are given by the Figure 6. The conquest of the solution to the problems of water has led in 2009 to the separation of both water and electricity, putting an end to the monopoly of the company energy of Mali (EDM - SA), which until now had in charge of the two sectors. It is as a result of this separation that the Government, by order N^o10-039/P-RM and 10-040/P-RM of August 5, 2010, decided the creation of Malian society of heritage of drinking water (Somapep: Société Malienne de Patrimoine de l'Eau Potable) and the Malian society's management of drinking water (Somagep: Société malienne pour la gestion de l'eau potable).

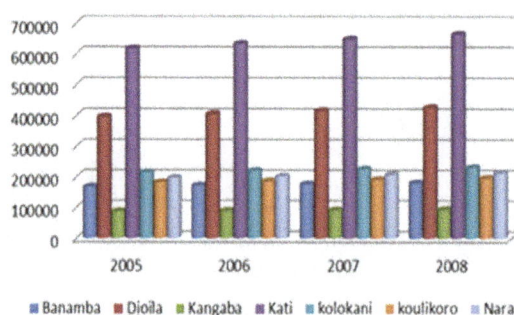

Figure 6. Water needs

2.2.7 Piezometer

Piezometric network is and will remain essential for establishing the water balance and determination of aquifer reserves. Piezometer is used for measuring the soil water pressure at the depth at which the slits lie. If water has a pressure greater than atmospheric pressure it will enter the piezometer. The piezometer is influenced by the rainfall, the runoff (surface water), the topography and the geology.

Koulikoro region has a total of 84 piezometers in boreholes. Only 17 piezometers are equipped and others are subject to manual measurements. The temporal evolution of the piezometric shows a general decline in levels, and some piezometers immediately respond to rainfall variations.

2.2.8 Aquifer Natural Recharge

Aquifer recharge is depending on many factors: climate, land use, vegetal covers …etc. Precipitation and river water percolate downward. In our study area, both (rainwater and river water) are essentially way to feed groundwater.

According to the authors of the Hydrogeological Synthesis of Mali (Robert, 1997), the relationship between infiltration (I) and average annual rainfall (P) is given by one of the following equations:

(a) rainfall regions between 300 mm and 700 mm, **I= 0.30P - 70**

(b) areas with greater than 700 mm annual rainfall, **I= 0.11P+ 63**

I and P are measured in millimetres (mm). Note that these equations were obtained after adjustment approach (approximation).

Thus, knowing the average annual rainfall, the infiltration rate of the study area is given by the Figure 7.

For estimating groundwater recharge by river, we used the Zero flux plane method. According to Amitha Kommadathit relies on the location of a plane of zero hydraulic gradients in the soil profile. Recharge over a time interval is obtained by summation of the changes in water contents below the plant. The position of the zero flux planeis usually determined by tensiometers.

The flow of water through the pores within a soil volume is generally considered as laminar, thus the velocity is

proportional to the hydraulic gradient. This is Darcy law given by the following formula: $q=VA=kiA=k\frac{\Delta h}{L}A$.

Where V: velocity of flow; K: coefficient of permeability of aquifer soil; A: cross-sectional area of the aquifer, and i hydraulic gradient.

3. Results and Discussion

We found that for the entire region of Koulikoro the average infiltration rate is less than 19.8% of the gross rainfall (Figure 7). This is approximately the same result like as the authors of the Hydrogeological Synthesis for all Mali (Average infiltration rate less than 20%). So this research allowed us to do some verification. On the other hand, it is well known groundwater and Rivers water feed each other. Interactions between groundwater and surface water basically proceed in two ways: groundwater flows through the stream bed into the stream (gaining stream), and stream water infiltrates through the sediments into the groundwater (losing stream).We got the goal which is to show the consequences of polluted Rivers on riverside groundwater. Indeed, it is very clear the polluted Rivers can easy reach its shallow groundwater.

Tableau 3 shows total coliforms >100 colonies per 100 ml. The coliform is very high. So River water is not allowed for consumption without treatment. The quality of surface water is deteriorating from year to year. Nitrogen in the form of nitrate (NO_3^-) indicates contamination with sewage. Excessive phosphorous concentrations in drinking water can cause health s' problems, while essential for growth.

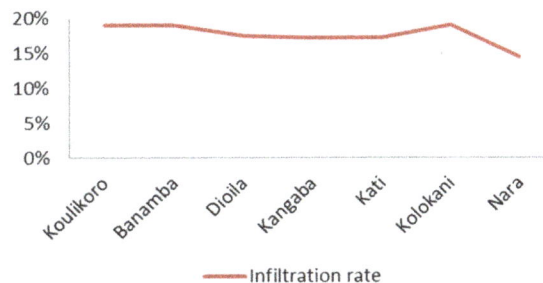

Figure 7. average infiltration rate

In most cases, we can say groundwater is cleaner than surface water. For Koulikoro aquifer, Information available suggests that groundwater is of good quality. However, increased salinity has been observed sporadically. Groundwater is heavily used for public supply. Around 55% of the population of the capital city of Bamako uses water from aquifer resources and more than 60% in Koulikoro region. We realized that: the information on the residence time of water in groundwater aquifer is necessary in the field of drinking water supply especially for risk analysis in cases of pollution and for the analysis of ecosystems. It is a commonplace that ecosystems depend on groundwater, vice versa. We know that the isotopic parameters of a stream vary very little. We have also rendered account that the knowledge of water balance is very important, which can be determined and managed by meteorological, hydraulic and hydrological studies. Those will take into account all the elements of the balance (rainfall, Evapotranspiration, runoff, infiltration, recharge and discharge).

In the closest flooding area to the River, the aquifer Continental terminal/Quaternary recharges by River would have estimated of order of 100,000 $m^3year^{-1}km^{-2}$(Anonym, 1990). An area of $510^{-3}km^2$ located a few km from koulikoroba (specified surface by the arrow Figure 1) the aquifer recharges by River would have estimated of order of 500 $m^3\,km^{-2}$.

Acknowledgments

The paper is supported by Chinese scholarship council and Malian government. The authors would like to thank very much anyone who made their contribution for realization of this paper in particularly Dr Li Peiyue and Mr. Qiao Liang.

References

Andersen, I., & Golitzen, K. G. (Eds.). (2005). *The Niger river basin: A vision for sustainable management.* World Bank Publications.

Cleary, R. W. (2003). *The groundwater pollution and hydrology course.*

Conway, D., Persechino, A., Ardoin-Bardin, S., Hamandawana, H., Dieulin, C., & Mahé, G. (2009). Rainfall and water resources variability in sub-Saharan Africa during the twentieth century. *Journal of Hydrometeorology, 10*(1), 41-59.

Diagnostic regional de Koulikoro situation de référence sur base d'analyse documentaire.

Gourcy L. (1994). *Fonctionnement Hydro Chimique De La Cuvette Du Fleuve Niger* (Mali).

International Water Management Institute. (2012). *Groundwater availability and use in subsaharan Africa a review of 15 countries.* Sri Lanka.

IWACO, Delft hydraulics. (1996). *le Niger supérieur.*

Kalbus, E., Reinstorf, F., & Schirmer, M. (2006). Measuring methods for groundwater–surface water interactions: a review. *Hydrology and Earth System Sciences, 10*(6), 873-887.

Karamouz, M., Szidarovszky, F., & Zahraie, B. (2003). *Water resources systems analysis.* Boca Raton, FL: Lewis Publishers.

Martin Jager and Sven Menge. (2012). *The Niger River Basin, An Assessment of Groundwater Need.*

Matthew, M. (2012). *Hydrogeology lecture notes* (2.3 ed).

Ministre De La Sante. (2010). *Analyse de la situation et estimation des besoins (ASEB) en Sante et Environnent au Mali dans le cadre de la mise en œuvre de la déclaration de Libreville.*

Programme des Nations Unies pour le Développement. (1990). *Synthèse hydrogéologique du mali, Direction de l'Hydraulique et de l'Energie.* Bamako, Mali.

Raghunath. H. M. (2006). Hydrology (revised 2nd ed.). Hydrology new age international publishers.

UN-WATER/WWAP. (2006). *Mali: La mise en valeur des ressources en eau.*

Valenza, A., Grillot, J. C., & Dazy, J. (2000). Influence of groundwater on the degradation of irrigated soils in a semi-arid region, the inner delta of the Niger River, Mali. *Hydrogeology Journal, 8*(4), 417-429.

Viessman Warren, Jr. L. L. G. (1997). *Introduction of hydrology* (4th ed.).

Zhao Y, Wang W, Bruen M, Zhao X. (2013). *International symposium on water resources and pollution control in arid/semi-arid regions.* 21-23 June 2013.

7

Geochemical Characteristics and Quality of Groundwater Around Okemesi Fold Belt, South Western Nigeria

O. A. Okunlola[1] & A. A. Afolabi[1]

[1] Department of Geology University of Ibadan, Ibadan, Nigeria

Correspondence: O. A. Okunlola, Department of Geology University of Ibadan, Ibadan, Nigeria. E-mail: o.okunlola@ui.edu.ng

Abstract

This study involves the determination of the hydrochemical character and quality of springs, shallow and deep wells around the Okemesi fold belt, south western Nigeria. This is with a view to elucidate their nature, type and evaluate portability, and suitability for agricultural and industrial purposes. The Okemesi fold belt is underlain by schistose rocks, mainly mica schist, quartz schist, and quartzite with minor amphibolites and gneisses.

The major hydrochemical water type are Na – (K) – Cl – (SO_4) and Ca – (Mg) – (Cl) – SO_4 water facies constituting about 52% and 47% respectively. The study showed that almost all of the physico-chemical parameters such as total dissolved solid TDS (18.75 - 790.50 mg/l), electrical conductivity EC (25 - 1054 µs/cm), pH (5.4 - 7.4), temperature (24.8 - 29.5°C), turbidity (2.2 - 40.5 N.T.U), total hardness (10 - 274 mg/l), total alkalinity (18 - 274 mg/l), dissolved oxygen (1.8 - 7 mg/l), chlorine demand (1.56 - 4.75 mg/l); and bacteriological analysis (5 - 80 MPN) results were within the World Health Organization (WHO) limits recommended for drinking water. However, some groundwater samples have Ni^+ and K^+ concentrations slightly above the recommended standard.

These physico-chemical parameters, especially cations: Ca^{2+} (1.6 - 72.8 mg/l), Mg^{2+} (0 - 4.39 mg/l), Na^+ (1.63 - 75.0 mg/l), K^+ (0 - 108.3 mg/l) , Si (4.0 - 10 mg/l), Cu (0 - 0.391 mg/l), Zn (0 - 0.29 mg/l), Cd (0 - 0 mg/l), Pb (0 - 0 mg/l), Fe (0.1 - 0.1 mg/l), Ni (0 - 0.043 mg/l) and anions such as HCO_3 (6.1 - 79.3 mg/l), Cl^- (5 - 109 mg/l), SO_4^{2-} (38 - 76 mg/l), NO_3^{2-} (0 - 0 mg/l) and PO_4^{3-} (0.005 - 0.03 mg/l) seem to reflect the chemical nature of the underlying rock units suggesting dissolution, weathering and water – soil / rock interaction processes.

Keywords: springs, wells, physico-chemical, hydrochemical, quality

1. Introduction

As water percolates through the rocks, some ionic exchanges take place and invariably the water takes into solution in different concentrations some elemental composition of the rocks. (Abimbola et al., 2002) Therefore, the chemical property, and also bacteriological composition of groundwater is important amongst other criteria in the assessment of the quality of waters. While it is generally accepted that water bodies, rivers, lakes, dams and estuaries are continuously subject to dynamic state of change with respect to the geological age and geochemical characteristics (Adefemi & Awokunmi, 2010), the availability of good quality water derived from these sources is an indispensable feature for preventing diseases and improving quality of life (Oluduro & Adeyowe, 2007). It is now generally recognized that the quality of water is just as important as its quantity (Fan & Steinberg, 1996; Gbodo & Ogunyemi, 1999; Abimbola et al., 2002, 2008; Tredoux et al., 2001; Adelana & Olasehinde, 2003; Adeyemi et al., 2003). Also apart from the fact that water quality is a function of physical, chemical and biological parameters, it is also subjective, as it depends on the particular intended use (Tijani, 1994). Therefore, water quality standards differ for various uses such as domestic (drinking), agriculture (irrigation) and industries.

The most significant geo hydraulic influence on groundwater chemistry arises from the source and circulation of groundwater itself (Amadi et al., 1989). Natural water contains some types of impurities whose nature and amount vary with source of water (Adeyemi et al., 2003). Metals for example, are introduced into aquatic system through several ways which include, weathering of rocks and leaching of soils, dissolution of aerosol particles from the atmosphere and from several human activities, including mining, processing and the use of metal based materials (Ipinmoroti, 1993; Adeyeye, 1994; Asaolu et al., 1997).

There has been some studies on rock composition, structure and association around Okemesi, southwestern Nigeria (Okunlola & Okoroafor, 2009; Ayodele & Odeyemi, 2010;) but little or none has been carried out on the quality of water in this region Hence the need for this study, which is aimed at elucidating the nature, origin and uses of the springs, shallow and deep well water in the study area..This involves the determination of the the the physical and chemical properties together with bacteriological load in the groundwater.

The petrogenetic features of the schistose rocks around the Okemesi fold belt area of investigation, which is part of the Ife-Ilesha schist belt has been described as one of the most complex lithological and structural frameworks amongst the Nigeria's metasedimentary belts (Olobaniyi, 2003; Okunlola & Okoroafor, 2009).

Okemesi is located in areas of high relief, (Figure 1) as a result, runoff is high and infiltration rates are low. Hence, some of the hand dug wells and boreholes are concentrated at the center of the fold belt, while the rivers and the springs follow a dendritic pattern (Figure 1).

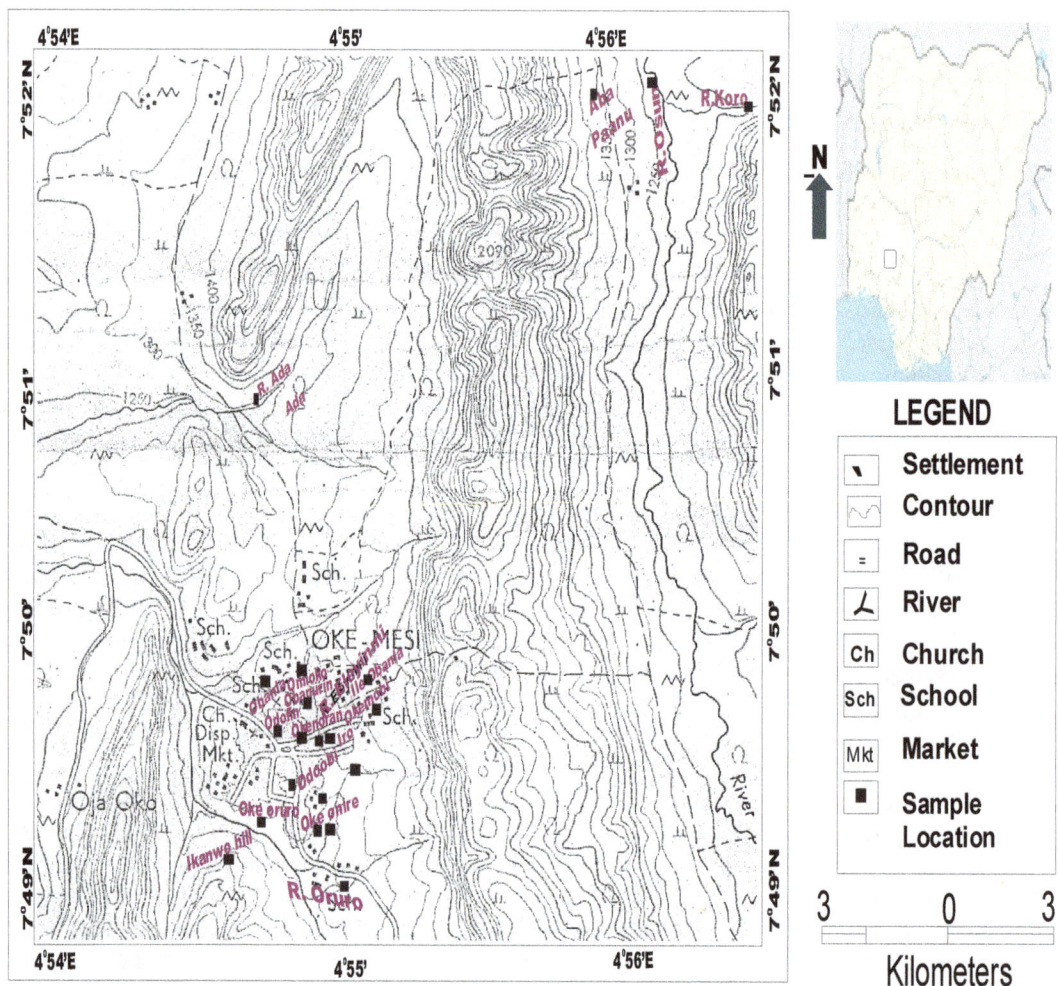

Figure 1. Map of sampling locations of surface and sub surface waters around Okemesi

The area of investigation lies completely within the 1: 50,000 sheet 253 (Ilesha NE) between latitude $7^0 49'$- $7^0 52'$ and longitude $4°54'$- $4°56'$, with the elevation ranging from 318 - 484m above the sea level. Okemesi is the largest town in the area. Other settlements in this area include Ajindo, Koro and Ada. The area is a planned settlement in spite of its rugged relief (Figure 1) and well serviced by a network of roads and footpaths (Figure 2).

A dendritic drainage system, upon which a trellis pattern has been super imposed mainly due to structural configuration characterize the area. Few large rivers and consequent streams that have their source from the nearby springs, together with deep and shallow wells present, are perennial, while small streams are seasonal with little or no meandering system (Figure 3). They flow westerly and a few northerly. The springs flow in a radial manner away from central range of schistose outcrops of Okemesi fold belt. Specifically, rivers like Oruro, Eleyinmi and Koro have utilized fractures in the area and cut across the divide.

Geomorphologically, the area forms an oval topographic feature with enclosing schistose and the surrounding gneissic complex (Figure 4). The range is broken in places by rivers, like Osun and Ada which have cut steep valleys and run almost at right angles to the trend of the range. The climate is typically equatorial, hot, dry and wet with mean monthly temperature around 27°C (80°F). Cloudiness and heavy precipitation help to moderate the daily temperature. It experiences about seven months (April-November) of torrential rain and associated high runoff and erosion, while December to March are often characterized by dry conditions.

1.1 Geological Setting:

Geologically, the study area is around the Okemesi fold belt of Ife – Ilesha schist belt within the Pre – Cambrian to Late Proterozoic Basement Complex rocks of southwestern Nigeria which in turn is part of the Pan – African mobile belt lying east of West African Craton (Jones & Hockey, 1964; Odeyemi, 1976; Rahaman, 1976). The major rock units in the area of investigation are ampibolite, banded gneiss, mica schist, quartz schist and quartzite (Okunlola & Okoroafor, 2009) (Figure 4). The details of the geological setting are contained in the aforementioned reference.

The basement rocks are commonly considered as poor aquifers because of their crystalline nature which leads to low porosity and permeability. However, appreciable porosity and permeability are developed through fracturing and weathering of the rocks (Davis & Deweist, 1966). This makes an otherwise barren rock to function as groundwater aquifer (Abimbola et al., 2002). The rocks in the area of investigation have undergone different tectonic episodes and have as a result responded differently to tectonic deformation that has affected the terrain, hence the numbers and sizes of lineaments (fractures) (Ayodele & Odeyemi, 2010) giving rise to the springs.

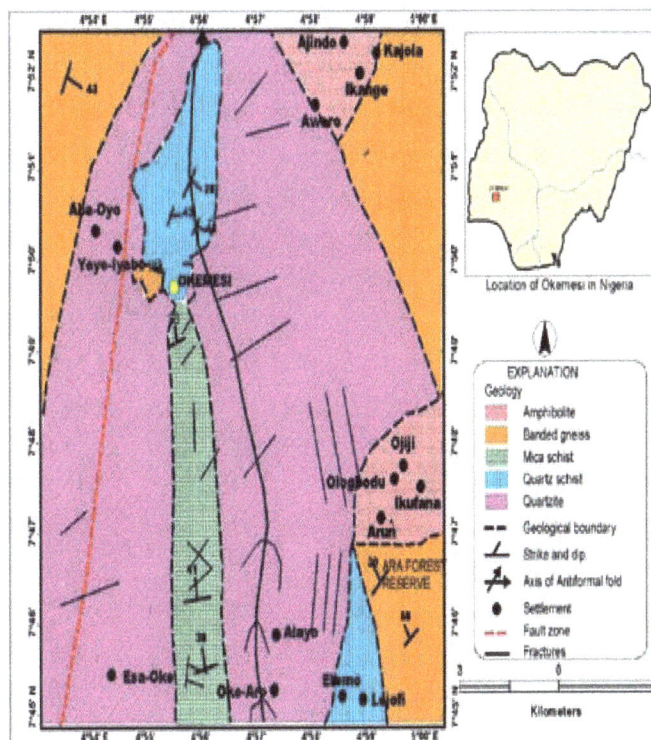

Figure 2. Geological map of Okemesi fold belt (after Okunlola & Okoroafor, 2009)

2. Materials and Methods

A total of twenty one representative samples of surface and groundwater were collected around Okemesi, Koro, Ajindo and Ada. Comprising 2 springs, 10 hand dug wells, 5 boreholes and 4 rivers which constitute the major water supply for the inhabitant. Sampling was done before the onset of heavy rain to minimize dilution effect. At each location, an aliquot of water to be sampled was used to rinse the plastic container to avoid contamination before collecting the water into 2 well drained separate clean new set of plastic containers or bottles, one for the determination of anions while the other set of samples were for cations and these were acidified immediately with two drops of HNO_3 in order to prevent loss of metals, bacterial and fungal growth. The first set was also stored in the refrigerator before bacteriological analysis.

The electrical conductivity, pH and temperature as a sensitive physico-chemical parameters were determined directly in the field using digital Hack pH meter while the location coordinates and elevation were determined by Ground Positioning System (GPS). Turbidity was determined by hack turbidity meter. Total hardness, total alkalinity, chlorine demand, dissolved oxygen, NO_3^{2-} and SO_4^{2-} Ca^{2+}, Mg^{2+}, Fe^{3+}, HCO_3 and silica were determined by titrimetry method at the Central Laboratory, Water Corporation of Oyo State, Nigeria. The laboratory method used for the detection and enumeration of total coliform count is the multiple tube method.

Cations such as K^+ and Na^+ together with the trace elements such as Cu, Zn, Cd, Pb, and Ni were analyzed using the Atomic Absorption Spectrometer (Bulk scientific 210/211 VGP AAS), 220GF Graphite Furnace and 220AS Autosampler at the Agronomy laboratory, University of Ibadan, Oyo state, Nigeria while PO_3^{2-} was analyzed by Ascorbic Acid method before determining its content in the solution by Spectrophotometer 70 at 882mμ. The detection limits of the metals are 0.001mg/l while for K^+ and Na^+ is 0.01mg/l

3. Results and Discussion

The data for physical, chemical and bacteriological parameters of both groundwater and surface water in the study area are presented in the Tables 1, 2, 3 and 4 respectively.

Table 1. The result of physical characteristics of surface water and groundwater around Okemesi

SAMPLE ID	NORTHINGS	EASTINGS	E (m)	TDS (mg/l)	EC μS/cm	pH	T (°C)	TU (N.T.U.)	TH (mg/l)	TA (mg/l)	DO (mg/l)	CD (mg/l)
RL1	N7°50′59″	E4°55′6.8″	318	42.0	56	7.4	25.5	15.2	20	30	6.7	2.296
WL2	N7°49′27″	E4°55′31.6″	402	114.75	153	6.5	29.5	6.8	68	68	4.4	2.169
RL3	N7°52′46″	E4°56′37.7″	380	19.35	129	7.4	25.8	7.5	32	66	3.8	4.75
WL4	N7°52′49″	E4°56′32.7″	398	51.0	68	6.0	26.3	14.2	50	38	4.3	1.585
RL5	N7°52′19″	E4°58′11.5″	441	401.25	535	6.7	25.4	14.2	252	274	1.8	2.196
WL6	N7°49′27″	E4°55′14.9″	395	179.25	239	6.5	27.6	5	40	18	3.2	2.308
RL8	N7°49′22″	E4°55′22″	396	50.25	67	6.5	25.0	2.2	30	24	4.0	2.558
SPL9	N7°49′15″	E4°55′3.8″	484	18.75	25	5.4	24.8	28.2	16	18	5.4	2.67
WL10	N7°49′20″	E4°55′21.8″	396	113.25	151	5.6	25.5	32.5	54	36	3.6	2.919
BL11	N7°49′21″	E4°55′13.6″	414	53.25	71	6.2	27.0	2.6	26	36	6.4	2.121
BL12	N7°49′12″	E4°55′19.7″	421	37.5	50	6.1	28.0	2.2	10	28	7.0	1.946
WL13	N7°50′04″	E4°55′27.3″	420	150.75	201	6.3	27.2	2.5	52	22	6.4	1.696
SPL14	N7°50′08″	E4°55′28.3″	443	43.4	62	6.0	25.3	8.4	16	32	3.7	1.946
WL15	N7°49′53″	E4°55′37.7″	431	123.75	165	5.7	27.5	2.2	38	20	5.5	2.162
RL16	N7°49′45″	E4°55′33.0″	399	104.25	139	6.1	25.8	17.6	46	138	6.9	2.296
BL17	N7°49′46″	E4°55′27.3″	408	642	856	6.0	26.8	7.5	274	144	3.4	1.56
BL19	N7°49′55″	E4°55′19.4″	414	790.5	1054	6.0	27.5	4.6	158	60	5.0	2.196
WL20	N7°49′54″	E4°55′20.3″	413	511.5	682	6.3	27.9	2.2	224	184	2.4	2.169
WL21	N7°50′06″	E4°55′27.9″	413	75	100	6.5	25.7	40.5	30	54	3.7	2.308
WL23	N7°49′57″	E4°55′23.3″	414	472.5	630	6.4	27.8	2.3	90	84	3.4	2.296
WL25	N7°49′44″	E4°55′25.2″	417	402.75	537	7.2	28.7	30.4	164	140	5.0	2.196

RL- river location   E- elevation   TU- turbidity   TH- total hardness

WL- well location   TDS- total dissolve solid   T- temperature   DO- dissolved oxygen

BL- borehole location   EC- electrical conductivity   CD- chlorine demand   SPL- spring location

SAR- sodium absorption ratio   TA- toltal alkalinity   S/N-serial number

Table 2. The result of chemical characterstics (cations and trace metals) of surface water and groundwater around okemesi

SAMPLE ID	Ca^{2+} (mg/l)	Mg^{2+} (mg/l)	Na^+ (mg/l)	K^+ (mg/l)	Cu (mg/l)	Zn (mg/l)	Cd (mg/l)	Pb (mg/l)	Fe (mg/l)	Ni (mg/l)
RL1	4.80	0.195	5.23	2.11	0.039	0.000	ND	ND	<0.1	0.015
WL2	16.0	0.683	6.86	8.49	0.021	0.000	ND	ND	<0.1	0.014
RL3	9.60	0.195	8.56	6.17	0.016	0.006	ND	ND	<0.1	0.004
WL4	12.0	0.488	6.95	8..27	0.391	0.080	ND	ND	<0.1	0.043
RL5	28.8	4.390	25.3	1.43	0.018	0.000	ND	ND	<0.1	0.006
WL6	9.60	0.390	22.00	16.20	0.016	0.006	ND	ND	<0.1	0.004
RL8	4.00	0.488	6.120	3.68	0.020	0.000	ND	ND	<0.1	0.025
SPL9	4.00	0.146	1.630	0.000	0.036	0.001	ND	ND	<0.1	0.007
WL10	21.6	0.000	4.100	3.860	0.037	0.027	ND	ND	<0.1	0.011
BL11	4.80	0.342	3.550	1.870	0.000	0.005	ND	ND	<0.1	0.014
BL12	1.60	0.146	2.760	4.660	0.002	0.005	ND	ND	<0.1	0.000
WL13	12.0	0.537	4.450	25.40	0.017	0.000	ND	ND	<0.1	0.018
SPL14	1.60	0.293	6.300	1.400	0.066	0.001	ND	ND	<0.1	0.000
WL15	28.8	0.439	17.70	4.92	0.032	0.013	ND	ND	<0.1	0.033
RL16	8.00	0.439	8.840	2..960	0.014	0.001	ND	ND	<0.1	0.018
BL17	72.8	2.245	34.00	23.60	0.019	0.020	ND	ND	<0.1	0.028
BL19	36.0	1.650	75.00	108.3	0.004	0.028	ND	ND	<0.1	0.029
WL20	64.0	1.562	32.10	48.20	0.048	0.000	ND	ND	<0.1	0.033
WL21	5.60	0.390	7.050	3..320	0.051	0.007	ND	ND	<0.1	0.016
WL23	20.8	0.927	41.70	86.60	0.000	0.018	ND	ND	<0.1	0.000
WL25	48.0	1.075	27.60	46.60	0.004	0.29	ND	ND	<0.1	0.026

ND - NOT DETECTED. Cd and Pb are below detection limits.

Table 3. The result of chemical characterstics (anions) of surface water and groundwater around okemesi

S/N	SAMPLE ID	SAR	SiO_2 (mg/l)	HCO_3^- (mg/l)	Cl^- (mg/l)	SO_4^{2-} (mg/l)	NO_3^{2-} (mg/l)	PO_4^{3-} (mg/l)
1	RL1	2.38	9.0	12.2	12.5	<35.0	0.000	0.011
2	WL2	1.68	9.0	24.4	11.5	<35.0	0.000	0.009
3	RL3	2.74	4.0	24.4	12.0	<35.0	0.000	0.014
4	WL4	1.97	<4.0	6.10	10.5	<35.0	0.000	0.020
5	RL5	4.39	10	42.7	19.0	38.0	0.000	0.016
6	WL6	6.96	4.0	12.2	29.5	<35.0	0.000	0.019
7	RL8	2.89	4.0	18.3	5.00	<35.0	0.000	0.012
8	SPL9	0.80	8.0	6.10	9.50	<35.0	0.000	0.005
9	WL10	0.88	8.0	6.10	15.0	<35.0	0.000	0.028
10	BL11	1.57	8.0	30.5	12.5	<35.0	0.000	0.016
11	BL12	2.08	6.0	12.2	8.50	<35.0	0.000	0.018
12	WL13	1.26	8.0	18.3	18.0	<35.0	0.000	0.030
13	SPL14	4.58	8.0	6.10	10.5	<35.0	0.000	0.020
14	WL15	3.27	8.0	12.2	20.0	<35.0	0.000	0.014
15	RL16	3.04	4.0	24.4	18.5	<35.0	0.000	0.019
16	BL17	3.92	8.0	36.6	67.0	70.0	0.000	0.020
17	BL19	12.22	6.0	24.4	109.0	<35.0	0.000	0.015
18	WL20	3.96	4.0	79.3	39.5	<35.0	0.000	0.016
19	WL21	2.88	8.0	12.2	16.5	76.0	0.000	0.020
20	WL23	8.95	4.0	30.5	51.0	53.0	0.000	0.015
21	WL25	3.94	<4.0	54.9	16.2	38.0	0.000	0.012

Table 4. The result of the bacteriological analysis around Okemesi

S/N	SAMPLE ID	CHLORINE RESIDUAL	COLONY COUNT (MPN)	MPN COLIFORM ORGANISM	LOCATION	ROCK TYPE
1	RL1	0	30	30	Ada	Quartz schist
2	WL2	0	80	80	Oke Onire 1	Mica schist
3	RL3	0	40	40	Aba Paanu	Quartzite
4	WL4	0	10	10	Aba Paanu 2	Quartzite
5	RL5	0	20	20	Ajindo/Koro	Amphibolite
6	WL6	0	20	20	Oke Onire 2	Mica schist
7	RL8	0	20	20	Oke Onire 3	Quartzite
8	SPL9	0	10	10	Ikanwo	Quartzite
9	WL10	0	40	40	Oke Oruro	Mica schist
10	BL11	0	20	20	Oke Onire 4	Quartzite
11	BL12	0	10	10	Oke Onire 5	Quartzite
12	WL13	0	20	20	Ile Obanla	Banded gneiss
13	SPL14	0	5	5	Omioko	Quartz schist
14	WL15	0	10	10	Okemobi	Quartz schist
15	RL16	0	20	20	Eleyinmi	Quartz schist
16	BL17	0	10	10	Iro	Mica schist
17	BL19	0	20	20	Odofin	Banded gneiss
18	WL20	0	40	40	Obanurin	Banded gneiss
19	WL21	0	30	30	Okenoran	Mica schist
20	WL23	0	10	10	Obanla	Quartz schist
21	WL25	0	30	30	Odoobi	Mica schist

The summary of the various parameters, their mean values as compared to the values of WHO (2004) standards is shown in table 5, while the result of the bacteriological analysis as compared with World Health Organization (W.H.O) 2004 Bacteriological Standard for Drinking Water is in Table 6.

The mean concentration of the cations is in the order $Ca^{2+} > K^+ > Na^+ > Mg^{2+}$ and the mean concentration of anions is in the order $SO_4^{2-} > Cl^- > HCO_3 > Silica > PO_4^{3-} > NO_3^{2-}$ while the trace metals mean concentration is in the order of $Fe > Cu > Zn > Ni > Cd > Pb$ and that of coliform mean concentration is 24ppm. Although all the analyzed parameters have concentration values less than the World Health Organization (WHO, 2004) recommended limits, K^+ in locations WL6 (16.20mg/l), WL13 (25.40mg/l), BL17 (23.6mg/l), BL19 (108.3mg/l), WL20 (48.2mg/l), WL23 (86.6mg/l), and WL25 (46.6mg/l) were observed to have the concentration value above WHO standard recommended limit (Table 5). These are all groundwater. Generally the concentration of calcium is much higher in the deep wells than the surface waters. for instance, sample BL17, BL19,WL20, and WL25 have calcium concentration of 72.8mg/l, 36.8mg/l, 64.0mg/l, and 48.0mg/l respectively while RL1, RL3, SPL9 and SPL14 have calcium concentration 4.8 mg/l, 9.6 mg/l, 4.0 mg/l, and 1.6 mg/l respectively.

Table 5. Summary of the physical and chemical parameters of waters around okemesi compared with WHO (2004) standard

PARAMETERS	N	MINIMUM	MAXIMUM	MEAN	STANDARD DEVIATION	HIGHEST DESIRABLE LEVEL (WHO)	MAXIMUM PERMISSIBLE LEVEL (WHO)
ELEVATION (m)	21	318.00	484.001	410.33	NC		
TDS (mg/l)	21	18.75	790.50	209.38	229.227	500	1000
EC (μS/cm)	21	25	1054	284.59	302 .015	1000	1400
Ph	21	5.4	7.4	6.32	0.532	7.0 – 8.5	6. 5 – 9. 2
TEMP (°C)	21	24.8	29.5	26.70	1.312	variable	variable
TURB (N.T.U.)	21	2.2	40.5	11.85	NC	5	NM
TH (mg/l)	21	10	274.00	80.48	82.197	NM	NM
TA (mg/l)	21	18	274.00	72.10	67.292	NM	NM
DO (mg/l)	21	1.8	7.00	4.57	NC	NM	NM
CD (mg/l)	21	1.56	4.75	2.30	NC	NM	NM
MPN (PPM)	21	5	80	23.57	NC	NM	NM
SAR	21	0.8	12.22	3.64	NC	NM	NM
Ca^{2+} (mg/l)	21	1.6	72.80	19.73	20.351	75	200
Mg^{2+} (mg/l)	21	0.00	4.39	0.84	0. 998	39	150
Na^+ (mg/l)	21	1.63	75.0	16.56	17.958	150	200
K^+ (mg/l)	21	0.00	108.30	19.43	29.614	10	15
SiO_2 (mg/l)	21	4.0	10.00	6.74	2.137	20	NM
HCO_3^- (mg/l)	21	6.1	79.30	23.53	18.227	500	1000
Cl^- (mg/l)	21	5.0	109	24.37	24.614	200	500
SO_4^{2-} (mg/l)	21	38	76	55	11.773	150	250
NO_3^{2-} (mg/l)	21	ND	0.00	0.00	NC	20	45
PO_4^{3-} (mg/l)	21	0.005	0.03	0.02	0.006	NM	NM
Cu (mg/l)	21	ND	0.391	0.04	0.082	1.0	1. 5
Zn (mg/l)	21	ND	0.29	0.02	0.064	0.2	5.0
Cd (mg/l)	21	ND	0.00	0.00	NC	0	0.005
Pb (mg/l)	21	ND	0.00	0.00	NC	0	0.05
Fe (mg/l)	21	0.10	0.10	0.10	NC	0..3	1. 0
Ni (mg/l)	21	0.00	0.043	0.016	0.012	0	0.02

NC – NOT CALCULATED   NM – NOT MENTIONED

Table 6. Bacteriological standard for drinking water compared with those around Okemesi area (WHO 2004)

S/N	CLASSIFICATION	MPN/100ml COLIFORM BACTERIA	SUMMARY FOR OKEMESI
1	Bacterial quality applicable to disinfection only	0-50	20
2	Bacterial quality requiring convectional methods of treatment (coagulation, filteration and di-infection)	50-5,000	1
3	Heavy pollutant requiring extensive type of treatment	5,000-50,000	-
4	Very heavy pollution, unacceptable unless special treatments designed for such water is used.	Above 50,000	-

In this study, the highest concentration of Ca^{2+} is found in borehole location BL17 at Iro, Okemesi having value of 72.8 mg/l. This may be attributed to dissolution of the water with the weathered calc-plagioclase feldspar, amphibole and pyroxene of the aquiferous zone by the groundwater. However, Mg^{2+} concentration in all the location is very low with mean value of 0.84 mg/l when compared to WHO highest desirable level of 39 mg/l for drinking water and could be attributed to generally lower dissolution of magnesium in water. Its values range from 0.146 mg/l to 2.245 mg/l in most places except in location RL5 at river Koro where the concentration is 4.390 mg/l. This river also has the highest concentration of Ca^{2+} of 28.8 mg/l for surface water. This may have incorporated the elemental component of the amphibolite in this area that the river passes through (Table 7).

Table 7. The statistical summary of average elemental composition of water sample control by rock type around Okemesi compared with WHO standard

ELEMENTS (mg / l)	AVERAGE FROM THE WATER SAMPLE OF AMPHIBOLITE (mg / l)	AVERAGE FROM THE WATER SAMPLE OF BANDED GNEISS (mg / l)	AVERAGE FROM THE WATEROF MICA SCHIST (mg / l)	AVERAGE FROM THE WATER SAMPLE OF QUARTZ SCHIST (mg / l)	AVERAGE FROM WATER SAMPLE OF THE QUARTZITE (mg / l)	W. H. O STANDARD 2004 (mg / l)
Ca^{2+}	28.8	37.33	29.07	12.80	10.13	75
Mg^{2+}	4.39	1.25	0.80	0.46	0.30	39
Na^+	25.3	37.18	16.94	15.85	4.93	150
K^+	1.43	58.97	17.01	19.598	4.11	10
SiO_2	10.0	6.0	7.17	6.60	5.70	20
HCO_3^-	42.7	40.67	24.73	17.08	16.27	500
Cl^-	19.0	55.5	25.95	33.8	9.67	200
SO_4^{2-}	38.0	35.0	48.17	38.6	35.0	150
NO_3^{2-}	0.00	0.00	0.00	0.00	0.00	20
PO_4^{3-}	0.016	0.02	0.02	0.02	0.01	
Cu	0.018	0.02	0.03	0.03	0.01	1.0
Zn	0.000	0.01	0.06	0.01	0.02	0.2
Cd	0.000	0.00	0.00	0.00	0.00	0.005
Pb	0.000	0.00	0.00	0.00	0.00	0.05
Fe	0.1	0.10	0.10	0.10	0.00	0.30
Ni	0.006	0.03	0.02	0.01	0.02	0.02

Generally, the concentration of K^+ and Na^+ varies for surface and groundwater. The highest K^+ and Na^+ concentration value of 108.3 mg/l and 75 mg/l respectively is found in borehole BL19 at Odofin while the lowest values of 0.00 and 1.63 for the two cations are found respectively at Ikanwo hills at location SPL9. This can be attributed to difference in the rock types and elevations of these two locations. While Odofin rock is banded gneiss with elevation of 414m, Ikanwo hills is quartzitic in composition with elevation of 484m above the sea level (Table 7). Hence, the borehole must have incorporated some minerals such as albite and orthoclase in the gneiss. The chloride concentration seems to have relationship with sodium and potassium in the area of investigation. This is because high concentration of chloride ion is found in area of high K^+ and Na^+. For example, sample BL19 with abnormal Cl^-, K^+ and Na^+ concentration values of 109 mg/l, 108.3 mg/l, and 75 mg/l, may have important geological control.

The source of HCO_3 with mean concentration of 23.53 mg/l can be attributed to dissolution of carbon dioxide of the air and organic matter. The lower concentration or absence of SO_4^{2-}, PO_4^{3-}, NO_3^{2-}, Fe, Cu, Zn, Cd and Pb in all the locations could be attributed to absent of industrial discharges, sewage discharge, landfill leachates and ore bodies. However, Ni concentration value is slightly above the maximum recommended value by WHO in 3 wells in location WL20, WL15 and WL4. Elevated trace metals concentrations in soils and shallow groundwater systems have been attributed to anthropogenic sources through agricultural practices (Abimbola et al., 1999: Mapanda et al., 2005; Tijani et al., 2009; Tijani, 2010) the possibility of geogenic contribution is high considering the presence of a basic rock unit associated with the gneisses in the area. The relatively higher electrical conductivity with a range of $50\mu s/cm$ to $1054\mu s/cm$ in the groundwater compared to a range of 25 $\mu s/cm$ to $535\mu s/cm$ in the surface water may also be due to increased water rock reaction and dissolution of solid components of the bed rock. However majority of the values are still within those for unpolluted waters

Statistical correlation using product moment coefficient and scattered plot indicates positive correlation between some pairs of parameters (Tables 8 & 9) and. Correlation studies show that calcium has positive correlation with cations: Mg ($r = 0.584$), Na ($r = 0.633$), K ($r = 0.481$) and some heavy metals: Zn (0.342) and Ni (0.505) indicating common source but show negative correlation with Cu ($r = -0.099$) suggesting different source. Also, potassium correlates positively

Table 8. Correlation coefficients for the physico-chemical parameters around Okemesi

	Ca	Mg	Na	K	HCO$_3$	Cl	SO$_4$	SiO$_2$	PO$_4$	Cu	Zn	Ni	TDS	EC	pH
Ca	1														
Mg	.584	1													
Na	.633	.531	1												
K	.481	.258	.893	1											
HCO$_3$	.739	.568	.455	.414	1										
Cl	.588	.407	.930	.832	.293	1									
SO$_4$	.294	.198	.173	.104	.063	.278	1								
SiO$_2$	.005	.251	-.202	-.327	-.168	-.140	.142	1							
PO4	.031	-.035	-.090	-.037	-.152	.049	.157	-.009	1						
Cu	-.099	-.107	-.190	-.167	-.252	-.184	-.064	-.220	.151	1					
Zn	.342	.049	.188	.258	.315	-.011	-.016	-.271	-.108	.138	1				
Ni	.505	.121	.258	.192	.273	.257	.032	-.157	.020	.486	.318	1			
TDS	.816	.671	.933	.815	.618	.889	.288	-.070	.013	-.216	.223	.301	1		
EC	.817	.668	.936	.817	.625	.891	.284	-.092	.006	-.224	.221	.287	.997	1	
pH	-.025	.121	.001	.022	.320	-.161	.008	-.091	-.232	-.174	.284	-.140	-.019	.015	1

Table 9. Correlations between some of the hydrochemical parameters around Okemesi

Variable	Correlation Coefficient
Mg^{2+} and Ca^{2+}	0.58
Mg^{2+} and EC	0.69
K^+ and Na^+	0.89
K^+ and EC	0.82
Na^+ and EC	0.94
TDS and Ca^{2+}	0.82
TDS and Mg^{2+}	0.67
Mg^{2+} and TH	0.87
TDS and HCO_3^-	0.62
Na^+ and Cl^-	0.93
Mg^{2+} and HCO_3^-	0.57
Na^+ and Ca^{2+}	0.63
K^+ and Cl^-	0.83
Ni and Cu	0.49
Ca^{2+} and Ni	0.51

with Na (r = 0.893), chloride (r = 0.832) and Carbonate (r = 0.414) indicating the same source and show negative correlation with silica (r = -0.327), phosphate (r = -0.037) and copper (r = -0.167). While correlation coefficient of sodium with carbonate (r = 0.455), chloride (r = 0.930), sulphate (r = 0.173), Zinc (r =0.188) and Nickel (r = 0.258) are positive, its correlation with silica (r = -0.202), phosphate (r = -0.090) and copper (r = -0.190) is negative. However, Ni has a weak positive correlation to all the ions; Calcium (r = 0.505), magnesium (r = 0.121), sodium (r = 0.258), potassium (r = 0.192), carbonate (r = 0.273), chloride (r = 0.257), sulphate (r = 0.032), phosphate (r = 0.020), and copper (r = 0.486), except silica (r = -0.070). There are relatively very strong correlation between total hardness and total alkalinity (r = 0.835), total dissolved solid and electrical conductivity (r = 0.997), total hardness and electrical conductivity (r = 0.849), calcium and total hardness (r = 0.893), magnesium and total hardness (r = 0.869).

3.1 Water Characterization

The plots of chemical parameters were carried out using trilinear Piper (Piper, 1944) and Scholler semi-log (Scholler, 1962) diagrams (Figures 3 & 4), to determine hydrochemical facies of the analyzed water sample collected around Okemesi. The Piper-Hill diagram is used to infer hydro-geochemical facies. These diagrams reveal the analogies, dissimilarities and different types of waters in the study area, which are identified and listed in Table 10. The concept of hydrochemical facies was developed in order to understand and identify the water composition in different classes. (Sadashivaiah et al., 2008). This water characterization by Piper's diagram has revealed 2 important water facies around Okemesi and these are Ca – (Mg) – (Cl) – SO_4 and Na – (k) – Cl – (SO_4) water facies.

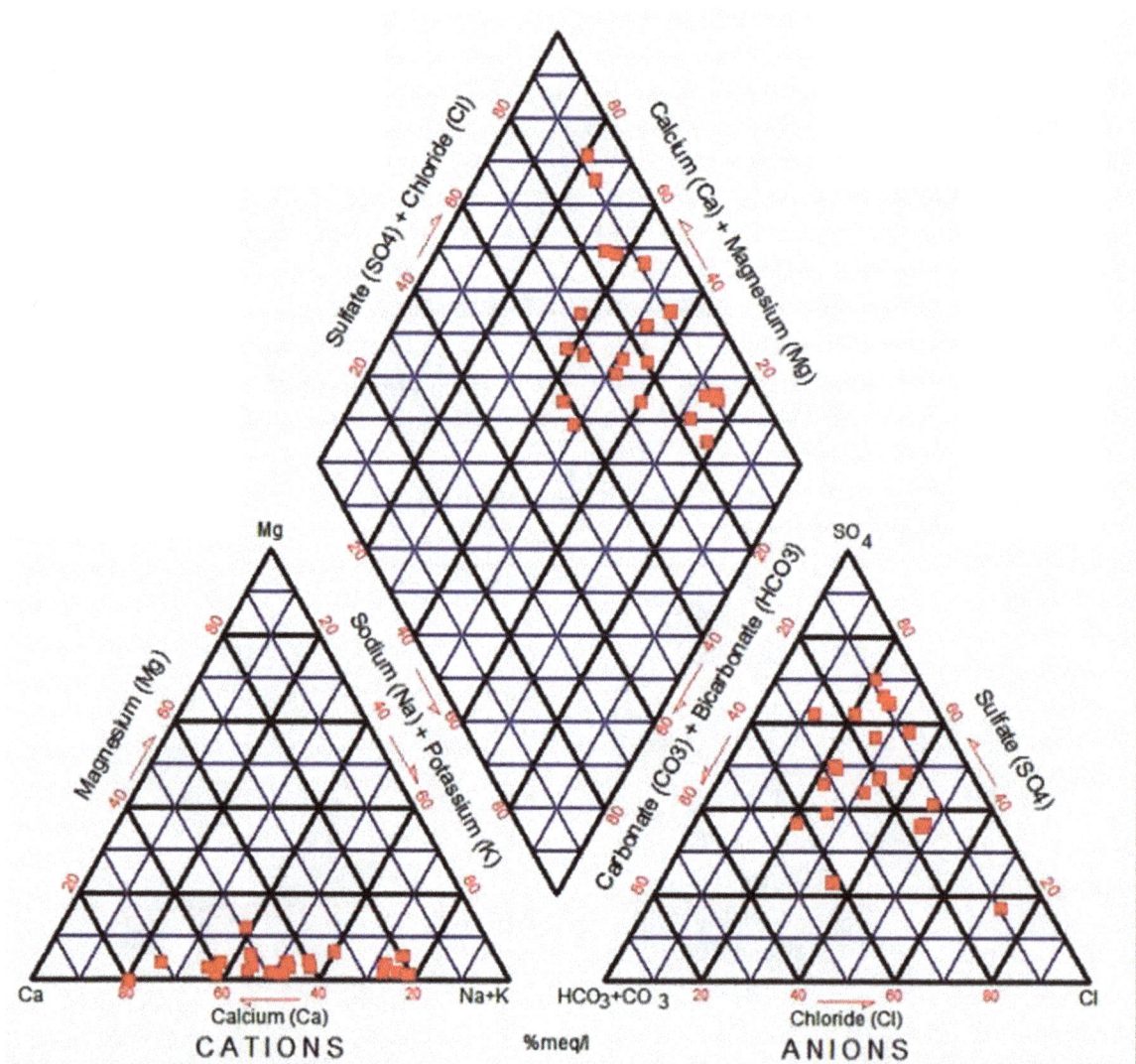

Figure 3. Piper Trilinear Diagram (1944) of hydrochemical analysis result around Okemesi

Table 10. Characterization of groundwater and surface water around Okemesi on the basis of Piper tri-linear diagram

Subdivision of the diamond	Characteristics of corresponding subdivisions of diamond-shaped fields	Percentage of samples in this category
1	Alkaline earth (Ca + Mg) Exceed alkali (Na + K)	50
2	Alkali exceeds alkaline earths	50
3	Weak acids (C03+HCO3) exceed Strong acids (SO4+Cl)	0
4	Strong acids exceeds weak acids	100
5	Magnesium bicarbonate type	0
6	Calcium-chloride type	10
7	Sodium-chloride type	52
8	Sodium-Bicarbonate type	0
9	Mixed type (No cation - anion exceed 50%)	48

$Ca - (Mg) - (Cl) - SO_4$ water facies constitute about 47.62% of the total water sample and is also referred to as normal alkaline earth with predominance of sulphate. This water suggests mixing process attributed to its interaction by dilution with weathered rock (Figure 5). It may also be due to anthropogenic input from inhabitant in the study area . $Na - (k) - (SO_4) - Cl$ water facies constitute about 52.38% of the total surface and groundwater samples and as a result is predominant water type in the study area. This refers to as alkaline water with high concentration of sodium and potassium.

This water type is common in groundwater in the area and likely to have originated from the dissolution of weathered gneissic rocks and schist, that is geogenic in origin.

Table 11. Modified Wilcox Quality Classification (1995) of irrigation water compared with number of water samples collected around Okemesi

WATER CLASS	ELECTRICAL CONDUCTIVITY (US/CM)	SALINITY HAZARD	SAR	NUMBER OF SAMPLES IN THIS CATEGORY
EXCELLENT	<250	LOW	0 - 10	20
GOOD	250 – 750	MEDIUM	10 – 18	1
PERMISSIBLE	750 – 2000	HIGH	18 – 26	
DOUBTFUL	200 -3000	VERY HIGH	26 - 30	

Where,

$$SAR = Na^+ / \frac{\sqrt{(Ca^{2+} + Mg^{2+})}}{2}$$

Table 12. Classification of water hardness modified after Todd (1980), compared with number of water samples collected around Okemesi

TOTAL HARDNESS RANGE (mg/L)	WATER CLASS	NUMBER OF WATER SAMPLES IN THIS CATEGORY
0 – 60	SOFT	14
61 – 120	MODERATELY HARD	2
121 – 180	HARD	2
> 180	VERY HARD	3

From Schoeller semi – log diagram (Figure 4), the water within the study area can be characterized as Ca –(Mg) - Na - (K) – (SO$_4$) – Cl water type because it constitutes about 99% of the total water types in the area. This is similar to the Piper characterization and is due to high percentage of chloride and sulphide ions in the water body, which is higher when compared to other ions within the study area.

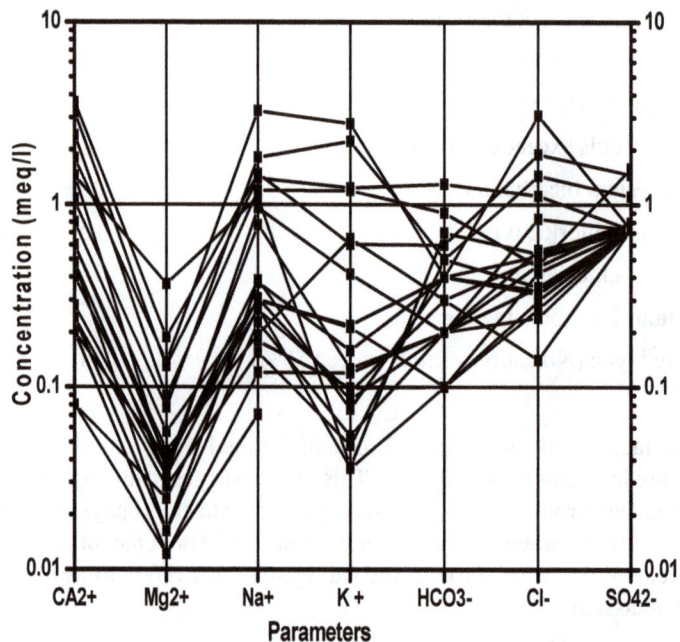

Figure 4. Scholler diagram showing chemical ions in the study area

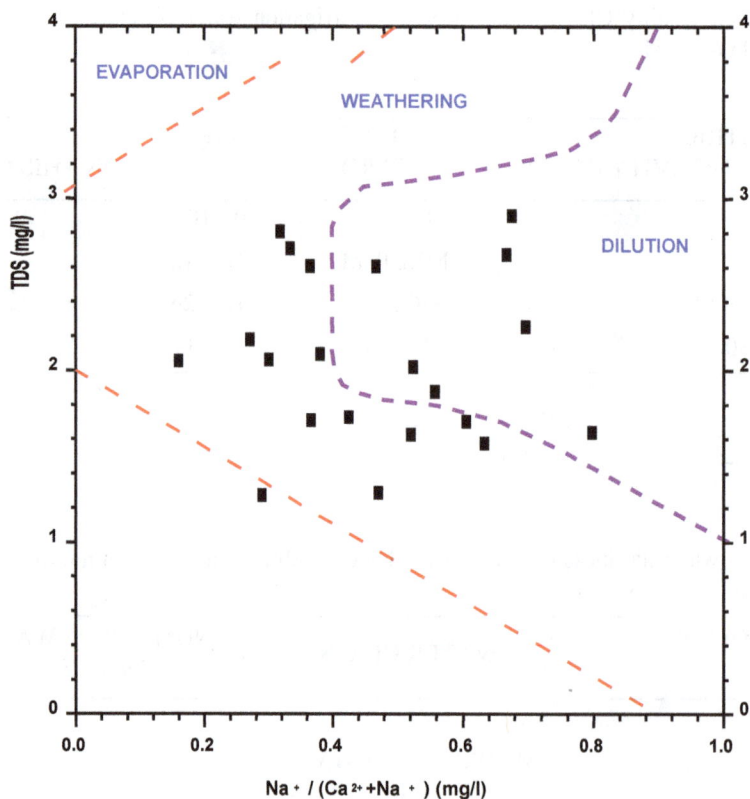

Figure 5. Modified Gibbs Diagram (1970) for the analysed water samples around Okemesi

It is also apparent that the water character around Okemesi are influenced greatly by weathering and dissolution activities of the bedrock. The higher concentration of relevant ions associated with the groundwater as compared to surface water is attributed to longer resident time of the contact between the water and the rock bodies. This leads to greater dissolution, increases concentration of solutes and enhanced weathering more in subsurface water than in surface water (Figure 5).

3.2 Water Quality and Usage

The chemical character of any water source determines to some extent its quality and usability. Geology and human activities through mining, industrialization, disposal of waste are liable to cause deterioration in the quality of surface water and groundwater of any environment especially when the groundwater is closer to the earth surfce. (Abimbola et al., 2002). Also, the standard or criteria for portable drinking water according to Davis and DeWeist (1966) include the absence of objectionable tastes, odour, colour, and substance of adverse effects. A quantitative measure of these criteria is stipulated by WHO (2004). Water quality requirements for different purposes differ; hence standards have been developed to appraise water usability for the various purposes. (Olobaniyi et al., 2007).

The result of the hydrochemical analysis shows that the surface and groundwater within the study area are chemically portable as compared with the WHO standard. Also, the quantity of trace metals present in the different water samples is low compared with the WHO Standard, this may have resulted from the low pH where most natural groundwater are mobile and the mass occur as charged metal ion which reach equilibrium with the solid phase usually a metal-hydroxide, metal carbonate or metal sulphide (Domenico and Schwartz, 1998; Oke, 2010). However, few of the samples have nickel values exceeding the highest permissible level in places especially those water sample from banded gneiss rich in amphibolitic bands (Table 2 and 7). The result of the total coliform count also show that only one of the samples has concentration above the WHO standard which may be as a result of faecal contamination in the water supply (Table 4) and (Figure 6).

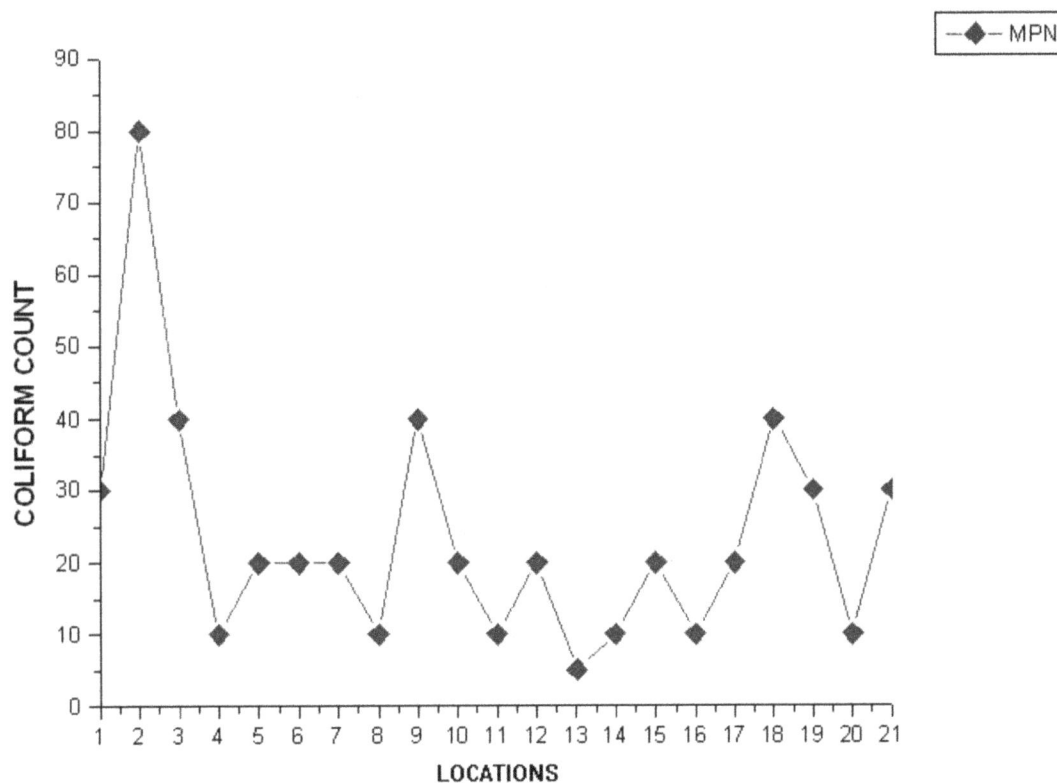

Figure 6. A plot of total coliform counts around Okemesi area against locations

3.3 Irrigation Use

The most important criteria for usage of water for irrigation purposes are total concentration of soluble salts, relative proportion of sodium to other principal cations, concentration of boron or other element that may be toxic, and under some condition, relationship of bicarbonate concentration with calcium and magnesium. These have been termed as the salinity hazard, sodium hazard, boron hazard and bicarbonate hazard (Wilcox, 1995). A better measure of the sodium hazard for irrigation is the SAR which is used to express reactions with the soil (Sadashivaiah et al., 2008). The classification of water samples from the study area with respect to SAR using Wilcox model is represented in Table 11. The SAR value of all the samples are found to be less than 10, and are classified as excellent for irrigation except in sample BL19 where SAR and electrical conductivity values are 12.22 and 1054 µs/cm respectively. This sample is within the permissible class and medium salinity hazard level. In general, all the water samples analysed are good for agricultural and irrigation purposes. However, other conditions necessary for plant growth such as amount of water, good soil type and favourable climate have to be met Also most of the samples (67%) have total hardness within 0 - 60mg/l,implying they are soft while while the rest are moderately hard (Table 12)

4. Conclusions

This study showed that most of the hydrochemical parameters in the groundwater around Okemesi are within the WHO limits for drinking water, posing no health threat to consumers. However, the levels of potassium, especially in some groundwater samples especially those of WL6 (16.20mg/l), WL13 (25.4 mg/l), BL17 (108.3 mg/l), WL20 (48.2 mg/l), WL23 (86.6 mg/l) and WL25 (46.6 mg/l), and, possibly, all the other areas such as WL4 (0.043mg/l), WL15 (0.033) and WL20 (0.033) where nickel levels exceeded the critical value of 0.02mg/l, should be closely monitored. Geological processes such as weathering and dilution of the water with rocks in this area have great impact on the quality and characteristics of the groundwater. Also a shallow well at WL 2 from Oke-Onire is found to have total coli form count of 80MPN, suggestive of heavy impact of human activity and is due to unnatural agents such as leakage of wastewaters and soak away.

Two major hydrochemical water facies were identified. These are $Ca - (Mg) - (Cl) - SO_4$ and $Na - (K) - Cl - (SO_4)$ water types. Forty eight per cent of the water sources are mixed water type having either HCO_3, SO_4^{2-} or Cl^- ions as the main anions predominating (Table 11).

Generally, the groundwater quality in this area is appropriate for drinking and agricultural purposes, except in few locations highlighted.

Acknowledgements

The authors appreciate the contribution of numerous colleagues and relations in the course of this study. The kings and chiefs of the study domains are specially acknowledged.

References

Abimbola, A. F., Ajibade, O. M., Odewande, A. A., Okunola, A. O., Laniyan, T. A., & Kolawole, T. (2008). Hydrochemical characterization of water resources around the semi-urban area of Ijebu-Igbo southwestern, Nigeria. *Water Resources. Journal of the Nigeria Association of Hydrogeologists, 18*, 10-16.

Abimbola, A. F., Odukoya, A. M., & Olatunjia, A. S. (2002). Influence of bedrock on the hydrogeochemical characteristics of bedrock of groundwater in Northern part of Ibadan metropolis. *Water Resources. Journal of the Nigeria Association of Hydrogeologists, 13*, 1-6.

Abimbola, A. F., Tijani, M. N., & Nurudeen, S. I. (1999). Some aspects of groundwater quality in Abeokuta and its environs Southwestern Nigeria. *Water Resources. Journal of the Nigeria Association of Hydrogeologists, 10*, 6-11.

Acworth, R. I. (1999). The development of crystalline basement aquifers in a tropical environment. *Quarterly Journal of Engineering Geology, 20*, 603-613.

Adefemi, S. O., & Awokunmi, E. E. (2010). Determination of physico-chemical parameters and heavy metals in water samples from Itaogbolu area of Ondo State, Nigeria. *African Journal of Environmental Science and Technology, 4*(3), 145-148p.

Adelana, S. M., & Olasehinde, P. I. (2003). High nitrate in water supply in Nigeria: Implication for Human Health. *Water Resources. Journal of Nigeria Association of Hydrogeologists, 4*, 12-18.

Adeyemi, G. O., Adesile, A. O., & Abayomi, O. B. (2003). Chemical characteristics of some well waters in Ikire, Southwestern Nigeria. *Water Resources. Journal of the Nigeria Association of Hydrogeologists, 14*, 12-18.

Adeyeye, E. L. (1994). Determination of heavy metals in Illisha Africana, associated water, soil sediments from some fish pond. *International Journal of Environmental Study, 45*, 231-240. http://dx.doi.org/10.1080/00207 239408710898

Amadi, P. A., Ofoegbu, C. O., & Morrison, T. (1989). Hydrogeochemical assessment of groundwater quality in parts of the Niger Delta, Nigeria. *Journal of Environmental Geology, 14*(3), 195-202.

Asaolu, S. S., Ipinmoroti, K. O., Adeyinowo, C. E., & Olaofe, O. (1997). Interrelationship of heavy metals concentration in water, sediment as fish sample from Ondo state coastal area, Nigeria. *African Journal of Science, 1*, 55-61.

Ayodele, O. S., & Odeyemi, I. B. (2010). Analysis of the lineaments extracted from LANDSAT image of the area around Okemesi, Southwestern Nigeria. *Indian Journal of Science and Technology, 3*(1), 31-36p.

Davis, S. N., & De-Wiest, R. M. J. (1966). *Hydrogeology.* New York: John Wiley and sons.

Domenico, P. A., & Schwartz, F. W. (1988). *Physical and chemical hydrogeology.* John Wiley and Son.

Fan, A. M., & Steinberg, V. E. (1996). Health implications of nitrate and nitrate in drinking water: An update on methaeomoglobinaemia occurrence and reproductive and developmental toxicology, Regulatory toxicology and pharmacology. Academy Press International.

Gbodo, V. O., & Ogunyemi, A. O. (1999). The quality and hydrogeochemistry of undergroundwater in Ago Iwoye/Ijebu Igbo as related to the underlying bedrocks. Unpublished B. Sc Geology Project, Olabisi Onabanjo University, Ago Iwoye.

Gibbs, R. J. (1970). Mechanisms controlling world chemistry. *Science, 170,* 1088-1090. http://dx.doi.org/10.1126/science.170.3962.1088

Hubbard, F. H. (1975). Precambrian crustal development in western Nigeria: indicators from the Iwo region. *Geo. Soc. Am. Bull, 86,* 548-554. http://dx.doi.org/10.1130/0016-7606(1975)86<548:PCDIWN>2.0.CO;2

Ipinmoroti, K. O. (1993). Determination of trace metals in fish associated water and soil sediments in fresh fish pond. *Discovery and innovations, 5,* 138p.

Jones, H. A., & Hockey, R. D. (1964). The geology of part of southwestern Nigeria. *Nigeria Min.Geol. and Metall. Soc., 1.*

Mapanda, F., Mangwayana, E. N., & Giller, K. E. (2005). The Effect of long term irrigation using waste water on heavy metal contents of soils under vegetables in Harare Zimbabwe. *Agri. Ecosyst. Environ., 107,* 151-165. http://dx.doi.org/10.1016/j.agee.2004.11.005

Odeyemi, I. B. (1993). A comparative study of remote sensing images of the structure of Okemesi fold belt Nigeria. *I. T. C. J. 1931*(1), 77-81.

Oke, S. A. (2010). Hydrogeochemical assessment of influence of bedrock weathering on chemical character of shallow groundwater system: A case study of Abeokuta. Unpublished M. Sc Project, Department of Geology, University of Ibadan.

Okunlola, O. A., & Okoroafor, E. R. (2009). Geochemical and petrogenetic features of Schistose rocks of the Okemesi fold belt, Southwestern, Nigeria. RMZ- *materials and geoenvironment, 56*(2), 148-162.

Olobaniyi, E. B., Ogala, J. E., & Nfor, N. B. (2007). Hydrogeology and bacteriological investigation of groundwater in Agbor area, Southwestern Nigeria. *Journal of Mining and Geology, 43*(1), 79-89. http://dx.doi.org/10.4314/jmg.v43i1.18867

Olobaniyi, S. O. (2003). Geochemistry of semi pelitic schist of Isanlu area, southwestern Nigeria: Implication for the geodynamic evolution of the Egbe-Isanlu schist belt. *Global Journal of Geological Science, 1*(2), 113-127.

Oluduro, A. O., & Adeyowe, B. I. (2007). Efficiency of moringa Oleifera seed extract on the microflora of surface and groundwater. *Journal of plant Science,* 6, 453-438.

Piper, A. M. (1944). A graphical procedure in geochemical interpretation of water analysis. *Trans American Geophysical Union,* 25, 914 – 923. http://dx.doi.org/10.1029/TR025i006p00914

Rahaman, M. A. (1976). Review of the basement geology of Southwestern Nigeria, in Geology of Nigeria. In C. A. Kogbe (Ed.) *Geology of Nigeria* (2nd Revised Ed., pp. 41-58). Elizabethan Pub. Co. Lagos.

Sadashivaiah, C., Ramakrishnaiah, C. R., & Ranganna, G. (2008). Hyrochemical Analysis and Evaluation of Groundwater in Tumkur Taluk, Karnataka State, India. *International Journal of Public Health, 5*(3), 158-164. http://dx.doi.org/10.3390/ijerph5030158

Schoeller, H. I. (1962). Les eanx souterraines, meson and paris.

Tijani, M. N. (1994). Hydrogeochemical assessment of groundwater in Moro area, Kwara State. Nigeria *Journal of Environmental Geology,* 24, 194 – 202. http://dx.doi.org/10.1007/BF00766889

Tijani, M. N. (2010). Groundwater system: A resource between the twin forces of nature and man. A faculty lecture delivered at the Faculty of Science, University of Ibadan. First Edition, Ibadan University Press.

Tijani, M. N., & Nton, M. E. (2009). Hydraulic, textural and geochemical characteristics of the Ajali Formation, Anambra Basin, Nigeria: Implication for groundwater quality. *Journal of Environmental Geology, 56,* 935-951. http://dx.doi.org/10.1007/s00254-008-1196-1

Todd, D. K. (1980). Groundwater Hydrology (2nd Ed., 315p). New York: Wiley.

Tredoux, G., Tama, A. S., & Engelbrecht, S. (2001). The increasing nitrate hazard in groundwater in the rural areas. Paper presented at Water Institute Conference. Sun City, South Africa.

WHO Guidelines for drinking water quality. *World Health Organization publication* (2004): Geneva, Switzerland.

Wilcox, L. V. (1995). Classification and use of irrigation waters, US Department of Agriculture, Washington Dc.

8

Development of Short-term Exposure Water Quality Standards for Cd, Cu, Pb and Zn in China

Zhen-guang Yan[1], Wei-li Wang[1], Xin Zheng[1] & Zheng-tao Liu[1]

[1] State Key Laboratory of Environmental Criteria and Risk Assessment, Chinese Research Academy of Environmental Sciences, Beijing, P. R. China

Correspondence: Zheng-tao Liu, State Key Laboratory of Environmental Criteria and Risk Assessment, Chinese Research Academy of Environmental Sciences, Beijing, P. R. China. E-mail: liuzt@craes.org.cn

Abstract

Environmental pollution sudden of heavy metals has posed serious ecological risks in China. However, local short-term exposure water quality standards (WQSs) are not yet established. In the present study, aquatic ecotoxicity data of Cd^{2+}, Cu^{2+}, Pb^{2+} and Zn^{2+} to local species were collected and screened. The suitability of species sensitivity distribution (SSD) methods assumed to be used to derive the WQSs was evaluated by data analysis. Then, the methodology of short-term WQSs was established with the principles of SSD and ecological risk assessment, and the tiered short-term WQSs values of Cd^{2+}, Cu^{2+}, Pb^{2+} and Zn^{2+} were derived with the established methodology. Finally, a case analysis was performed with the developed cadmium short-term WQSs and risk grades. The results may provide technical references for response to sudden environmental pollution.

Keywords: short-term exposure, water quality standard, sudden environmental pollution, ecological risk assessment, heavy metals

1. Introduction

Water quality standards (WQSs) play important roles in protection of ambient water environment quality. They can be divided into long-term exposure WQSs and short-term exposure WQSs. The latter meant to estimate severe effects and to protect most species against lethality during intermittent and transient events (e.g., spill events to aquatic-receiving environments, infrequent releases of short lived/ non persistent substances.). In contrast, long-term exposure guidelines are meant to protect against all negative effects during indefinite exposures (CCME, 2007). The technical system of long-term exposure WQSs have been established maturely in developed countries, such as the criterion continuous concentration (CCC) of the United States (USEPA, 1985), the water quality guidelines issued by the Canadian Council of Ministers of the Environment (CCME, 1991), the predicted no effect concentration of the European Union (ECB, 2003), the trigger values of Australia and New Zealand (ARMCANZ & ANZECC, 2000) and the negligible concentration (NC), the maximum permissible concentration (MPC), the serious risk concentration (SRC) issued by the Netherland (Traas, 2001).

The short-term exposure WQSs were studied earlier in the United States (US). In the American water quality criteria (WQC) document issued in 1968, also called "Green Book" (National Technical Advisory Committee to the Secretary of the Interior, 1968), the criterion maximum concentration (CMC) was proposed to deal with the acute exposure of pollutants, and the concept was still in use today in the US (USEPA, 2009). In recent years, many countries strengthen the study on the short-term exposure WQSs. Fox example, the Netherland issued the revised guidance for the derivation of environmental risk limits in 2007. In the guidance (van Vlaardingen and Verbruggen, 2007), in addition to the NC, MPC and SRC, a new concept of maximum acceptable concentration for ecosystem (MAC_{eco}) was proposed to protect the aquatic ecosystem against acute toxic effects exerted by exposure to short-term peak concentrations or against acute effects of transient exposure peaks.

The intense efforts were conducted to evaluate the acute toxicity of heavy metals on aquatic organisms (Redeker and Blust, 2004; Johnson et al., 2007; Priel and Hershfinkel, 2006; Birungi et al., 2007; Karntanut and Pascoe, 2002). Lots of acute toxicity data are available and the present CMCs were mainly derived from standardized acute toxicity data. Heavy metals such as Cd, Cu, Pb and Zn can damage freshwater organisms. For example, for zinc, the 96-h LC_{50} to invertebrate range from 0.1 mg/L to 14 mg/L and that of vertebrate range from 0.654 mg/L to 46.5

mg/L (Wu et al., 2011). Recently, Rachel, Andrew, Claudia, Pereira, and John (2014) ranked metals according to the threat they pose to freshwater organisms in the UK.

After decades of development, China has established relatively mature long-term exposure WQSs to protect the quality of surface water, ground water, marine water and so on. For instance, surface waters can be divided into five classes based on the surface water specific function classification and protection target, Class I and Class V belong to the most rigorous and the least rigorous WQS (GB3838-2002). However, short-term exposure WQSs are not yet developed, not even studied. On the other hand, at present, China has entered the period of high risk of pollution accident, and unexpected environment pollution events of various pollutants, especially heavy metals, occurred often. Fox example, recently, the serious sudden accident of cadmium pollution taking place in the Longjiang River in Guangxi Province has caused tens of tons of adult fish and more than one million fish fry death (Xinhua News Agency reported). The short-term exposure WQSs is needed urgently in China to assess the ecological risk posed by heavy metals in sudden pollution accidents. This study collected and screened the acute ecotoxicity data of Cd, Cu, Pb and Zn, and established the methodology of short-term WQSs with the principle of species sensitivity distribution (SSD) and ecological risk assessment. And, the tiered short-term exposure WQSs for the four heavy metals was derived. The results can provide valuable information to the environmental management of sudden pollution accident of heavy metals.

2. Method

The Method section describes in detail how the study was conducted, including conceptual and operational definitions of the variables used in the study, Different types of studies will rely on different methodologies; however, a complete description of the methods used enables the reader to evaluate the appropriateness of your methods and the reliability and the validity of your results, It also permits experienced investigators to replicate the study, If your manuscript is an update of an ongoing or earlier study and the method has been published in detail elsewhere, you may refer the reader to that source and simply give a brief synopsis of the method in this section.

2.1 Collection of Published Acute Ecotoxicity Data of Cd, Cu, Pb and Zn

The published acute toxicity data of Cd^{2+}, Cu^{2+}, Pb^{2+} and Zn^{2+} to aquatic animals were collected from the ECOTOX database (http://cfpub.epa.gov/ecotox), TOXNET Database (http://toxnet.nlm.nih.gov), the China National Knowledge Infrastructure (www.cnki.net) and other open literatures. The data were screened according to the guidelines for deriving WQC for the protection of aquatic organisms in the US (USEPA, 1985). Unqualified data with unsuitable exposure time (for daphnia and midge, 2 days, and for other organisms, 4 days are suitable), unusual diluted water (such as distilled water), unscientific experimental design and relatively insensitive life stages were not selected. Data of non-Chinese species were also abandoned. As for the test endpoints, the 48 h-LC_{50} or EC_{50} for daphnia or larvae of midge, and 96 h-LC_{50} or EC_{50} for fish, mollusks, shrimp and other organisms were chosen.

2.2 Evaluation of the Suitability of Four SSD Methods

In order to obtain the optimal model, the suitability of several SSD methods assumed to be used to develop the methodology of short-term WQSs were evaluated. The hazardous concentrations for 5% of the species (HC_5) were calculated according to the four SSD methods that based on log-triangle (USEPA, 1985), log-normal (Van Vlaardingen, Traas, Wintersen, & Aldenberg, 2004), log-logistic (Aldenberg & Solb, 1993) and BurrIII function (Hose & Van den Brink, 2004), respectively. The model that gained a suitable HC_5 value was chosen to derive the pollutant concentration corresponding to different affected fractions of species.

2.3 Establishment of Methodology of Tiered Short-Term Exposure WQSs

The methodology of tiered short-term exposure WQSs was developed with the principle of SSD and ecological risk assessment. The SSD curve was fitted by the desirable SSD method that screened out through the above procedure. The tiered ecological risks were defined according to different affected fractions of species, and the corresponding pollutant concentrations were calculated by the fitting function. Then, the tiered short-term exposure WQSs were developed according to the tiered pollutant concentration and a correction factor.

2.4 Data Analysis and Development of Short-Term WQSs for Cd, Cu, Pb and Zn

The data were analyzed using the PASW statistics 18. The normality of the data was checked by Kolmogorov-Smirnov test. Statistical significances were considered to be significant at $p \leq 0.05$. The species acute toxicity data was used to generate the SSD curve. If a species has more than one toxicity datum, the species mean acute value (SMAV) was used instead, and it equal to the geometric average of all the qualified toxicity data of the species. According to the methodology that established above, the short-term exposure WQSs for the four heavy metals were developed. Finally, a case analysis of sudden cadmium pollution in Longjiang River, Guangxi Province in 2012 was performed with the derived short-term WQSs of cadmium.

3. Results

3.1 Freshwater Species Sensitivity of Cd, Cu, Pb and Zn

Published acute toxicity data of Cd, Cu, Pb and Zn to freshwater organisms were collected and screened, and the results were shown in the supplementary materials. Qualified toxicity data of 45 species for Cd^{2+}, 54 species for Cu^{2+}, 26 species for Pb^{2+} and 26 species for Zn^{2+} were obtained. The normality of these ecotoxicity data were analyzed by Kolmogorov-Smirnov test and the results showed that they are all acceptable.

The statistic characteristics of the qualified data were analyzed and the results were shown in Table 1. We can see that the data of Cu^{2+} is sufficient and the data of Pb^{2+} and Zn^{2+} is relatively insufficient. Fortunately, they all meet the minimum toxicity data requirement of SSD generation (ten data for fitting of one SSD curve (Wheeler, Grist, Leung, Morritt, & Crane, 2002). In term of the average value, Cu^{2+} has higher toxicity, while Pb^{2+} has lower toxicity to freshwater organisms.

Table 1. Statistic characteristics of toxicity data of Cd, Cu, Pb and Zn

Heavy metals	Sample number	Minimum/(µg/L)	Maximum/(µg/L)	Average/(µg/L)[*]	SD
Cd^{2+}	45	0.15	4.76	2.85	1.07
Cu^{2+}	54	-0.80	4.46	1.82	0.89
Pb^{2+}	26	1.80	5.84	3.78	1.23
Zn^{2+}	26	1.93	4.85	3.40	0.91

Note: The minimum value have been transformed by common logarithm.

As for the sensitivity of freshwater organism to the four heavy metals, the most sensitive and insensitive species to Cd^{2+} are *Salmo trutta* (LC_{50} = 1.40 µg/L) and *Branchiura sowerbyi* (LC_{50} = 58 020 µg/L), respectively. Except 1 fish and 1 rotifer, in the 10 most sensitive organisms to Cd^{2+}, the other 8 are all crustaceans. The most sensitive and insensitive species to Cu^{2+} are *Tubifextubifex* (LC_{50} = 0.16 µg/L) and *Sinopotamon henanense* (LC_{50} = 28 610 µg/L), respectively. Except the tubificid worm *Tubifex tubifex*, in the 10 most sensitive organisms to Cu^{2+}, the other 9 are all crustaceans. The most sensitive and insensitive species to Pb^{2+} are *Ceriodaphnia dubia* (LC_{50} = 63.8 µg/L) and *Sinopotamon henanense* (LC_{50} = 692 090 µg/L). The 10 most sensitive organisms to Pb^{2+} contain 5 crustaceans, 4 fish and 1 shrimp, indicating the diversity of species sensitivity to this heavy metal. The most sensitive and insensitive species to Zn^{2+} are *Ceriodaphnia reticulate* (LC_{50} = 85.4 µg/L) and *Rana catesbeiana* (LC_{50} = 70 000 µg/L), respectively. Except 1 fish and 1 rotifer, in the 10 most sensitive organisms to Zn^{2+}, the other 8 are all crustaceans. So, the SSD of the four heavy metals suggested that crustaceans may be the most sensitive species to heavy metals.

3.2 The Suitability of the Four SSD Methods

In order to screen the desired constuction method of SSD curve, four SSD fitting methods were adopted to analyze the acute toxicity data of the four heavy metals to acquire the HC_5 values. The results were shown in Table 2, and it suggested that different SSD method produced different HC_5 values. Generally, the values derived with "SSD-RIVM" and "SSD- AU & NZ" methods were relatively higher, while the "SSD-EU" method was relatively stringent.

Table 2. Comparison of the derived HC_5 values of the four SSD fitting methods

Fitting Methods	Fitting Functions	HC_5/(µg/L)			
		Cd^{2+}	Cu^{2+}	Pb^{2+}	Zn^{2+}
SSD-USA	Log-triangular	4.32	3.32	88.54	90.14
SSD-EU	Log-logistic	0.72	0.26	10.28	23.19
SSD-RIVM	Log-normal	12.09	2.19	54.26	75.71
SSD-AU & NZ	BurrIII	9.42	3.22	110.26	115.31

Note. The derivation of the fitting methods: SSD-USEPA (USEPA, 1985), SSD-EU (Aldenberg & Solb, 1993), SSD-RIVM (Aldenberg &Jaworska, 2000), SSD-AU & NZ (Hose & Van den Brink, 2004).

The "SSD-AU & NZ" method was chosen as a desirable fitting method for its moderate derived values to develop the methodology of the short-term WQSs.

3.3 Establishment of the Methodology of Tiered Short-term WQSs

The methodology of tiered short-term exposure WQSs were developed with the principle of SSD and ecological risk assessment. Different affected fraction of the aquatic organism corresponds to different ecological risk level caused by the pollutant. In the Netherlands, if the affected fraction of the aquatic species reaches 50%, the risk level caused by the pollutant is considered serious (Traas, 2001). On the other hand, if the affected fraction of the aquatic species less than 5%, the ecological risk posed by the pollutant can be ignored generally (USEPA, 1985). Taking the above principles as references, four grades of risk levels were designed in this study according to different affected fractions of the aquatic species, and the corresponding four grades of short-term WQSs were derived. As shown in Figure 1, they are "4th grade (IV)" (serious risk, the affected fraction is greater than 50%), "3rd grade (III)" (apparent risk, the affected fraction is greater than 30%), "2nd grade (II)" (some risk, the affected fraction is greater than 15%) and "1st grade (I)" (potential risk, the affected fraction is greater than 5%). The X value in Figure 1 indicates the hazardous concentration (Shcheglov, Moiseichenko, & Kovekovdova, 1990) of the pollutant, and the WQS equals the HC value divided by the correction factor that was generally assumed to be 1 to 5 (van Vlaardingen & Verbruggen, 2007). Because the uncertainty of the ecological risk rise with the increasing of concentration of pollutant, the correction factor value was set to be 5, 4, 3 and 2, corresponding to the four grades of WQSs, IV, III, II and I, respectively.

In addtion, the short-term duration time and frequency of the derived WQSs in this study were designated "3 hrs" and "not more than one time pre three years" according to the technical guidelines of the US (USEPA, 1985). They were proposed according to the results of related sciencitific research that concerned the toxic effects of pollutant to individual species and ecosystem.

3.4 Derivation of the Tiered Short-term WQS for Heavy Metals

The tiered short-term WQS for the four heavy metals were derived according to the established methodology. The calculated SSD parameters were shown in Table 3, and the results were shown in Figure 2 and Table 4.

Table 3. The SSD parameters calculated with "SSD-AU & NZ" method

Heavy Metals	Fitting Functions	Calculated SSD parameters		
Cd^{2+}	BurrIII	1122.381 (b)	0.745 (c)	0.834 (k)
Cu^{2+}	BurrIII	24.957 (b)	0.791 (c)	1.662 (k)
Pb^{2+}	ReWeibull	21.158 (α)	0.415 (β)	
Zn^{2+}	ReWeibull	38.224 (α)	0.536 (β)	

Table 4. The four-grade short-term exposure WQSs of the heavy metals

Grades	Risks	Affected Fraction	Short-term exposure WQSs/(µg/L)			
			Cd^{2+}	Cu^{2+}	Pb^{2+}	Zn^{2+}
IV	Serious	50%	158	11.5	746	353
III	Apparent	30%	58.1	5.80	247	157
II	Some	15%	20.5	3.20	110	90.1
I	Potential	5%	4.71	1.61	55.1	57.7

3.5 Case Analysis with the Developed Short-term WQSs of Cadmium

A sudden cadmium environmental pollution was occurred in Longjiang River, Guangxi Province in China in 2012. The reported peak values of cadmium concentration exposed in the river in the accident was 400 µg/L, and after emergency disposal, it decreased to about 125 µg/L (Xinhua News Agency reported). In term of the short-term WQSs of cadmium developed in this study, the 400 µg/L of cadmium may pose serious ecological risks (risk grade: IV), and most of the regional aquatic species are threatened, while the 125 µg/L of cadmium may pose apparent ecological risks (risk grade: III) and the affected aquatic species is greater than 30% (Table 4 and Figure 2). The affected organisms contained shrimps, some sensitive fish, e.g. the bighead fish, and some aquatic invertebrates, such as hydras, daphnia and rotifers. Basically, after emergency disposal, there were no risks to some insensitive species, such as common carp, loach, amphibians, oysters and crabs with the time limitation of 3 hrs.

4. Discussion

WQSs play important role in water environment management. They can be divided into long term exposure WQS and short term exposure WQS (USEPA, 1985). Generally, the value of the former are lower, and they are determined according to the results of WQC study, risk assessment and cost-benefit analysis etc (USEPA, 1994), while the value of the latter are higher, and they are developed according to the results of WQC study and risk assessment (van Vlaardingen & Verbruggen, 2007; Sloof, 1992). Compared with WQC from USEPA (2009), WQS of grade I for the four heavy metals in our study is in the same order of magnitude. Grade I for Cd^{2+} is slightly higher than CMC in America, and grade I for Pb^{2+} is slightly lower. CMC of Zn^{2+} in America was between 90.1μg/L of Grade II and 157μg/L of Grade III in our study. However, grade IV for Cu^{2+} is still lower than that in America, indicating that Chinese native species were more sensitive to Cu^{2+}.

As for the minimum toxicity data requirement for deriving the WQC, the USEPA prescribed that at least three phyla and eight families should be used in the calculation (USEPA, 1985). The short term exposure WQS were divided into two sorts, A and B, in the WQS guidelines issued by the Canadian Council of Ministers of the Environment (CCME, 2007). The WQS of sort A should be developed by the SSD method with sufficient toxicity data that at least include data of 3 fish (including 1 Salmonidae fish and 1 non-Salmonidae fish), 3 aquatic or semi-aquatic invertebrate (including 1 pelagic crustacean) and if have, aquatic plants and amphibian. When the toxicity data is insufficient, the assessment factor (AF) method was recommended to derive the WQS of sort B. In the guidance for derivation of environmental risk limits issued by the Netherlands (van Vlaardingen & Verbruggen, 2007), according to different situations, three methods, the AF method, the SSD method and simulation of ecosystem were recommended to develop the short term WQS, respectively. In the guidance, at least three acute toxicity data of three different trophic level (for example, algae, daphnia and fish), are required to be used in the AF method, and at least eight acute toxicity data of aquatic organisms, including 6 aquatic animals, 1 algae and 1 higher plant, are required to be used in the SSD method. Two kinds of vertebrate, including 1 fish should be contained in the 6 aquatic animals. In the present study, sufficient toxicity data were collected for all the four heavy metals, and the data quantity is qualified for all the SSD methods.

There are several popular SSD methods for derivation of the WQSs, and all of them were accepted by the Organization for Economic Co-operation and Development (OECD, 1992; Posthuma, Suter, & Traas, 2002). These SSD methods were evaluated in the present study with calculation of the HC_5 value. Through comparison analysis, the "SSD-AU & NZ" method produced relatively moderate results for all the four heavy metals and was chosen to derive the WQS in this study. The derived short term exposure WQS was set to four grades (I, II, III and IV) in this study. The correct factors for the HC values were set to be 2, 3, 4 and 5 for the four grades of risks due to the increasing risks with the higher pollutant concentrations, and they still need to be validated in field study or management of sudden environmental pollution accident.

According to the methodology of the short-term WQSs, the duration time is 3 hrs. In an accident of sudden environmental pollution, when the exposure time beyond 3 hrs, the posed ecological risk could be increased. How to assess the increased risks in an emergency is worth study. Before a perfect theory being proposed, at least some aquatic organisms can be taken as biological indicators for risk assessment. For example, in the above case analysis of the accident of the Longjiang River cadmium pollution, according to the Table S1 (supplementary materials) and figure 2, we can know that when the sensitive freshwater shrimps were hard to survive, the pollutant can be considered to have posed some ecological risks, and the death of bighead fish and amphibians indicate apparent and serious risks, respectively.

Generally, the ecotoxicity of heavy metals can be affected by some water quality parameters, such as hardness, temperature, pH, etc. So a perfect WQS should be developed according to different regional water conditions. Moreover, the water quality conditions in different basins or regions in China are of high diversity. WQSs should be developed according to different ecoregions to facilitate risk assessment, ecoregion protection and environmental management in pollution accident. The derived WQSs in the present study may be improved in these aspects in the future.

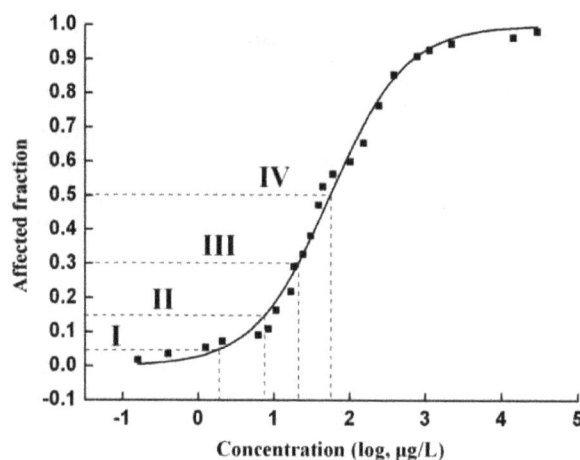

Figure 1. Sketch map of SSD curve fitting

Note. "I", "II", "III" and "IV" in the figure represents the four grade WQSs, and is corresponding to the four grade ecological risk levels: I (potential risk, the affected fraction is greater than 5%), II (some risk, the affected fraction is greater than 15%), III (apparent risk, the affected fraction is greater than 30%) and IV (serious risk, the affected fraction is greater than 50%).

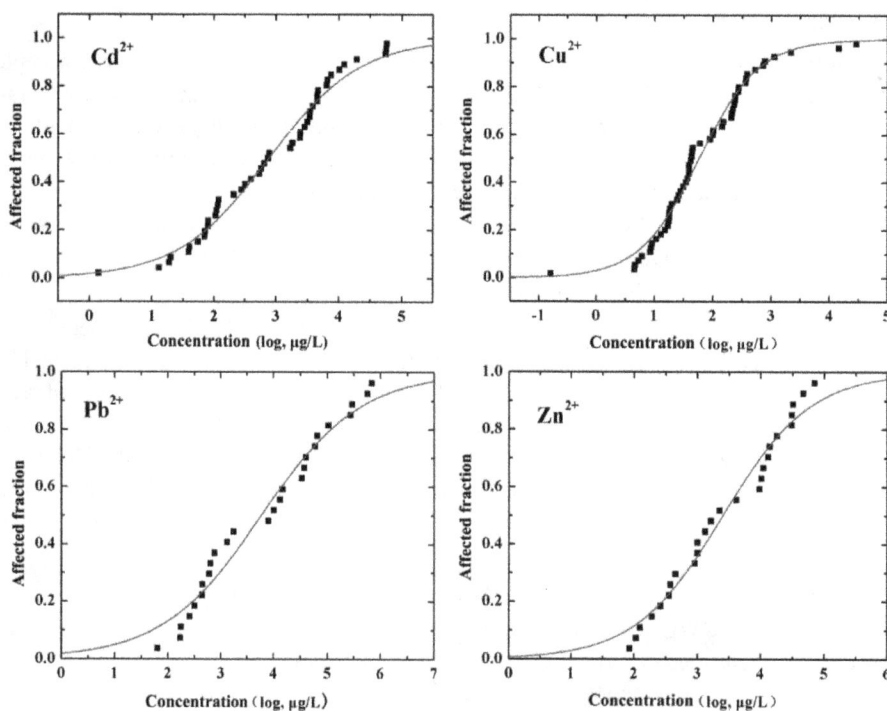

Figure 2 SSD fitting of four heavy metals

Note. The solid square in the figure indicate the acute ecotoxicity data of different heavy metals.

Acknowledgements

This work was financially supported by the National Science and Technology Project of Water Pollution Control and Abatement of China (Grant No. 2012ZX07501-003-06).

References

Aldenberg, T., & Jaworska, J. S. (2000). Uncertainty of the hazardous concentration and fraction affected for normal species sensitivity distributions. *Ecotoxicology and Environmental Safety, 46*(1), 1-18. http://dx.doi.org/10.1006/eesa.1999.1869

Aldenberg, T., & Solb, W. (1993). Confidence limits for hazardous concentrations based on logistically distributed NOEC toxicity data. *Ecotoxicology and Environmental Safety, 25*(1), 48-63. http://dx.doi.org/10.1006/eesa.1993.1006

ARMCANZ, & ANZECC. (2000). *Australia and New Zealand guidelines for fresh and marine water quality.* Canberra, Australia: Agriculture and Resource Management Council of Australia and New Zealand and the Australian and New Zealand Environment and Conservation Council.

Birungi, Z., Masola, B., Zaranyika, M. F., Naigaga, I., & Marshall, B. (2007). Active biomonitoring of trace heavy metals using fish (*Oreochromis niloticus*) as bioindicator species. The case of Nakivubo wetland along Lake Victoria. *Physics and Chemsitry of the Earth, 32*, 1350-1358. http://dx.doi.org/10.1016/j.pce.2007.07.034

CCME (1991). *A protocol for the derivation of water quality guidelines for the protection of aquatic life.* Winnipeg, Manitoba: Canadian Council of Ministers of the Environment.

CCME (2007). *A protocol for the derivation of water quality guidelines for the protection of aquatic life.* Winnipeg, Manitoba: Canadian Council of Ministers of the Environment.

ECB (2003). *Technical guidance document on risk assessment in support of commission directive 93/67/EEC on risk assessment on new notified substances, commission regulation (EC) No. 1488/94 on risk assessment for existing substances and directive 98/8/EC of the European parliament and of the council concerning the placing of biocidal products on the market. part II. environmental risk assessment.* Ispra Italy.

Hose, G. C., & Van den Brink, P. J. (2004). Confirming the species sensitivity distribution concept for endosulfan using laboratory, mesocosm, and field data. *Archives of Environmental Contamination and Toxicology, 47*(4), 511-520. http://dx.doi.org/10.1007/s00244-003-3212-5

Johnson, A., Carew, E., & Sloman, K. A. (2007). The effects of copper on the morphological and functional development of zebrafish embryos. *Aquatic Toxicology, 84*, 431–438. http://dx.doi.org/10.1016/j.aquatox.2007.07.003

Karntanut, W., & Pascoe, D. (2002). The toxicity of copper, cadmium and zinc to four different Hydra (*Cnidaria: Hydrozoa*). *Chemosphere, 47*, 1059-1064. http://dx.doi.org/10.1016/S0045-6535(02)00050-4

National Technical Advisory Committee to the Secretary of the Interior. (1968). *Water quality criteria.* Washington DC.

OECD. (1992). *Report of the OECD workshop on the extrapolation of laboratory aquatic toxicity data to the real environment.* OECD environment monograph No. 59, Paris.

Posthuma, L., Suter II, G. W., & Traas, T. P. (2002). *Species sensitivity distributions in ecotoxicology.* Boca Raton, CRC: Lewis Publishers.

Priel, T., & Hershfinkel, M. (2006). Zinc influx and physiological consequences in the β-insulinoma cell line, Min6. *Biochemical and Biophysical Research Communications, 346*, 205–212. http://dx.doi.org/10.1016/j.bbrc.2006.05.104

Rachel L. D., Andrew C. J., Claudia M., Pereira M. G., & John P. S. (2014). Using risk-ranking of metals to identify which poses the greatest threat to freshwater organisms in the UK. *Environmental Pollution, 194*, 17-23. http://dx.doi.org/10.1016/j.envpol.2014.07.008

Redeker, E. S. & Blust, R. (2004). Accumulation and toxicity of cadmium in the aquatic Oligochaete *Tubifex tubifex*: A kinetic modeling approach. Environmental Science & Technology, *38*(2), 537-543. http://dx.doi.org/10.1021/es0343858

Shcheglov, V. V., Moiseichenko, G. V., & Kovekovdova, L. T. (1990). Effect of copper and zinc on embryos, larvae and adult individuals of the sea urchin *Strongylocentrotus intermedius* and the sea cucumber *Stichopus japonicus. Biological Morya, 3*, 55-58.

Sloof, W. (1992). RIVM documents. *Ecotoxicological effect assessment: deriving maximum tolerable concentrations (MTCs) from single-species toxicity data.* Report No. 719102018. National Institute for Public Health and the Environment (RIVM), Bilthoven, the Netherlands.

Traas, T. P. (2001). *Guidance document on deriving environmental risk limits*. Report No. 601501012. Bilthoven, the Netherlands: National Institute of Public Health and the Environment.

USEPA. (1985). *Guidelines for deriving numerical national water quality criteria for the protection of aquatic organisms and their uses*. PB 85-227049 Washington D C.

USEPA. (1994). *Water quality standards handbook.* Washington D C.

USEPA. (2009). National recommended water quality criteria. Washington D C.

van Vlaardingen, P. L. A., & Verbruggen, E. M. J. (2007). *Guidance for the derivation of environmental risk limits within the framework of 'international and national environmental quality standards for substances in the Netherlands' (INS)*. National Institute for Public Health and the Environment (RIVM), Bilthoven, the Netherlands.

Van Vlaardingen, P. L. A., Traas, T. P., Wintersen, A. M., & Aldenberg, T. (2004). *ETX2.0 - A program to calculate hazardous concentration and fraction affected, based on normally distributed toxicity data*. Report 601501028. National Institute for Public Health and the Environment (RIVM), Bilthoven, the Netherlands.

Wheeler, J. R., Grist, E. P. M., Leung, K. M. Y., Morritt, D., & Crane, M. (2002). Species sensitivity distributions: data and model choice. *Marine Polluttion Bulletin, 45*, 192-202. http://dx.doi.org/10.1016/S0025-326X(01)00327-7

Wu F. C., Feng C. L., Cao Y. J., Zhang R. Q., Li H. X., Liao H. Q., & Zhao X. L. (2011). Toxicity characteristic of zinc to freshwater biota and its water quality criteria. *Asian Journal of Ecotoxicology, 6*(4), 367-382.

9

Trends in Illegal Killing of African Elephants (*Loxodonta africana*) in the Luangwa and Zambezi Ecosystems of Zambia

Vincent R. Nyirenda[1], Peter A. Lindsey[2], Edward Phiri[3], Ian Stevenson[4], Chansa Chomba[5], Ngawo Namukonde[1], Willem J. Myburgh[6] & Brian K. Reilly[6]

[1] Department of Zoology and Aquatic Services, School of Natural Resources, Copperbelt University, Kitwe, Zambia

[2] Department of Zoology and Entomology, Mammal Research Institute, University of Pretoria, Pretoria, South Africa

[3] Lusaka Agreement Task Force, Nairobi, Kenya; Directorate of Conservation and Management, Zambia Wildlife Authority, Chilanga, Zambia

[4] Conservation Lower Zambezi, Chirundu, Zambia

[5] Disaster Management Training Centre, School of Agriculture and Natural Resources, Mulungushi University, Kabwe, Zambia

[6] Department of Nature Conservation, Faculty of Science, Tshwane University of Technology, Pretoria, South Africa

Correspondence: Vincent R. Nyirenda, Department of Zoology and Aquatic Sciences, School of Natural Resources, Copperbelt University, Kitwe, Zambia, Email: nyirendavr@hotmail.com

Abstract

The resurgence in African elephant (*Loxodonta africana*) poaching for ivory and bushmeat threatens the persistence of elephant populations, continent wide. In addressing the scourge, monitoring of illegal killings of elephants plays a key role in effectively directing counter measures. This study evaluated spatiotemporal trends and patterns in elephant poaching. Illegal killing of elephants occurred mostly along major rivers, mainly in late dry season during which period elephants were more vulnerable to illegal exploitation. However, during the wet season, retaliatory killings of "problem elephants" marauding crop fields also took place. Elephant poaching was attributed to socio-economic and ecological drivers such as high poverty levels, weak governance, high demand for elephant ivory, and low social capital. These drivers are likely to apply to other elephant range states as well. We propose that local strategies that empower communities economically, build broad-based law enforcement capacity in stakeholders to counter illegal killing of elephants, and which positively shift the risk/reward ratio for ivory poachers trade, be urgently developed and implemented.

Key words: CITES, elephant poaching , ivory trade, resource monitoring

1. Introduction

The risk of local extinction of African elephants (*Loxodontaafricana*) in the 37 range states is a growing concern. Illegal wildlife trade is driving global extirpation of populations of commercially valuable species (Wittemyer, Northrup, Blanc, Douglas-Hamilton, & Omondi, 2014). African elephants are illegally killed for ivory and bushmeat in increasingly unsustainable numbers (Bennett et al., 2007; Lindsey et al., 2013; Maiselset al., 2013; Nellemann, Formo, Blanc, Skinner, Milliken, & De Meulenaer, 2013; Wittemyer, Daballen, & Douglas-Hamilton, 2013). Though wildlife poaching results in modest returns for entire communities, such illegal activity is often attractive to individuals because of the high associated private returns and limited benefits from legal wildlife based land uses (Lindsey et al., 2011; Lindsey et al., 2013; Rentsch & Damon, 2013). Furthermore, elephants often impose significant costs to communities via damage to crops (Lamarque et al., 2009). Lastly, commercial poaching is often conducted by individuals who are from areas distant from the wildlife areas, whose motives are unaffected by local benefits to communities derived from wildlife. This suite of issues creates challenges to reduce illegal killing of elephants.

There has been a major resurgence in the illegal elephant killing in the last few years, resulting in the decline and even local extinction of some elephant populations (Douglas-Hamilton, 2009; Poulsen, Clark, Mavah, & Elkan, 2009; CITES et al., 2013; Wittemeyer et al., 2014). Subsequently, range states and international non-governmental organisations (NGOs) held the first African Elephant Summit in Gaborone, Botswana from 2-4 December 2013 to agree on urgent measures to counteract this resurgence. These measures combine market and non-market based strategies to address demand and supply of elephant parts and derivatives (IUCN, 2013). Non-market based approaches deal with issues such as governance and collaborative action in law enforcement.

In the Luangwa Valley, Zambia, African elephant populations experienced excessive illegal exploitation in the 1970's and 80's, primarily for ivory and later for both ivory and bushmeat (Leader-Williams, Albon, & Berry, 1990; Jachmann & Billiouw, 1997; Dalal-Clayton & Child, 2003) (Table 3). Lower Zambezi ecosystem also experienced excessive elephant poaching during the late 1970s and early1980s, and the illegal harvest of elephants has continued until present (Table 4). Despite the ban by the Convention on International Trade in Endangered Species of Wild Fauna and Flora (CITES) in 1989 (Lausanne, CoP7) on international trade in elephants, their parts and derivatives, illegal killing of elephants continues unabated.

Monitoring of the Illegal Killing of Elephants (MIKE) programme was established in 60 selected sites of the elephant range in Africa and Asia by CITES (Bangkok, Conf. 10.10, Rev. CoP16). One of the 31 MIKE sites in Africa is South Luangwa National Park in Zambia. The MIKE programme provides information for making management and enforcement decisions, improving ability to monitor elephant populations, detecting changes in levels of illegal killing, and assessing the effectiveness of law enforcement. Information collection centres on elephant population sizes and trends, incidence and trends in illegal killing, and measures of the effort employed for the detection and prevention of illegal killing. The methods applied in MIKE sites for monitoring illegally killed elephants have also been replicated for non-MIKE sites such as Lower Zambezi ecosystem. In addition to MIKE programme related monitoring, assessing and identifying underlying local drivers of elephant poaching remain critical (Jachmann, 2012). This study attempts to achieve site-based understanding on elephant poaching to provide bases from which to make management decisions.

2. Methods and Materials

2.1 Study Sites

The study was conducted across three sites in Zambia: a) Lower Zambezi National Park (NP) (4,092 km^2) and Chiawa Game Management Area (GMA) (2,344 km^2) (hereafter collectively referred to as the 'Lower Zambezi', Figure 1); b) North Luangwa NP (4,636 km^2), Mukungule GMA (6,500 km^2), Musalangu GMA (17,350 km^2) and Munyamadzi GMA (3,300 km^2) (hereafter referred to as 'North Luangwa') (Figure 2); c) South Luangwa NP (9,050km^2) and Lupande GMA (4,840km^2) (which together comprise the MIKE site) (hereafter referred to as 'South Luangwa' - Figure 3). More than 72% of Zambia's elephant population occur in the Luangwa Valley, being elephant stronghold (McIntyre, 2004; Chomba, Simukonda, Nyirenda, & Chisangano, 2012). Unlike in the GMAs, human habitation is excluded from national parks except for park management purposes (CSO, 2012). Inhabitants of the GMAs are mainly subsistence farmers, though some derive income from largely illegal bushmeat hunting.

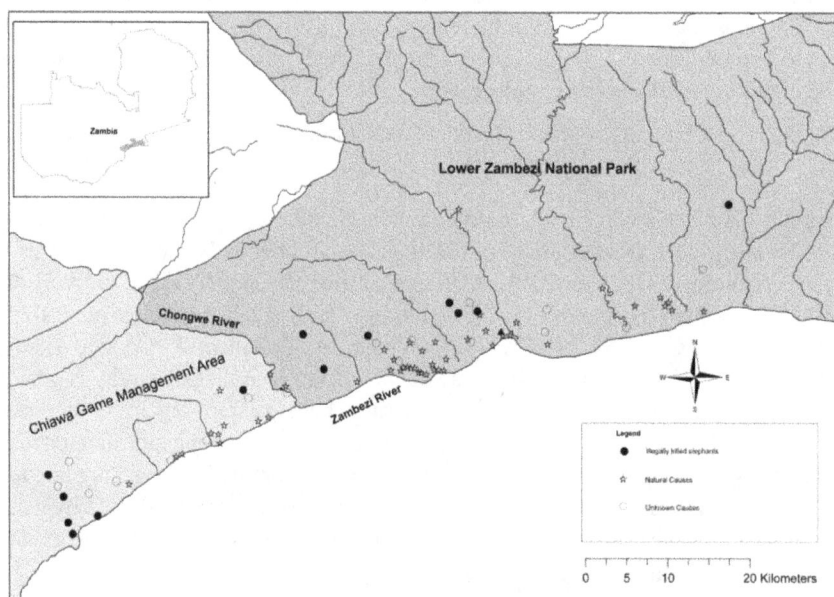

Figure 1. Geographical locations of illegal killings of elephants encountered in the Lower Zambezi ecosystem, Zambia, 2012-2013

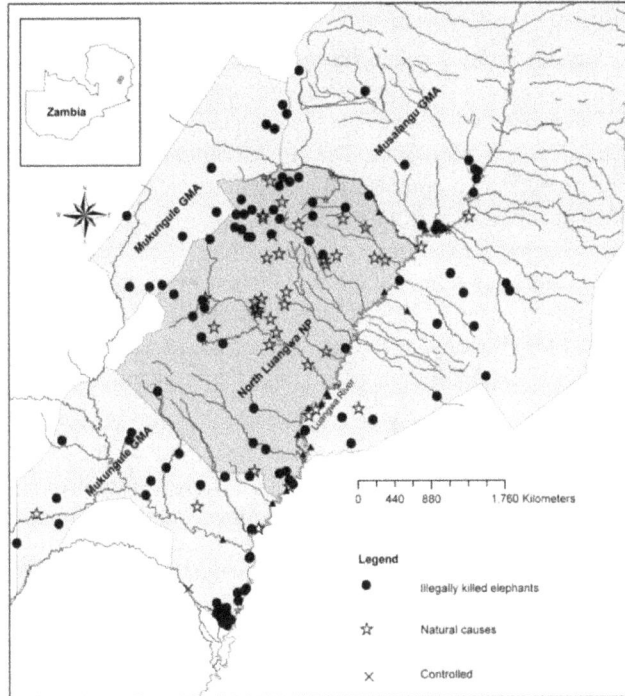

Figure 2. Geographical locations of illegal killings of elephants encountered in the North Luangwa ecosystem, Zambia, 2007-2013

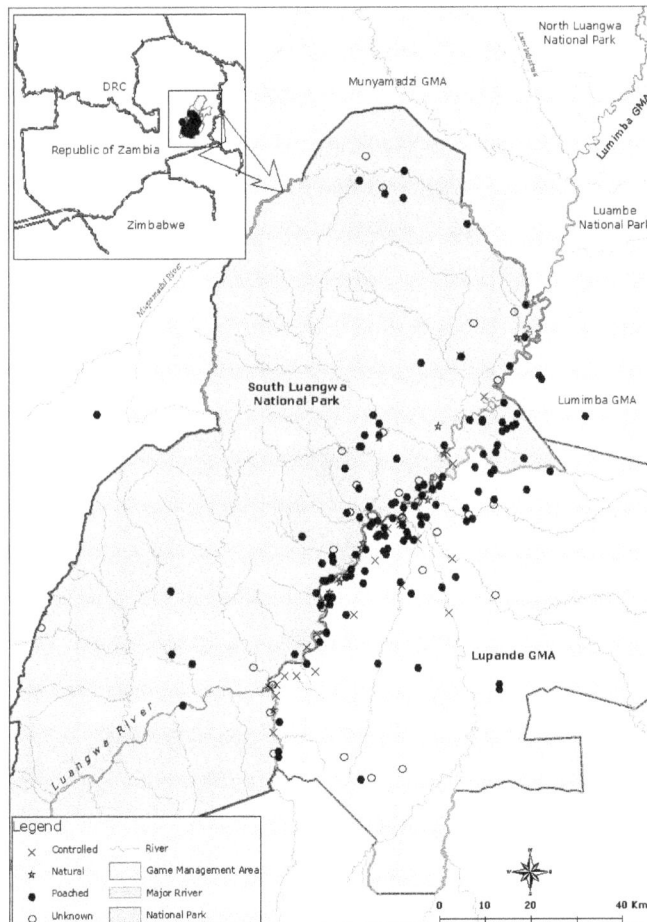

Figure 3. Geographical locations of illegal killings of elephants encountered in the South Luangwa MIKE site, eastern Zambia, 2010-2013

2.2 Assessing Elephant Mortalities

Trained patrol teams collected elephant mortality data during January 2012 to December 2013 in the Lower Zambezi, January 2007 to December 2013 in North Luangwa, and January 2010 to December 2013 in the South Luangwa MIKE site. There were 6-8 members in each patrol team. Occasionally, the patrol teams were assisted by use of aircraft in the detection of elephant mortalities. During surveillance flights, any sightings of elephant mortalities were georeferenced, recorded and communicated to patrol teams for physical inspection. Patrol teams searched the sites for evidence of elephant poaching. We assumed that mortality incidences and the detection rates of carcasses were unaffected by the cause of death. For each carcass, the geographical location, age of the elephant, sex, cause of death, and whether poachers had removed ivory and meat, were recorded. Data from legal elephant sports hunting, which occurs on a small scale CITES voluntary quota of 60 individuals per annum, were not included. There were four age classes for allotment of individual carcass: fresh (≤1 month), recent (>1 month and ≤1 year), old (>1 year) or very old (≤10 years), following Douglas-Hamilton and Hillman (1981).

The age and sex determination of elephant carcasses were conducted following Moss (1996). Sex was determined by tusk shape and weight (where tusks were available), head shape and genitalia. The cause of death was evaluated following Roffe, Friend, and Locke (1996) and was classified as: Poached by poisoning; Poached by snaring; Poached by shooting; Natural mortality; Management problem animal control; Accident or Unknown. Sex or cause of death were not denoted for elephant carcasses in the Old and Very Old carcass classes as it was impossible to identify. 'Intact'; 'Removed'; 'Naturally absent'; or 'Recovered' formed status of elephant ivory. 'Found'; 'Recovered'; 'Removed'; or 'Parts removed' constituted the elephant meat status. The season in which the carcass was found was recorded as: 'Late wet season' (January-March); 'Early dry season' (April-June); 'Late dry season' (July-September); and 'Early wet season' (October-December). Proportion of Illegally Killed Elephants (PIKE), defined as the total number of illegally killed elephants found divided by the total number of carcasses found per year (CITES et al., 2013; Nellemann et al., 2013), depicted severity of elephant poaching. Mean PIKE values were calculated within a 95% confidence interval.

We determined the impact of mortalities on the elephant population of the Luangwa Valley, Zambia, where the mortalities due to poaching were relatively high (Figure 4). Aerial survey reports (Tables 3, 4 and 5) gave data on elephant populations. D-tests (Norton-Griffiths, 1978) were used to test the significance of population changes between episodes of comparable population estimates. Calculations of carcass ratios provided further insights into the impact of elephant mortalities on populations. Carcass Ratio=(Number of dead elephants)/(Number of dead elephants+Number of live elephants)*100 (Douglas-Hamilton & Burrill, 1991).

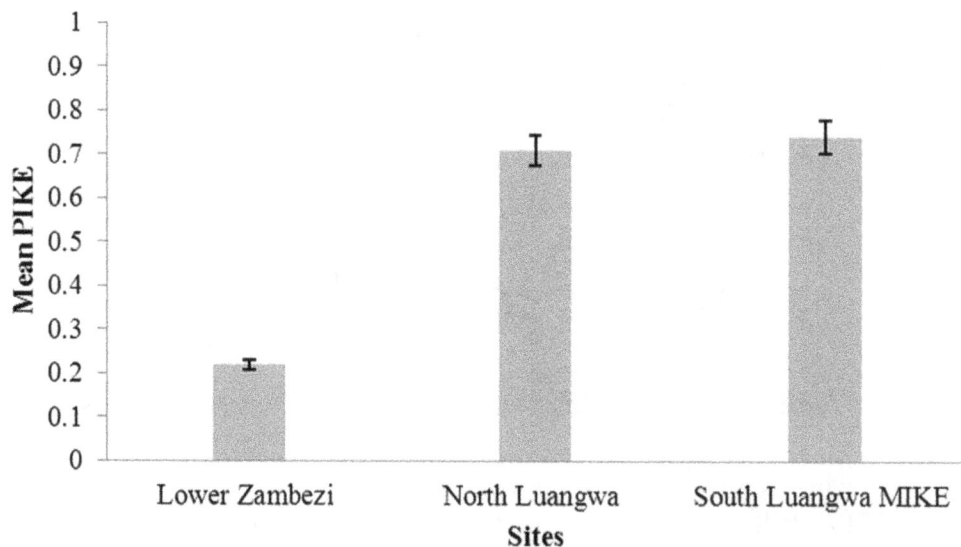

Figure 4. Trends in proportional illegal elephant mortalities encountered in Luangwa and Lower Zambezi ecosystems of Zambia, between 2007-2013

3. Results

3.1 Spatiotemporal Distribution of Elephant Mortalities

At Lower Zambezi, there were 87 elephant mortalities during 2012-2013. Most mortalities occurred within riparian zones of Zambezi River and its tributaries (Figure 1). The PIKE value declined from 32.1% (n=9) in 2012 to 17.0% (n=10) in 2013, and averaged 21.8% (Figure 4). Management control operations accounted for 1.1% (n=1), natural causes of death contributed more than half of mortalities at 55.2% (n=48) and unknown causes comprised 21.8% (n=19) of total mortalities. These mortalities were detected largely in late dry season and early wet season (Figure 5).

During 2007-2013, North Luangwa experienced 202 elephant mortalities. Many elephant mortalities occurred along the river systems, particularly during the dry season. However, there were also substantial mortalities away from rivers (Figure 2). This finding contrasted with the situations in South Luangwa and Lower Zambezi. In North Luangwa, the PIKE value increased from 50.0% in 2007 to 76.9% in 2013 and averaged 70.9% (Figure 4). Illegally killed elephants totalled 70.3% (n=142) of the total mortalities. Management control operations accounted for 0.99% (n=2) and natural causes of death contributed 28.7% (n=58) of mortalities. Elephant poaching was detected most commonly in the late dry season (Figure 5).

There were 215 mortalities recorded in South Luangwa during 2010-2013. Most elephant poaching occurred close to the Luangwa River and its tributaries (Figure 3). The PIKE value was 58.5% in 2010, increased to 81.3% in 2013, and averaged 74.1 % (Figure 4). Illegally killed elephants comprised 64.7% (n=139) of total mortalities. Other mortality causes included: management control operations (15.3%; n=33); natural deaths (5.6%; n=12); and unknown causes (14.4%; n=31). Most elephant poaching was detected in late dry season (Figure 5).

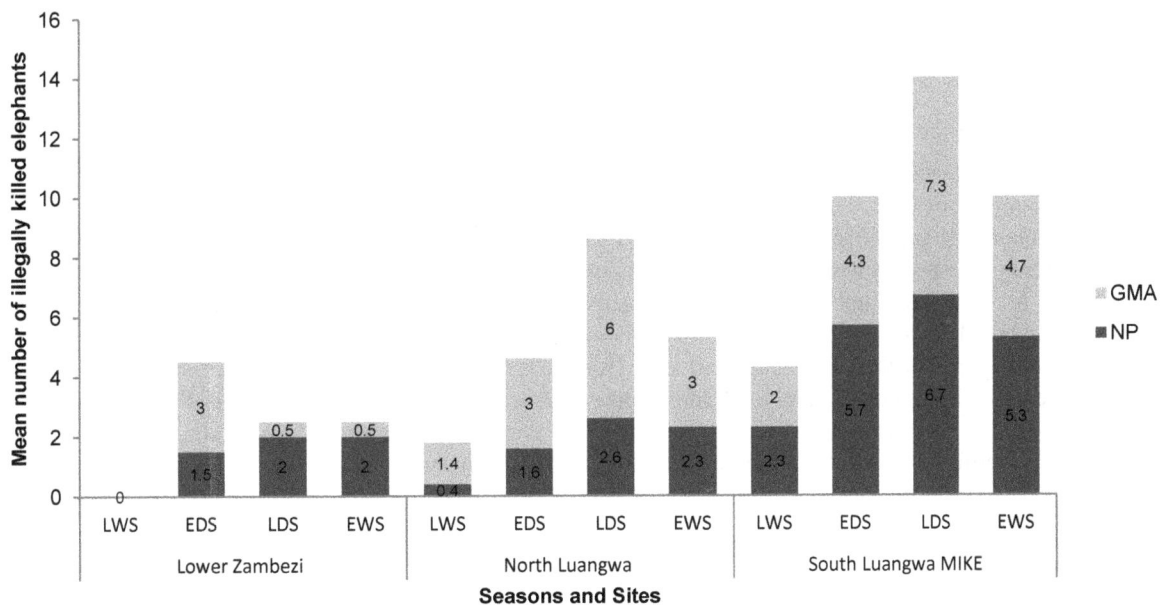

Figure 5. Mean number of illegally killed elephants in Luangwa and Zambezi ecosystems of Zambia, 2007-2013

3.2 Elephant Age and Carcass Status

Adult males were the most impacted upon age-sex category (Table 1). Poached elephants were typically adults or sub-adult, though young elephants were also snared or shot by local farmers and poachers in some cases.

Table 1. Elephant carcass status and age by gender in Lower Zambezi and South Luangwa ecosystems, Zambia, 2010-2013

Carcass Status-Age	Lower Zambezi, 2012-2013				South Luangwa MIKE, 2010-2013			
	Male	Female	Unknown	Total	Male	Female	Unknown	Total
Fresh-Adult	18 (5)	8 (2)	5 (3)	31(10)	106(73)	19(15)	0	125(88)
Fresh-Sub~adult	7 (1)	4	5	16(1)	21(16)	7(7)	0	28(23)
Fresh-Young	4		6	10	3(3)	3(2)	0	6(5)
Fresh-Unknown	0	0	0	0	0	0	0	0
Recent-Adult	6 (2)	(2)	1	7(4)	21(18)	4(3)	0	25(21)
Recent-Sub~adult	1	5	2	8	2(2)	0	0	2(2)
Recent-Young	0	0	4 (1)	4(1)	0	0	0	0
Recent-Unknown	1	2(2)	2(1)	5(3)	0	0	0	0
Old-Adult	3	0	0	3	0	0	20	20
Old-Sub~adult	0	0	0	0	0	0	4	4
Old-Young	1	0	0	1	0	0	1	1
Old-Unknown	0	0	3	3	0	0	0	0
Very Old-Adult	0	0	0	0	0	0	2	2
Very Old-Sub~adult	0	0	0	0	0	0	2	2
Unknown	0	0	0	0	0	0	0	0
Total	**41(8)**	**19 (6)**	**28 (5)**	**88(19)**	**153(112)**	**33(27)**	**29**	**215(139)**

N.B: In brackets in Table 1 are the numbers of elephants illegally poached and without brackets are total elephant mortalities.

3.3 Illegal Meat and Ivory Extractions

The motivation for elephant poaching was the acquisition of meat and ivory in all the three monitoring sites (Table 2). In North Luangwa however, elephants were killed primarily for ivory (Table 2). In some cases, illegally killed elephants had their tails cut off (e.g. 15.8% in North Luangwa, and 42.1% in Lower Zambezi). In other cases, neither meat nor ivory was removed from elephant carcasses, yet the elephants had evidence of having been either snared or shot. In the Lower Zambezi, however, elephants suspected to have died from anthrax, a disease caused by a bacterium called *Bacillus anthracis* had meat also intact. In 2013, some meat was found intact in North Luangwa, where three poached elephants and over 300 vultures of different species were associated with poisoning. The wildlife was poisoned by toxic carbamate pesticides called Carbofuran, traded as Furidan. In 2010, Aldicarb traded as Temik poison was used by poachers to kill an unknown number of elephants. The culprit was arrested by the wildlife authority and confessed to using Temik in elephant poaching.

Table 2. Percentage of meat and ivory extracted and a total number of illegally killed elephants in Luangwa Valley, eastern Zambia, 2007-2013

Parameter	Lower Zambezi, 2012-2013			North Luangwa, 2007-2013								South Luangwa MIKE site, 2010-2013				
	Year		Total No. Killed	Year							Total No. Killed	Year				Total No. Killed
	2012	2013		2007	2008	2009	2010	2011	2012	2013		2010	2011	2012	2013	
Meat and ivory poached	50.00	0	5	75.00	33.33	16.67	33.33	31.58	18.18	9.30	27	54.17	60.98	0	29.27	50
Meat and ivory intact	0	33.33	3	0	0	0	16.67	5.26	0	11.63	8	8.33	14.63	37.50	12.20	25
Meat intact but ivory poached	40.00	22.22	6	25.00	66.67	58.33	33.33	36.84	81.82	58.14	75	20.83	12.20	3.13	29.27	23
Meat poached but ivory intact	10.00	44.44	5	0	0	16.67	16.67	15.79	0	18.60	15	12.50	2.44	59.38	29.27	35
Meat poached but ivory absent (tusklessness)	0	0	0	0	0	8.33	0	10.53	0	2.33	4	4.17	4.88	0	0	3
Meat intact but ivory naturally absent (tusklessness)	0	0	0	0	0	0	0	0	0	0	0	0	4.88	0	0	3
Total %	100.00	100.00	-	100.00	100.00	100.00	100.00	100.00	100.00	100.00	-	100.00	100.00	100.00	100.00	-
Total number of illegally killed elephants	10	9	19	4	6	12	12	19	33	43	129	24	41	33	41	139

3.4 Impacts of Mortalities on Elephants Populations

Elephant populations declined remarkably in the Luangwa Valley ecosystem during 2009-2012, primarily due to high levels of poaching (Frederick, 2013). A number of elephant carcasses is believed to have been underestimated in the Luangwa Valley during that period, hence, low carcass ratios recorded (Table 3). The detection levels were low compared to data sets generated by field patrol teams in Lower Zambezi and Luangwa ecosystems (Table 4 and 5). Field patrol data also incorporated data handed over from aerial surveillances for physical mortality verification. In addition, the large differences in aerial population estimates are due to variations in methods, resulting in high variances and wide confidence limits. There is varied carcass ratios sequel to high variances such that they became incomparable.

Table 3. Trends of Luangwa Valley System elephant populations and the carcasses detected from aerial surveys, 2009-2012

Year	Pop. Est.	Carcass	CR	*d*-test (comparison with 2012)
2009	12,352	49	0.40	-3.97
2011	10,649	35	0.33	-3.05
2012	6,361	104	1.61	-

Adapted from the Fredrick (2012) unpublished report data

Table 4. Comparison of African elephants in the Lower Zambezi National Park (old and fresh not separated)

Year	Elephant Pop. Est.	Carcasses	Carcass Ratio	Sources
1970	564	-		Zyambo & Simwanza, 2003
1991	374	60	13.82	Zyambo & Simwanza, 2004
1994	26	13	33.33	Zyambo & Simwanza, 2005
1995	112	66	37.08	Zyambo & Simwanza, 2006
1996	116	6	4.92	Zyambo & Simwanza, 2007
2003	1303	14	1.06	Simwanza, 2003
2005	1710	7	0.41	Simwanza, 2005
2008	289	65	18.36	Simukonda, 2008
2013	2200	15	0.68	Viljoen, 2013

Table 5. Comparison of African elephants in the North Luangwa National Park (Only fresh carcasses included)

Year	NLNP	Carcasses	Carcass Ratios	Sources
1973	17700	-	-	Caughley & Goddard, 1975
1979	7360	-	-	Douglas-Hamilton et al., 1979
1985	5282	-	-	Lewis, 1985
1989	5077	-	-	Owens, 1993
1989	3170	-	-	Owens, 1993
1992	1650	3	0.18	Owens, 1993
1993	2282	5	0.22	Owens, 1993
1994	2000	5	0.25	Owens et al., 1994
1998	767	0	0.00	Aucamp, 1998
1999	1167	2	0.17	Aucamp, 1999
2000	1599	0	0.00	Aucamp, 2000
2001	3750	-	-	Aucamp, 2001
2003	3235	8	0.25	Aucamp, 2003
2009	3749	0	0.00	WCS, 2010

3.5 Illegal Activities Correlates With Elephant Mortalities

The Pearson correlation test for the five serious offences associated with the elephant poaching indicated that there was insignificant correlation between illegal killings of elephants and gunshot heard, footprint pairs, drying racks, poachers' camps or snares encountered during the foot surveillance patrols in the Lower Zambezi ecosystem. Gunshots heard, footprint pairs, drying racks, poachers camps and snares encountered were insignificantly correlated with poached elephants (Pearson correlation=0.002, $p < 0.995$; -0.248, $p < 0.413$; -0.216, $p < 0.479$; 0.004, $p < 0.990$; 0.105, $p < 0.732$ respectively). There were 93 gunshots heard on foot patrol per annum (median), with a range of 31 – 198 gunshots during the period between 2001 and 2013. Footprint pairs encountered on patrols ranged from 4 – 58 footprint pairs per annum, with a median of 17 footprint pairs. The drying racks encountered ranged from 0 – 38, with a median of 5 drying racks per annum while poachers camps encountered were between 0 – 49, with median of 4 poachers camps per annum. Encounters of snares ranged from 0 – 122, with a median of 15 snare encounters per annum. Figure 6 summaries the annual correlates between 2001 and 2013.

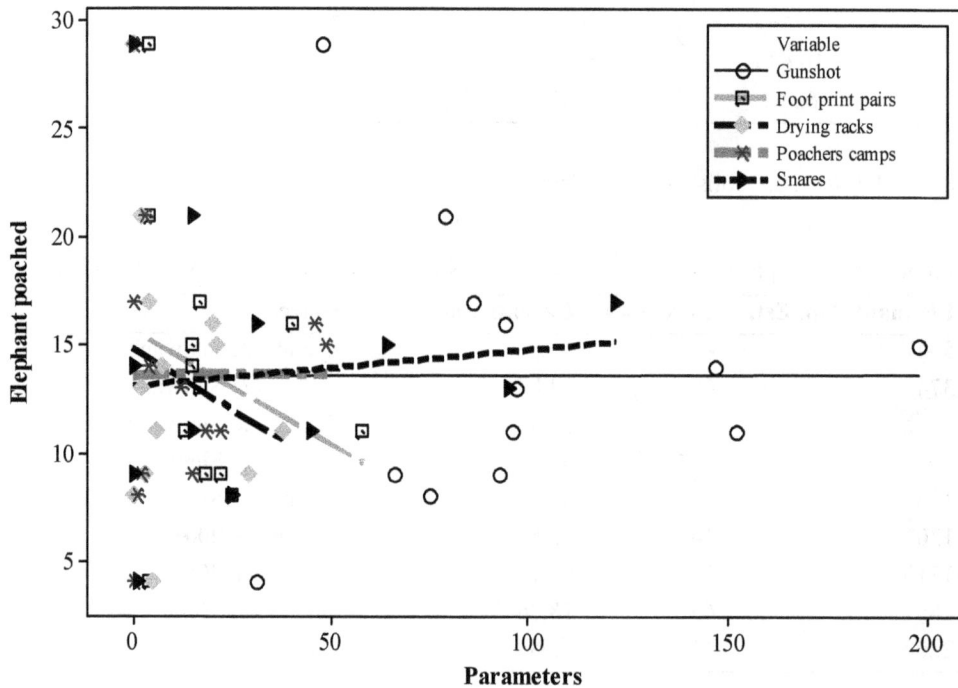

Figure 6. Annual correlates between poached elephants and five serious offences in the Lower Zambezi ecosystem, Zambia, 2001-2013

4. Discussion

The data from the three sites provided insights into elephant poaching. The ground-based method of monitoring illegally killed elephants; though labour intensive and covering a vast elephant range, proved useful as it provided key insights into trends in the elephant poaching. The underlying motives for the illegal killing of elephants appear to have been primarily ivory trafficking, the bushmeat trade and retaliatory killings.

Simukonda(2008) assessed that elephant populations in Luangwa Valley were either stable or increasing, based on 1.57% elephant carcass ratio for the Luangwa ecosystem. However, more recently, Frederick (2013) raised concerns over the increase in illegal killings of elephants in the Luangwa Valley. Based on 2012 population estimates by Frederick (2013) combined with the MIKE data, we derive 5.17% elephant-carcass ratio in South Luangwa MIKE area. This maybe unsustainable as it exceeds the natural elephant population growth rate of 5% (Dunham, 2012; CITES et al., 2013). Based on the 2012 aerial survey results of the Lower Zambezi (Viljoen, 2013), the calculated elephant carcass ration is 2.68%. Our study re-emphasises elephant poaching threats as there is a growing trend of poaching (Figure 4 and Table1). For instance, high numbers of fresh and recent carcasses in Table 1 offer evidence of increasing threat of poaching to the Luangwa Valley elephant populations. PIKEs of greater than 0.5-0.54 or 50-54% are precursors of severe poaching threats to elephant populations, resulting in declining populations (Nellemann et al., 2013; Wittemyer et al., 2014). Despite low ranger (scout) density of approximately 1 scout to 208 km^2 (1:208) in Lower Zambezi in comparison to 1:50 and 1:36 in North and South Luangwa respectively, the PIKE for Lower Zambezi exceptionally remained lower than 54% (Figure 4).

Though there were comparatively fewer illegally killed elephants annually in the three sites evaluated than some sites in Africa, they were higher than the average recorded in Southern Africa (CITES et al., 2013; Nellemann et al., 2013). Most elephant poaching occurred along the river systems in late dry seasons, as earlier stated. Elephant distribution was dependent on the water availability in dry and semi-dry areas (Chamaillé-Jammes, Valeix, & Fritz, 2007; de Beer & van Aarde, 2008). In dry season, elephants concentrate along the riparian environments in the Luangwa Valley (Simukonda, 2008; Chomba et al., 2012) and during the same season, local farmers rest from their main activity of subsistence agriculture and engage more in other economic activities, including in some cases illegal hunting of wildlife (Gibson & Marks, 1995; Lewis, 2007). During the late dry season, mangoes (*Mangifera indica*) are ripe and elephants tend to devour the mangoes in the villages but also forage in vegetable gardens. Mangoes are common in villages in the Luangwa and Lower Zambezi, and invading elephants pose a threat to human life, and usually trigger retaliatory killings of elephants by inhabitants. However, a greater threat of poaching from other parts of the Luangwa Valley, also tend to confine elephants to the Luangwa Valley floor even during wet season, resulting in 'compression effect' (Lewis, 1986). In the Zambezi ecosystem, seasonal elephant movements between the escarpment and the valley floor have been well documented and are attributed to inter-seasonal pasture pressures (ZAWA, 2001).

Presumably due to their relatively larger body and ivory sizes (Moss, 1996), adult and sub-adult male elephants are targeted by poachers more than female elephants. Further, the status of tusklessness of a proportion of elephant population in the Luangwa Valley may have positive impact on containing levels of poaching of female elephants for ivory (Jachmann, Berry, & Imae, 1995). During the wet season, local people either snare or shoot problem elephants that enter their fields and damage crops. Elephants that are not killed instantly retreat and eventually die in secluded places along the Luangwa River and its tributaries. Relatively high levels of poached elephants found with both meat and ivory intact (Table 2) are attributed to retaliatory killings and, in some cases, to intervening action from law enforcement officers. Further, the motives for collection of tails are unclear but probably for traditional rituals, and this threat requires forensic investigations. The recent practice of use of poison in elephant poaching also needs to be decisively addressed.

Despite being relatively well managed (Mwima, 2007), the study sites have a number of socio-economic and ecological drivers for elephant poaching that need to be addressed. These are characterized as: (1) high poverty levels among the rural populace, (2) inadequate legal benefits from wildlife utilisation, (3) weak governance, low social capital and prevalent corruption, (4) high demand for elephant ivory locally and abroad, and (5) inadequate natural resource monitoring and law enforcement.

With high human population growth of over 3.8% (regional mean: 2.6%) per annum in Lupande GMA (CSO, 2012), lack of protein and low rate of employment, elephant poaching is accentuated. Some key features exacerbating illegal killing of elephants in the Luangwa Valley include retaliatory killings of elephants by local farmers for crop raiding by marauding elephants (Nyirenda, Chansa, Myburgh, & Reilly, 2011) and high incidences (60 %) of droughts (Gilvear, Winterbottom, & Sichingabula, 2000), leading to food insecurity. Human influx in GMAs is common and poses a threat to wildlife persistence (Watson, Becker, McRobb, &Kanyembo, 2013). Poverty levels in Zambia's rural areas, measured by UNDP Human Development Index remain high and typical of several other elephant range states (UNDP, 2011; World Bank, 2013). In such situations where black markets exist and non-poaching wages are low and poverty is prevalent, elephant poaching occurs. The benefits are greater than the costs and risks associated with poaching (Messer, 2010; Knapp, 2012). Proximity to urban markets increases demand for elephant products (Leader-Williams, Albon, & Berry, 1990). Therefore, there is little incentive for skilled hunters to desist from poaching, which provides potentially high private benefits, for the sake of wildlife conservation. In this case, transformation of local hunters to conservationists insinuated by Gibson and Marks (1995) is put to test. The legal benefits acquired from wildlife are often inadequate to sufficiently provide incentives against illegal hunting of elephants (Dalal-Clayton & Child, 2003).

Weak governance in traditional structures of local communities, low social capital and perceived and actual corruption (Nyirenda, Chansa, Myburgh, & Reilly, 2010), undermine strategies to counter elephant poaching. Further, rampant human settlements and elephant habitat encroachments in GMAs of the Luangwa Valley facilitate elephant by-catch from wire snaring (Watson et al., 2013; Becker et al., 2013). Similarly, human encroachment in the form of human habitation, clearings, charcoal production and cultivation are widely spread in the Zambezi ecosystem (Viljoen, 2013). Local communities are also tempted to poach elephants through the increasing prices for elephant products, which are driven by increasing demand in consumer countries such as China (Nellemann et al., 2013). The rising price of the elephant and other poached products has also attracted organized crime to wildlife trading. Partly as a result of the high potential rewards, elephant poaching continues despite the harsh penalties associated with the illegal killing of the species and trading in elephant parts. Based on Zambia's wildlife legislation, illegal killing and trafficking of elephants attract penalties not less than 5 years and 7 years respectively and not exceeding a term of imprisonment of 20 years and 25 years respectively, without the option of a fine. Law enforcement efforts are undermined by funding shortages for protected area management (Lindsey et al. 2014), and by low salaries for law enforcement officers, which makes them vulnerable to being coerced into assisting ivory traders. Such factors may compromise the effectiveness of enforcement systems designed to save charismatic species (Bennett, 2011).

There is a need for improving and intensifying law enforcement efforts through implementation of rigorous systems to ensure that field officers are motivated trained, fittest, strongest people possible and to improve the working conditions of field officers. In the national parks, this would require elevated funding for Zambia Wildlife Authority (ZAWA) or co-management through partnerships between ZAWA, Non-Governmental Organisations and private sector. Wildlife populations in protected areas in Zambia with co-management agreements in place appear to be faring better than in parks without such support (Lindsey et al. 2014). There is currently co-management in place for North Luangwa and Lower Zambezi National Parks, and support is provided to ZAWA for law enforcement by an NGO in South Luangwa National Park. Other NGOs also involved provide support for law enforcement in some of the GMAs around those parks. Containing poaching of wildlife species requires financial and technical support towards local alternative and beneficial livelihoods in the adjacent GMAs.

In addition to elevated funding for law enforcement and stricter penalties for poaching and trafficking in elephant products, there is a need for other tools and regulations such as anti-money laundering and asset forfeiture legislation to be employed in support of wildlife legislation. There is a need to sustain the monitoring of illegal killings of elephants in protected areas and encourage use of resultant data by field staff to form the scientific bases for interventions. Efforts are required to tackle ivory and bushmeat traders in addition to the people who poach elephants. Collectively, such efforts would shift the risk/reward ratio for the would-be elephant poachers and make the illegal activity less attractive.

In addition to 'stick' approaches, there is a need for strategies that increase support for conservation. In the areas neighbouring the protected areas, there is a need for strategies to enable communities to benefit from tourism in the parks, and to own and benefit from wildlife that occurs in the GMAs (Lindsey et al., 2013). Such measures should ideally be combined with allocation of land ownership to local communities and land use planning, to limit on-going unplanned human immigration to areas adjacent to national parks. In the GMAs, there is a need for durable and equitable joint ventures between local communities and the private sector or NGOs which allow for emergence of elevated anti-poaching and wildlife management. Efforts to gather intelligence from local communities should be increased and are likely to be more effective if they receive a fairer share of the benefits from wildlife-based land uses in the area.

5. Conclusion

Resurgence of elephant poaching is a great concern at sites, national and international levels. The emerging patterns from monitoring of illegal killings of elephants on parameters such as spatiotemporal patterns, proportion of elephants illegally killed to other types of mortality and gender aspects of illegally killed elephants are critical for decision-making. Corresponding specific and multiple counteractive strategies are required to be put in place to urgently address the scourge at sites' levels along with those at national and international levels. Such strategies would target the supply chain of elephant products and key drivers for illegal killing of elephants (Keane, Jones, Edwards-Jones, & Milner-Gulland, 2008; Challender & MacMillan, 2014).

Acknowledgements

This study was conducted with the support of Zambia Wildlife Authority. We are grateful to all the field staff, too many to itemise, who took part in the data collection. The geographical map (Figure 3) was produced by Benson Kabungo.

References

Becker, M., McRobb, R., Watson, F., Droge, E., Kanyembo, B., Murdoch, J., &Kakumbi, C. (2013). Evaluating wire-snare poaching trends and the impacts of by-catch on elephants and large carnivores. *Biological Conservation, 158*, 26-36.

Bennett, E. L. (2011). Another inconvenient truth: the failure of enforcement systems to save charismatic species. *Oryx, 45*(4), 476-479.

Bennett, E. L., Blencowe, E., Brandon, K., Brown, D., Burn, R. W., Cowlishaw, G., ...Wilkie, D. S. (2007). Hunting for consensus: reconciling bushmeat harvest, conservation and development policy in West and Central Africa. *Conservation Biology, 3*, 884-887.

Central Statistical Office (CSO). (2012). *Zambia 2010 Census of population and housing: population summary report.* Lusaka: Central Statistical Office.

Challender, D. W. S., & MacMillan, D. C. (2014). Poaching is more than an enforcement problem. *Conservation Letter.* http://dx.doi.org/10.1111/conl.12082

Chamaillé-Jammes, S., Valeix, M., & Fritz, H. (2007). Managing heterogeneity in elephant distribution: interactions between elephant population density and surface-water availability. *Journal of Applied Ecology, 44*, 625-633.

Chomba, C., Simukonda, C., Nyirenda, V., & Chisangano, F. (2012). Population status of the African elephant in Zambia. *Journal of Ecology and the Natural Environment, 4*(7), 186-193.

CITES, IUCN, SSC African Elephant Specialist Group, & TRAFFIC International. (2013). Status of African elephant populations and levels of illegal killing and the illegal trade in ivory: a report to the African Elephant Summit.

Dalal-Clayton, B., & Child, B. (2003). *Lessons from Luangwa: the story of the Luangwa Integrated Resource Development Project, Zambia.* Wildlife and Development Series 13. London: International Institute for Environment and Development.

de Beer, Y., & van Aarde, R. J. (2008). Do landscape heterogeneity and water distribution explain aspects of elephant home range in southern Africa's arid savannah? *Journal of Arid Environment, 72*, 2017-2025.

Douglas-Hamilton, I. (2009). The current elephant poaching trend. *Pachyderm, 45*, 154-157.

Douglas-Hamilton, I., & Burrill, A. (1991). Using elephant carcass ratios to determine population trends. *African Wildlife Research and Management, 1*, 98-105.

Douglas-Hamilton, I., & Hillman, A. K. K. (1981). Elephant carcasses and skeletons as indicators of population trends.From low-level aerial survey techniques workshop. African Monograph No. 4, Addis Ababa.

Dunham,K. M. (2012). Trend in populations of elephant and other large herbivores in Gonarezhou National Park, Zimbabwe, as revealed by sample aerial surveys. *African Journal of Ecology, 50*(4), 476-488.

Frederick, H. (2013). Aerial survey report: Luangwa Valley, 2012. COMACO, Lusaka.

Gibson, C. C., & Marks, S. A. (1995).Transforming rural hunters into conservationists: An assessment of community-based wildlife management programs in Africa. *World Development, 23*(6), 941-957.

Gilvear, D., Winterbottom, S., & Sichingabula, H. (2000). Character of channel platform change and meander development: Luangwa River, Zambia. *Earth Surface Processes and Landforms, 25*(4), 421-436.

IUCN (The World Conservation Union). (2013). African Elephant Summit, Gaborone, Botswana, 2-4 December 2013: urgent measures. Retrieved December 4, 2013, from http://cms.iucn.org/downloads/africa_elephant_summit_final_urgent_measures_3_dec_2013.pdf

Jachmann, H. (2012). Pilot study to validate PIKE-based inferences at site level. *Pachyderm, 52*, 72-87.

Jachmann, H., & Billiouw, M. (1997). Elephant poaching and law enforcement in the central Luangwa Valley, Zambia. *Journal of Applied Ecology, 33*, 1241-1250.

Jachmann, H., Berry, P. S. M., & Imae, H. (1995). Tusklessness in African elephants: a future trend. *African Journal of Ecology, 33*(3), 230-235.

Keane, A., Jones, J. P. G., Edwards-Jones, G., Milner-Gulland, E. J. (2008). The sleeping policeman: understanding issues of enforcement and compliance in conservation. *Animal Conservation, 11*(2), 75-82.

Knapp, E. J. (2012). Why poaching pays: a summary of risks and benefits illegal hunters face in western Serengeti, Tanzania. *Tropical Conservation Science, 5*(4), 434-445.

Lamarque, F., Anderson, J., Fergusson, R., Lagrange, M., Osei-Owusu, Y., & Bakker, L. (2009). *Human-wildlife conflict in Africa: causes, consequences and management strategies.* Rome: FAO.

Leader-Williams, N., Albon, S. D., & Berry, P. S. M. (1990). Illegal exploitation of black rhinoceros and elephant populations: pattern of decline, law enforcement and patrol effort in Luangwa Valley, Zambia. *Journal of Applied Ecology, 27*, 1055-1087.

Lewis, D. M. (1986). Disturbance effects on elephant feeding: evidence for compression in Luangwa Valley, Zambia. *African Journal of Ecology, 24*, 227-241.

Lewis, D. M. (2007).Opportunities and constraints for protected area management through increased connectivity to local livelihood needs in surrounding border areas: lessons from Luangwa Valley, Zambia. In: Redford K. H. and Fearn E, eds., *Protected areas and human livelihoods*. Working Paper No. 32. pp. 38-49. New York: Wildlife Conservation Society.

Lindsey, P. A., Nyirenda, V. R., Barnes, J. I., Becker, M. S., McRobb, R., Tambling, C.J., ... t'Sas-Rolfes, M. (2014) Underperformance of African protected area networks and the case for new conservation models: insights from Zambia. *PLoS ONE, 9*(5), e94109. http://dx.doi.org/10.1371/journal.pone.0094109

Lindsey, P., Balme, G., Becker, M., Begg, C., Bento, C., Bocchino, C., ...Zisadza, P. (2013). The bushmeat trade in African savannas: Impacts, drivers, and possible solutions. *Biological Conservation, 160*, 80-96.

Lindsey, P., Romañach, S., Matema, S., Matema, C., Mupamhadzi, I., &Muvengwi, J. (2011).Dynamics and underlying causes of illegal bushmeat trade in Zimbabwe. *Oryx, 45*(1), 84-95.

Maisels, F., Strindberg, S., Blake, S., Wittemyer, G., Hart, J., Williamson, E. A., ... Warren, Y.(2013). Devastating decline of forest elephants in central Africa. *PLoS ONE, 8*(3), e59469. http://dx.doi.orrg/10.1371/journal.pone.0059469

McIntyre, C. (2004). *Zambia: the Bradt travel guide*. Third edition. St Peter: Chalfont, Bradt.

Messer, K. D. (2010). Protecting endangered species: when are shoot-on-sight policies the only viable option to stop poaching? *Ecological Economics, 69*, 2334-2340.

Moss, C. (1996). Getting to know a population. In: Kangwana K, ed., *Studying elephants*. Handbook series 7.Nairobi: African Wildlife Foundation.

Mwima, H. K. (2007). *Management effectiveness tracking tool for protected areas*. Lusaka: Ministry of Tourism, Environment and Natural Resources.

Nellemann, C., Formo, R. K., Blanc, J., Skinner, D., Milliken, T., & De Meulenaer, T. (2013). *Elephants in the dust: the African elephant crisis. A rapid response assessment*. GRID, Arendal, Birkel and Trykkeri AS: United Nations Environment Programme. Retrieved March 4, 2014, from www.grida.no

Norton-Griffiths, M. (1978). *Counting animals*. Nairobi: African Wildlife Foundation.

Nyirenda, V. R., Chansa, W. C., Myburgh, W. J., & Reilly, B. K. (2010). Social capital and community responses to natural resource management in the Luangwa Valley, Zambia. *Journal of Sustainable Development in Africa, 12*(8), 158-180.

Nyirenda, V. R., Chansa, W. C., Myburgh, W. J., & Reilly, B. K. (2011).Wildlife crop depredation in the Luangwa Valley, Zambia. *Journal of Ecology and the Natural Environment, 3*(15), 481-491.

Poulsen, J. R., Clark, C. J., Mavah, G., & Elkan, P. W. (2009). Bushmeat supply and consumption in a tropical logging concession in northern congo. *Conservation Biology, 23*, 1597-1608.

Rentsch, D., & Damon, A. (2013). Prices, poaching, and protein alternatives: an analysis of bushmeat consumption around Serengeti National Park, Tanzania. *Ecological Economics, 91*, 1-9.

Roffe, T. J., Friend, M., & Locke, L. N. (1996). Evaluation of causes of wildlife mortality. In: Bookhout TA., ed., *Research and management techniques for wildlife and habitats*(5th ed., revision, pp. 324-348). MA.: The Wildlife Society.

Simukonda, C. (2008). *National-wide large mammal surveys and population estimates*. Lusaka:New Horizon Printing Press.

United Nations Development Programme (UNDP). (2011). *Zambia Human Development Report 2011: service delivery for sustainable human development*. Lusaka: UNDP.

Viljoen, P. (2013). *Lower Zambezi National Park, Rufunsa and Chiawa Game Management Areas: aerial wildlife survey*. Chirundu: Conservation Lower Zambezi.

Watson, F., Becker, M. S., McRobb, R., & Kanyembo, B. (2013). Spatial patterns of wire-snare poaching: implications for community conservation in buffer zones around National Parks. *Biological Conservation, 168*, 1-9.

Wittemyer, G., Daballen, D., & Douglas-Hamilton, I. (2013).Comparative demography of an at risk African elephant population. *PLoS One, 8*(1), e53726. http://dx.doi.org/10.1371/journal.pone.0053726

Wittemyer, G., Northrup, J. M., Blanc, J., Douglas-Hamilton, I., & Omondi, P. (2014). Illegal killing for ivory drives global decline in African elephants. *PNAS Early Edition*, 1-5. http://dx.doi.org/10.1073/pnas.1403984111

World Bank.(2013). *Country Partnership Strategy for the Republic of Zambia, year 2013-2016*. Lusaka: World Bank.

Zambia Wildlife Authority (ZAWA).(2001). *Lower Zambezi National Park, and Chiawa and Rufunsa Game Management AreasGeneral Management Plan*. Chilanga: ZAWA.

10

Current Livelihood Condition of and Futurity of Tea Farming for Marginal Small Tea Farm Holders (MSTH) of Sri Lanka: Case Study From Badulla and Matara District

Indika R. Palihakkara[1,2], Abrar J. Mohammed[2] & Makoto Inoue[2]

[1] Faculty of Agriculture, University of Ruhuna, Sri Lanka

[2] Graduate School of Agricultural and Life Sciences, The University of Tokyo, Japan

Correspondence: Palihakkara R. Indika, Graduate School of Agricultural and Life Sciences, The University of Tokyo, Tokyo, Japan. E-mail: irpalihakkara@gmail.com.

Abstract

With an area of 120,955 ha, Small Tea Farm Holders (STH) constitute about forty percent of the total tea land area of Sri Lanka. In addition to possessing small tea land area, portion of small holders face another serious problem, low productivity of the tea land. With the aim of filling the knowledge gap on the livelihood of Marginal Small Tea Farm Holders (MSTH) and guide future policies and other interventions based on existing realities, this research analyzes the livelihood of MSTH in Badulla and Matara district using Sustainable Livelihood (SL) Framework and discusses futurity of tea production for MSTHs. Since land is the most limiting factor for their livelihood, the MSTH were first divided into four land categories ranging from very small to large holder. Then, their livelihood capital was measured using the five livelihood capitals, i.e., natural, human, social, physical and financial. The result showed inconsistent relationship between land size of MSTH and other capitals such as number of trees planted and income from tea production. The district with smaller average land size (Matara) and the very small and small land holders within district were found to generate more income from tea reduction. Improving human capital through education was also found to contribute negatively toward labor contribution for tea production. Between districts, weather and elevation, two forms of natural capital which are mostly neglected in rural studies using SL approach, were found to play important role in determining outcome from tea based livelihood. Overall, however, majority of the farmers are in the view that tea generates low benefit and should look for other alternatives. Considering this, possibility of other livelihood activities such as conversion of the marginal tea land to fuelwood planation should be considered. This, in addition to improving farmers' livelihood through enhanced income from their land, will also contribute for overcoming the ongoing shortage of fuelwood in Sri Lanka while also improving the environmental service from the MSTHs' land.

Key words: marginal tea plantations, smallholders, land use conversion, Tea Small Holding Development Authority

1. Introduction

Tea (*Camellia sinensis* L.) has been playing an important role in economic development of Sri Lanka from its inception as a plantation production system during the European colonial expansion period in the middle of 19th century by Sir James Taylor in 1867 (Ganewatta, 2000; TRI, 2006).Initially, tea was cultivated in central highlands of Sri Lanka. With growing demand in international markets for this cash crop, the tea area expanded further into untouched forest as well as coffee plantations in central Sri Lanka, and finally to Uva (includes Badulla district) and low lands of south western Sri Lanka (includes Matara District). The current extent of tea plantations of the country accounts for 11% of total agricultural land area that is about 212,700 ha (Ministry of Plantation Industries, 2012), which is managed as either a private estates, government estates or Small Tea farm Holders (STH).

Small Tea farm Holders (STH) are defined as farmers with tea plantations area of less than 20.2 hectares (50 acres) and without their own processing facilities. This definition was operationally established during the 1972 and 1976 land reform laws when all the tea farms larger than 50 acres were expropriated (Deepananda, 2009). Currently, there are about 390,346 STH with a total extent of tea farm area of 120,955 ha representing higher percentage tea land extent out of three plantation management categories (TSHDA, 2012). The number of small holders in the country increased by 70% for the previous 10 year before the last censes taken in 2005 (TSHDA, 2012).

In addition to having small tea land area, portion of the STH face another serious problem, or low productivity of the tea land. These low productive farmlands are characterized as marginal tea land. Marginal tea land are tea production lands with a yield below the national average tea yield which is 1615 kg/ha (Ministry of Plantation Industries, 2012). With Sri Lanka's tea industry claiming the highest cost of production among other competition countries (Ganewatta et al., 2000), Marginal Small Tea Holders' (MSTH) ' interest in tea based livelihood as well as land development is being further reduced. For example, tea bushes belonging to most of the tea plantations were planted more than 40 to 60 years ago (Dissanayake, 2013). Moreover, according to censes taken in 2005, about 7310 hectares of the small holder plantations were abandoned (TSHDA, 2012).

Totally or partially abandonment of MSTH's tea plantations create many problems. In addition to the socio-economic drawbacks it case to the farmers, it also contributes for environmental problems of the area as well as to the country. Sri Lanka is facing environmental problems such as irregular rainfall, drying up of natural streams, soil erosion and associated loss of fertility etc., especially in tea producing areas (Mungai, 2004), Therefore, understanding the prevailing reality of marginal tea small holders is important to tackle the socio-economic as well as environmental consequences of existing tea-based land use systems (FAO, 2012). By utilizing the livelihood capital approach, this research is aimed at exploring the current livelihood condition of STH farmers and the futurity of tea farming. Specifically, it explores natural, financial, human, physical and social capital of the small holders and their perspective towards tea farm production.

The remaining part of the paper is organized as follows. The next section elaborates on the analytical framework utilized for this paper followed by the research method section. In the result and discussion section, the current livelihood condition of the MSTH farmers, measured in terms of the five livelihood capital, as well as their perception towards tea based livelihood strategy is presented and discussed. The pearls of the paper as well as possible remedies to improve the existing situation of MSTH farmers is finally detailed in conclusion and policy recommendation section.

2. Analytical Framework

During the last decades, with the renewed international commitment to poverty reduction together with the ever increasing environmental concern, there have been significant theoretical and practical advances in the way poverty-environment linkages are considered. Rural poverty has been accepted as both a major cause and result of degraded soils, vegetation, forests, water and natural habitats (DFID 2000; Mohammed, 2014). Although poor people in developing countries are particularly dependent on natural resources and its ecosystem services for their livelihoods, they live in areas of high ecological vulnerability and relatively low levels of resource productivity (Mohammed & Inoue, 2013; Mohammed & Inoue, 2014). On the other hand, rapid deforestation and biodiversity losses are depriving people of valuable forest resources, such as fuel wood, food and medicine. Soil degradation is a major threat to the livelihoods of 1 billion people, mostly the poor who are more likely to live in degraded or fragile areas (Baumann, 2002). Considering such linkage between nature resource degradation and poverty, understanding the livelihood of poor living in degraded environment is vital for designing sustainable resource management or prescribe alternatives.

The Sustainable Livelihoods Approach (SLA) emerged partly as a result of this rethinking of poverty-environment linkages and has since become a driving force in its evolution (Brocklesby & Fisher, 2003). One of the key attribute of SLA is the livelihood capital of household as capital serve as vehicles for making a living, making living meaningful and challenging the structures under which one makes a living (Bebbington, 1999).The livelihood framework identifies five core capital which sometimes are called livelihood building blocks upon which livelihoods are built. These are natural, social, human, physical and financial capital (Scoones, 1998).

Natural capital is the natural resource stocks including plants, trees, soil, water, land with its characteristics etc. while financial capital refers to monetary capital bases such as incomes, savings and other (Sherbinin et al., 2008). The skills, knowledge, ability to labor etc. are the human capital. Physical capital includes productive assets held by the household like vehicles, tools, oxen etc. (Scoones, 1998; Sherbinin et al., 2008). The social capital include networks, social claim, social relations, affiliations, association upon which people draw when pursing different livelihood strategies requiring individual or coordinated action (Scoones, 1998).

Improving capital assets is a primary strategy for improving rural people's livelihood from a given livelihoods strategy. Rural poor usual require a range of capitals to pursue their strategies. Core concepts for an understanding of local decision-making to follow a given livelihood strategy, therefore, is the notions of asset substitution and trade-offs. Such understanding enables framing of strategic questions for development planning (Baumann, 2002). It particularly helps understand local people perceptions towards a given livelihood strategy (Figure 1).

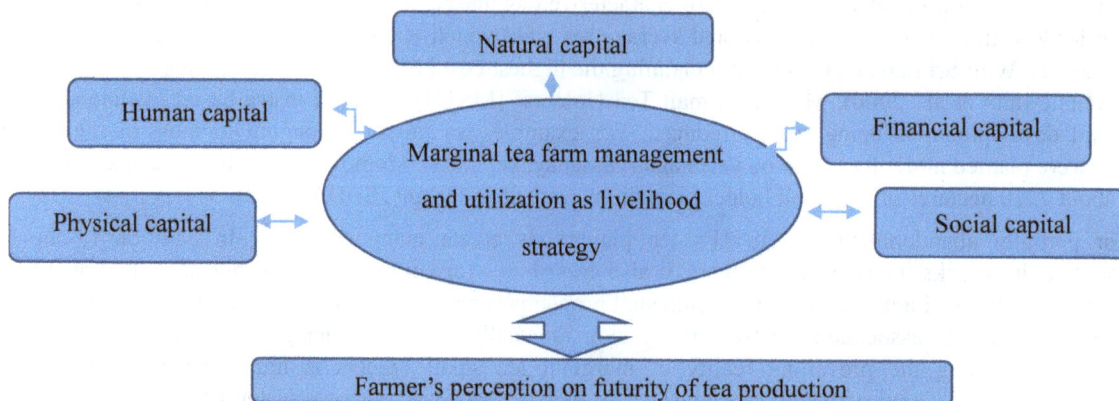

Figure 1. Analytical framework of the study

3. Research Method

3.1 Study Site Description

The study is conducted in Matara and Badulla districts which represent lowland and highland tea plantation of the country respectively. After preliminary discussion with staffs from Tea Small Holding Development Authority (TSHDA) regional officers of the two districts, MSTH from Urubokka, Haliella, Lunugala, and Ella TSHDA sub divisions were selected for the study as they have relatively large number of MSTH.

Urubokka is a TSHDA sub division in Matara district, while Haliella, Lunugala, Ella are TSHDA sub divisions in Badulla district. Urubokkais situated, North East border of Matara district, adjoining Hambantota and Rathnapura districts. Haliella is situated south western part of Badulla district while Lunugala and Ella are found in the border of Badulla district adjoining to Moneragala district (Figure 2).

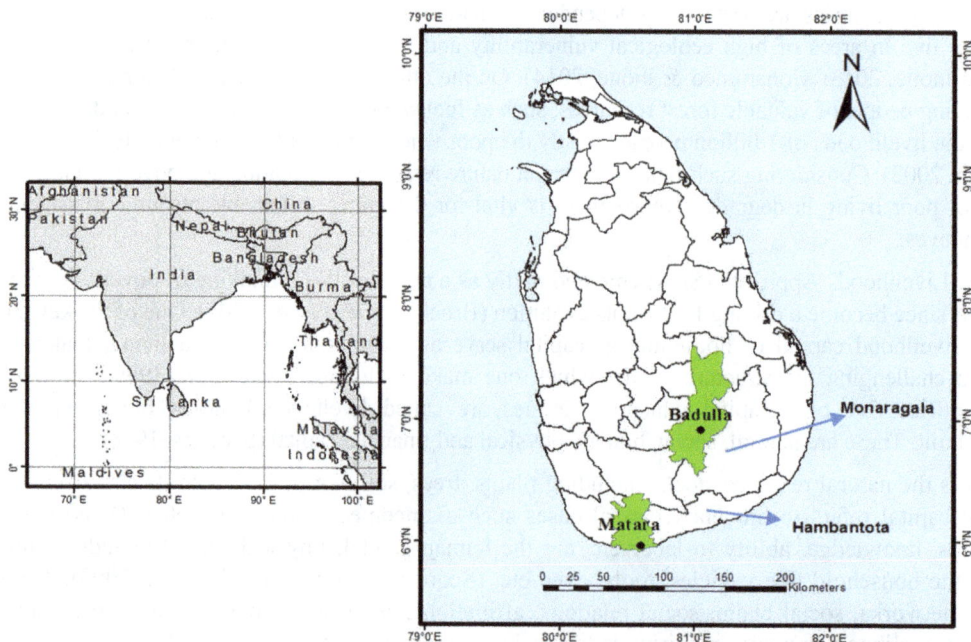

Figure 2. Location of Sri Lanka and the study sites Matara and Badull districts in Sri Lanka

3.2 Data Collection and Analysis

Data collection was conducted in two stages. In the first stage, review of archival records such as field records as well as reconnaissance survey of the research site together with TSHDA officer in charge of the area and chairman of tea farmer organizations was undertaken. By using the archival record review, a total of 648 and 177 MSTH were identified from Matara and Badulla district respectively. After consultation with TSHDA officers, a total 50 MSTH for Matara and 31 MSTH for Badulla were randomly selected for data collection.

Based on land size classification of TSHDA, the selected farmers were categorized into four land class, i.e., very small land holders (VSLH) (<1 ha), small land holders (SLH) (1-2 ha), middle land holders (MLH) (2-3 ha), large land holders (LLH) (3-20 ha) (Table 1). Direct observation was conducted to MSTHs' plantation and land conditions such as slope, maturation, type of management. The visit was also used as an opportunity to get familiarized with the village and the farmers.

Table 1. Sampled MSTH categorized according to their land size

Category	Land extent	Number of farmers	
		Matara	Badulla
Very small land holders (VSLH)	Less than 1hectare	32	8
Small land holders	Between 1 and 2 hectare	8	8
Middle land holders (MLH)	Between 2 and 3 hectare	5	5
Large land holders (LLH)	Between 3 and 20 hectare	5	10

In the second stage, detailed quantitative and qualitative data were collected using interviews and archival records. A pre-tested structured and semi structured questionnaire was used to collect data on the five capitals of MSTH as per the analytical framework developed. The collected data includes information on human capital (number of family members, their contribution to the plantation, number of formal trainings, and experience on tea industry); natural capital (plantation extent, % slope of the plantation land), Physical capital (possession of bike, motorbike, car, tractor); social capital (organization membership, linkage with external actors) and financial capital (farmer's income, green leaf productivity, income from crops other than tea) of the respondents. The data that requires farmers opinion such as their way of evaluating the plantations and, their opinion about futurity tea based livelihood was measured using a Likert scale ranging from 1 = strongly disagree to 5 = strongly agree. Qualitative information was collected by using key informant interviews and focus group discussions. Two group discussions were conducted, one in each district. Information was written down and cross-checked by reading them for the attendant/ respondents. The overall situation of tea planting, the problems they are facing as well as reason for farmers keeping uncultivated lands in their plantations were among the topic discussed during group discussion. The data was collected from beginning of July to end of August, 2013.

For analyzing family members' contribution to the plantation, rational scoring weight method was utilized (Jayamannna et al., 2002). Five selected activities, i.e. harvesting or tea plucking; weed management; shade tree management; fertilizer application, and other routine work such as, pest and disease control, maintenance of drains; were identified from previous works (Herath et al., 2009). Rationality of giving weight for each category was decided by the importance of each work. For example, tea harvesting claim 60% of the total green leaf production cost (Wijerathne et al., 2007) therefore, highest marks were allocated for harvesting. Slope angle of the plantations was calculated using the equation: $v = sin-1(h/l)$, where v is the slope angle, h is the measured height, and l the length along the slope. The length (l) was fixed at two meters for ease of calculations (Yang et al., 1997). For the other data, deceptive statistics and One-way ANOVA was conducted using SAS software to find out averages as well as statistical significances.

4. Result and Discussion

4.1 MSTHs' Livelihood Capitals

4.1.1 Natural Capital

In Matara, large proportion of the randomly sampled households belong to the VSLH category which has an average land area of 0.7 ha. In Badulla, on the other hand, most of the farmers are from the LLH category with an average land holding of 15.9 ha. In addition, the average land size of farmers in the first category is smaller in Matara compared to Badulla (Table 2). These may imply that the Badulla district is relatively well endowed in terms of the most important natural capital, i.e. farmland. The reason for this disparity can be traced back to the origin of current land tenure system of the two regions. In Badulla district, tea plantation was established during the British colonial era with large scale tea plantations (Wenzlhuemer, 2008; De Silva, 1981). After the independence, government of Sri Lanka introduce new land reform act in 1972 declaring that individual farmer can hold up to 20 ha (50 ac) land and the rest were acquired and managed as government property. These 20 ha plantations with the time were further divided into smaller portions among the family members (Samaraweera, 1982). In the case of Matara District, however, small holder tea plantations were popularized after the independence with smaller land extents using other agricultural lands or clearing forest lands (De Silva, 1981).

While limitation of the slope of moderately suitable farmland recommended by TRI for Matara is in between 25% to 70%, and Badulla is in between 25% to 55%, actual average slope ranges from 48%-53% in Matara and from 49% -61% in Badulla. Farmers belong to MLH category with an average land area of 2.4 and 2.1 have the highest percentage slope at 53%, and 61%, in Matara and Badulla respectively (Table 2). The slope of all categories of both districts are in the upper levels of moderately suitable lands of TRI's land suitability classification for tea except for MLH in Badulla that is in sever category with a slope of 61%. In general, however, the percentage slopes were not significantly different between the districts and among the four land categories in both districts.

Results on percentage of uncultivated land has showed that LLH category of the Badulla, which represented 32% of the sampled farmers had the largest uncultivated land proportion at 30%. Highest uncultivated land percentages were also observed in the SLH and LLH of Matara which is equal to 22.2%. Overall, the difference in percentage of uncultivated land was significant between VSLH and LLH at $p<0.1$ level. The relatively high percentage of uncultivated land from LLH mean that increase in large size may not guarantee improve in livelihood for this farm holds.

Another important capital for the tea growers is shade trees. Well maintaining shade trees provides not only shade for tea, but also a cheap source of firewood to meet energy requirements, timber, and nutrients for the soil in which tea bushes grow and reduce soil erosion. Since Sri Lanka has no oil or natural gas reserves, biomass from shade trees and other sources is of central importance to the overall energy supply of the country as identified by Forestry Sector Master Plan-Sri Lanka (FSMP, 1995).

Table 2. Natural capital of MSTH in the two districts

Natural & bio-physical capital	Matara District				Badulla District			
	VSLH	SLH	MLH	LLH	VSLH	SLH	MLH	LLH
Farm size (ha)	0.7	1.3	2.4	13.7	0.8	1.3	2.1	15.9
% uncultivated	11	22	13	22	15	6	7	30
% Slope	48	48	53	49	49	54	61	52
Number of tree	1806	202	861	476	459	309	848	1484
(high shade, medium shade)	(178, 1628)	(42, 160)	(141, 720)	(26, 450)	(119, 340)	(104, 205)	(193, 655)	(384, 1100)
location	Low elevation				High and mid elevation zone			
weather	Higher temperature				Low temperature			

Interestingly, the result showed that VSLH farmers in Matara have the highest average number of trees, or 1806 trees of which 178 are high shade trees. In Badulla district, however, LLH has higher average number of trees (Table 2). In addition to providing shade, these trees have vital importance for the farmers. Usually, pollarding and periodic lopping of high and medium shade respectively are practiced to ascertain the optimal shade levels of 10-40 %. In addition, when the tree life span is over, it is removed from the plantation and replanted. For example, the average life span of Gravillea and Albizia is 30 and 12 years respectively. These pollarded and lopped materials as well as the entire trees that have been removed after finishing their life span are used for fuel wood. Such wood/fuel wood production from non-forest tree resources is highly significant in the Sri Lankan context. It is vital especially for tea growers to minimize the fuelwood shortage they commonly face. Due to fuelwood shortage, for example, the price of Gliricidia, increased to Rs 1500 per cubic meter at present from Rs 1000 per cubic meter two years ago (TRI Sri Lanka, 2012). In addition to fuelwood, these trees also provide diverse goods for the framers: from providing shade to and mulch for tea plantation, to providing edible food, medicine, fuelwood and timber to local people. *Gliricidia sepiumis* the dominant species in both district followed by *Milia azedarach, Albizzia moluccana* and *Calliandra calothrysus, Erythrina lithosperma* in Matara and Badulla respectively.

The last two major natural capital and the most determinant in terms of affecting the productivity as well as income of tea production are elevation and weather. Matara belong to low elevation wet zone of Sri Lanka with higher temperature. On the other hand, Badulla is located in high to mid elevation zone of Sri Lanka with relatively low temperature. Due to this higher temperature, Matara tea yield /unit area is higher with higher number of harvesting rounds per year than Badulla (Wijerathna, .2007). In addition, low elevation of Matara gives its green leaf higher price than mid and higher elevation tea leaf of Badulla (Sri Lanka tea board 2014).

4.1.2 Financial Capital

Financial capital of the farmers, which is measured by income from four major sources , i.e., from tea, other agriculture products (other than tea), salary /pension/rental of property other than tea plantation and income from non-regular labor work in Sri Lankan Rupees(Rs) per celender year is presented in Table 3. Compared to Badulla, tea was found to contribute to major portion of the income in Matara. This is despite Badulla having comparatively large tea farmland than Matara. Moreover, the percentage of farmers having tea as the only income source were high for Matara with 43%,50%, 80% ,40% from VSLH to LLH respectively. Tea income was found to be relatively less important in Badullla where salary/pension/ renting contributed the major source of income.

Table 3. Financial capital of MSTH in the two districts (Rs/year)

Category		Matara District				Badulla District			
		Tea crop	Agriculture	Salary/ Pension/ Rent	Labor	Tea crop	Agriculture	Salary/ Pension/ Rent	Labor
VSLH	amount	57471 (45)	17800 (14)	46908 (37)	4333 (3)	54462 (27)	16133 (8)	126000 (64)	2000 (1)
	%	43 *	43	34	9	38*	38	38	12.5
SLH	amount	54336 (52)	21000 (20)	30000 (28)	0 (0)	59276 (32)	26766 (14)	100800 (54)	0 (0)
	%	50 *	37	13	0	50 *	38	25	0
MLH	amount	73419 (31)	8000 (3)	156000 (66)	0 (0)	36572 (23)	14400 (9)	111000 (68)	0 (0)
	%	80 *	20	20	0	0 *	20	80	0
LLH	amount	28023 (8)	75000 (21)	259992 (71)	0 (0)	26681 (14)	14800 (8)	144000 (78)	0 (0)
	% Farmers	40 *	20	60	0	30 *	30	50	0

*Indicates %farmers getting income only from tea.

Especially LLH farmers, despite the expectation due to their large land, got only 14% of their income from tea with 78% of it coming from salary and pension. Labor work was found to be important income source only for the VSLH farmers (Table 4). This probably is because the labor force of relatively large land holders utilized their time and energy within their plantation. Overall, the total income from tea land seemed to be irrespective of the total land area of the farmers with small and middle land area holders generating much income than the farmers with large land holding.

4.1.3 Human Capital

For effective management of MSTH farmland and get good income, farmer's individual performances as well as their group work is vital. Hence number of family members and their labor contribution for tea planting, experience on tea farming, formal training received and their level of education play an important role (TSHDA Sri Lanka 2008). As shown in Table 5, the average number of family members ranged from 2.8 for MLH of Matara to 4.4 of VSLH in Badulla. Badulla district average number of family members is always above 4 for all categories. This is larger than the 2011 average household size of Sri Lanka which is 3.9.

Table 4. Human capital of MSTH

Human capital	Matara District				Badulla District			
	VSLH	SLH	MLH	LLH	VSLH	SLH	MLH	LLH
No of family members	4.3	4.4	2.8	3	4.4	4	4.2	4
Score/FM contribution	7	7.6	1.3	0.5	6.4	5.4	3.2	0.1
Experience/ tea farming (Number of Years)	30	26	40	43	49	52	53	57
No of formal training (Total trainings)	1.4	2.3	6.8	9.2	2	1.1	1.6	3.7
Farmer Education (Number of years in formal education)	6.5	7.6	13.8	13.6	8.4	10.9	9.6	12.9

Table 5. Social capital of MSTH (Average. Number of members in social organizations)

Social organizations	Matara District				Badulla District			
	VSLH	SLH	MLH	LLH	VSLH	SLH	MLH	LLH
TSHDA	1	1	1	1	1	1	1	0.8
Government supported	0.5	0.3	1	0.7	0.25	0.25	0.4	0.5
Local autonomous	0.2	0.1	0.2	0.6	0	0.4	0.4	0.4
NGO supported	0.2	0.1	0	0.2	0	0	0	0

Family member contribution decreased with the increasing land size, the lowest being for MLH and LLH of Matara. The reason for this, according to respondents, is that household members in MLH & LLH have higher social recognition and education levels which made them reluctant to contribute for assisting their parents in the tea farm works. Their argument also supported by data on education level (Table 4). The presence of leaches (blood sucking small creature) in the tea farm that is said to be responsible for black scare in their skin was another justification for the educated youth to be unwilling to help their family. Such scars will make them easily recognized by other fellows that they are working in tea plantations which, for them, is socially degrading.

Experience in tea production for the two district varied from 26 for SLH of Matara to 57 for LLH of Badulla. Overall, Badulla district farmers have more experience than Matara farmers in all 4 categories (Table 4). The main reason for this is Badulla farmers either inherited or purchased the plantations under the Land Reform Act introduced in 1972 and with most of them having working experience with tea either on their own plantations or plantations managed by the British companies. While in Matara, majority of small holder farmers in the Matara District established in late 1970s and early 1980s after establishment of TSHDA and introduction of open economy system to Sri Lanka in 1977 (TSHDA 2012).

The education level of family members is good for both districts. Although there is not as such significant inter-district difference in education level, the intra district distribution was found to be quite different. Unlike Badulla, farmer's education level is higher at for MLH and LLH (13.8 and 13.6 respectively) and was significantly higher than VSLH and SLH which have 6.5 and 7.6 respectively. The possible reason for this is number of schools functioning in the Badulla by large scale tea plantations which established in the colonial period. Number of formal training received, however, is significantly higher for Matara than Badulla as well as for large holders compared to small holders except for VSLH of Badulla. (Note): To calculate score of FM contribution, rational scoring weight method was used

4.1.4 Social Capital

Owing to the labor intensive nature of Tea farming, MSTH can acquiring more benefits by working collectively and forming groups or networks that can be recalled when labor is demanded. This is quite advantageous even if there are opportunity costs such as time allocated to participate in meetings, payments needed for membership etc (Barrett 2001). In this study site, four types of societies were identified: TSHDA, Organization with government support and involvement, autonomous local organization formed and maintain by the local people, and NGO supported organization. MSTH from all land categories in both districts were found to belong to TSHDA (Table 5). Next to TSHDA, farmers were found to be more affiliated to government supported organization than autonomous organization and NGO supported ones. The more interest in local autonomy organization over NGO supported one is due to the diverse goods and services provided by the former. The roles of these local organizations are providing temporally constructing materials, cooking utensils, financial loans and labor during special social occasions such as funeral and wedding. In Matara few farmers attached to organization under type 3, where it mainly focused on pipe born water project for house hold consumption and in Badulla farmers were not involved in any NGO activity in the study area.

4.1.5 Physical Capital

Availability and mode of transportation in the rural farming areas plays key role in rural economy (Holden, 1998). Hence, analysis of physical capital involved assessment of possession of different transportation means. Different type of vehicles, including bicycles, motor bicycles, three weelers, tractors, trucks and cars were possessed by the MSTH, but neither of them used animal drafted vehicles. Bicycles and motor bicycles were common among all the four categories in two districts, they use it as transportation and carrying goods and bicycle usage was significantly lower among LLH in both districts(table 07). Usage of tractors were extremely lower than trucks and it was higher among LLH in both districts, which they used as a dual purpose uses like human and material transportation. Car usage is extremely luxurious and used by the LLH in two districts where Matara has the higher value, 0.6 % than 0.2% Badulla, results the comparatively better road network in the district (Table 6).

Table 6. Physical capital of MSTH

physical capital	Matara District				Badulla District			
	VSLH	SLH	MLH	LLH	VSLH	SLH	MLH	LLH
Bicycle	0.5	0.5	0.4	0.2	0.5	0.5	0.6	0.2
Motor bicycle	0.4	0.8	0.6	0.6	0.5	0.8	0.6	0.6
Three wheelers	0.06	0.2	0.2	0	0.1	0.1	0.2	0.1
Tractors	0	0	0	0.2	0	0.1	0	0.1
Trucks	0.1	0.1	0.4	0.8	0	0.1	0.2	0.9
Cars	0.03	0	0	0.6	0	0	0	0.2

4.2 Futurity of Tea Production Based Livelihood Strategy

Owing to the less productivity and profitability of the tea farming livelihood strategies, most of the STH are in view that they should look for other alternative livelihood strategy. Especially the MLH and LLH of Matara district and SLH and LLH of Badulla district were in agreement with the idea that it is proving low benefit and they need another option (Table 7). The perception of these category of farmers is also in line with the outcome from the livelihood analysis in the previous section. For MLH and LLH in Matara, for example, only 31% and 8% of their income are obtained from tea as compared to the 45% and 52% of VSLH and SLH respectively (Table 3). Moreover, owing to small number of family members and also low family member contribution (Table 4), they have to utilize hired labor that increases the cost of tea production and hence reduce its profitability.

Consequently, farmers are forced to abandon portion of all of their tea production land, exacerbating the already exited social (poverty) and environmental (soil erosion, land degradation etc.) problems. Overall, the prominent reason for farmers leaving portion of their land uncultivated (Table 2) are poor natural capital (unproductivity of land, slope of land), low financial return (low profit from the land) and lack of human capital (difficult to find labor to manage the land) and problem of dying out of the plants. Among these, unproductive and eroded land as well as dying of plants were very important factors of VSLL of Matara district while the latter being vital for all categories of Badulla as well as SLH and LLH of Matra district.

Labor shortage is another important factor forcing farmers to leave their farm uncultivated. It is particularly important for MLH of Matara and LLH of Badulla. Labor shortage, however, is found to be less important problem for the VSLH in both district (Table 3). This is because smaller size farmers can efficiently use their family labor as has also been witness by other researches such as Basnayake et al. (2002) and Eastwood,et al. (2010). Overall, for MSTH, tea farming seemed to be compatible livelihood strategy only for VSLH, farmers with land area less than 1ha, as compared to the other land categories, albeit all categories would seem to prefer shifting to other livelihood strategy, if they get the opportunity to do so.

Table 7. MSTH's general attitude about tea production based livelihood strategy

Perception	Matara District				Badulla District			
	VSLH	SLH	MLH	LLH	VSLH	SLH	MLH	LLH
Generate low profit (High COP, Input prize)	3.5	3	3.8	4.4	3.3	3.8	3.2	3.7
Have to find another alternative	2.8	2.6	4	4.2	3	3	3.2	3.9

5. Conclusion and Policy Implication

This paper explores livelihood capitals of Marginal Small Tea Holder farmers (MSTH) and futurity of tea based livelihood strategies for this category of farmers in Matara and Badulla districts, two districts representing the upper and lower tea production areas of the country respectively. The average landholding ranged from 0.7ha for VSLH in Matara to 16ha for LLH in Badulla and the percentage of uncultivated land from 7% for MLH in Badulla to 30% for LLH. The major reason for having uncultivated were un-productivity and slope of the farming land, low profitability, labor shortage and dying of tea plants. The income generation from Tea crop was found to have reciprocal relationship with size of land. Matara, the district with relatively small size of tea farmland had generated large proportion of income from tea production. Within district too, the VSLH and SLH were found to generate large proportion of their income from tea crop production compared to the MLH and LLH. Although the human capital was found to be good for all, formal education was found to contribute negatively to household member's contribution for tea production. With the fear of being recognized by peers at school of their household role as tea farmer, the educated members were found to be reluctant to help their family in the tea farm activities. Important livelihood capitals, which mostly is neglected in suitable livelihood approach, but found to be salient in this study are elevation and weather. These two, categorized under natural capital in this study, were found to play important role in affecting the livelihood outcome from tea production.

Despite its important contribution for household income for most of the samples households, famers were found to be pessimistic to continue tea production and in favor of alternative livelihoods. The high input requirement of this livelihood strategy together with the low productivity of land and the weather condition seemed to contribute for their attitude. One possible alternative for farmers is to convert the Tea plantation in to fuelwood plantation. Doing so will minimize the cost of production that is discouraging the farmers to continue tea farming. Tree species such as *Gliricidia septum* and *Calliandra calothrysus* that are important source of fuel wood have been growing well in the existing topography and weather condition in the marginal lands as witnessed by current dominance in the landscape. Hence they can be considered for fuelwood planation establishment. In addition to logs from these trees when they reach to harvestable age, twigs and branches from lopping and pruning during their development can also serve as an important source of fuelwood for subsistence and commercial uses. With appropriate advertisement and negotiation, these plantations can also be incorporated in to Payment for Environment Service (PES) for their local (such as flood and soil erosion protection) and global (carbon sequestration) environmental services to generate intermediate income for local people. Market for the fuelwood will also continue to be promising considering the existing severe shortage of fuelwood, especially in the tea production industry.

In order to facilitate the aforementioned land use conversion that will benefit both the environment and local people, future detail participatory feasibility studies are vital. It is also crucial to evaluate the existing policy environment such as Land policy, Forest policy, Environmental policy, Agriculture policy and Energy policy of the country with respect to their implications on farmers' decision to combine their tea based livelihood with that of fuelwood production.

Acknowledgments

The authors thank and acknowledge partial financial support from Grants-in-Aid for Scientific Research (A) supported by the Government of Japan (No. 24248026, Project leader: Makoto Inoue). The finding, however, does not represent the views of the Japanese Government.

References

Amerasinghe, Y. R. (1993). Recent Trends in Employment and Productivity in the Plantation Sector of Sri Lanka: with special reference to the Tea Sector. ILO (pp 3-25) Geneva, Switzerland.

Barret, H. R., Browne, A. W., Harris, P. J. C., & Cadoret, K. (2002). Organic certification and the UK market: organic imports from developing countries. *Food policy, 27*(4), 301-318. http://dx.doi.org/10.1016/S0306 -9192(02)00036-2

Barrett, H. R., Browne, A. W., Harris, P. J. C., & Cadoret, K. (2001). Smallholder farmers and organic certification: Accessing the EU Market from the developing world. *Biological Agriculture & Horticulture: An International Journal for Sustainable Production Systems, 19*(2), 183-199. http://dx.doi.org/10.1080/01448765.2001.9754920

Basnayake, B. M. J. K., & Gunaratne, L. H. P. (2002). Estimation of Technical Efficiency and Its Determinants in the Tea Small Holdings Sector in Mid Country Wet zone of Sri Lanka. *Sri Lankan Journal of Agricultural Economics, 4*, 21-37. http://dx.doi.org/10.4038/sjae.v4i0.3488

Baumann, P. (2002). Improving access to natural resources for the rural poor: A critical analysis of central concepts and emerging trends from a sustainable livelihoods perspective. *FAO LSP Working Paper No. 1. Rome.*

Bebbington, A. (1999). Capitals and capabilities: a framework for analyzing peasant viability, rural livelihoods and poverty. *World Development, 27*(12), 2012-44. http://dx.doi.org/10.1016/S0305-750X(99)00104-7

Brocklesby, M. A., & Fisher, E. (2003). Community development in sustainable livelihood approaches-an introduction. *Community Development Journal, 38*(3), 185-198. http://dx.doi.org/doi: 10.1093/cdj/38.3.185

De Jong, R., Ariyaratne, M. G., & Ibrahim, M. N. M. (1999). Performance of dairy farming on abandoned marginal tea lands in the mid country of Sri Lanka. *Tropical Agricultural Research and Extension, 2*(1), 55-66.

De Silva, K M. (1981). A history of Sri Lanka. Berkeley, CA: University of California Press.

De, P. K., & Ratha, D. (2012). Impact of remittances on household income, asset and human capital: Evidence from Sri Lanka. *Migration and Development, 1*, 163-179. http://dx.doi.org/10.1080/21632324.2012.719348

DFID. (1999). *Sustainable livelihoods guidance sheets.* Department for International Development, London

Dissanayake, D. R. R. W, Udugama J. M. M., & Jayasinghe-Mudalige, U. K. (2013). Development of an Alternative Microfinance Scheme to finance in the Tea Small Holding sector: A success story. *Journal of Food and Agriculture, 3*(1-2), 31-40. http://dx.doi.org/10.4038/jfa.v3i1-2.5168

FAO. (2012). *Report of the Intercessional Session of the IGG on Tea.* FAO: Colombo.

Forestry Sector Master Plan [FSMP]. (1995). *Sri Lanka Forestry Sector Master Plan.* Forestry Planning Unit, Ministry of Agriculture, Lands and Forestry.

Ganewatta, G., & Edwards, G. W. (2000). The Sri Lanka Tea Industry: Economic Issues and Government Policies. *44th Annual Conference of Australian Agricultural and Resources Economics Society*, University of Sydney, Australia, 23-25 January 2000.

Herath, D., & Weersink, A. (2009). From plantations to smallholder production: the role of policy in the reorganization of the Sri Lankan tea sector. *World Development, 37*(11), 1759-1772. http://dx.doi.org/10.1016/j.worlddev.2008.08.028.

Jayamanne V. S., Wijeratne, M., & Wijayaratna, C. M. (2002). Adoptability of New Technology in the smallholdings Tea Sector. *Journal of Agriculture and Rural Development in the Tropics and Subtropics, 103*(2), 125-131.

Korf, B., & Oughton, E. (2006). Rethinking the European countryside - can we learn from the South? *Journal of Rural Studies, 22*, 278-28. http://dx.doi.org/10.1016/j.jrurstud.2005.09.005

Markelova, H., & Mwangi, E. (2010). Collective action for smallholder market access: evidence and implications for Africa. *Review of policy research, 27*(5), 621-640. http://dx.doi.org/10.1111/j.1541-1338.2010.00462.x

Mendis, P. (1992). *Human environment and spatial relationship in agricultural production: The case study of Sri Lanka and other tea producing countries.* Peter Lang Publishing: NY.

Mohammed, A. J. (2014). *Decentralization, Forest and Poverty: Framework and Case Studies from Ethiopia.* Nova Science Publishers: NY.

Mohammed, A. J., & Inoue, M. (2013). Forest-dependent communities' livelihood in decentralized forest governance policy epoch: case study from West Shoa zone, Ethiopia. *Journal of Natural Resource Policy Research, 5*(1), 49-66. http://dx.doi.org/10.1080/19390459.2013.797153

Mohammed, A. J., & Inoue, M. (2014). Linking outputs and outcomes from devolved forest governance using a Modified Actor-Power-Accountability Framework (MAPAF): Case study from Chilimo forest, Ethiopia. *Forest Policy and Economics, 39*, 21-31. http://dx.doi.org/10.1016/j.forpol.2013.11.005

MungaI, D. N., Ong, C. K., Kiteme, B., Elkaduwa, W., & Sakthivadivel, R. (2004). Lessons from two long-term hydrological studies in Kenya and Sri Lanka. *Agriculture, Ecosystem and Environment, 104*(1), 135-143. http://dx.doi.org/10.1016/j.agee.2004.01.011

Samaraweera, I. (1982). Land Reform in Sri Lanka. *Third Word Legal Studies* (Vol.1, Article 7).

Scoones, I. (1998). *Sustainable rural livelihoods: a framework for analysis* (Working Paper 72). Brighton: Institute for Development Studies.

Sherbinin, A., Vanwey, L. K., McSweeney, K., Aggarawal, R., Barbieri, A., Henry, S., & Walker,R. (2008). Rural household demographics, livelihoods and the environment. *Global Environment Change, 18*, 38-53. http://dx.doi.org/10.1016/j.gloenvcha.2007.05.005

Sivapalan.P, Kulasegaram, S., & Kathiravetpillai, A. (2006). *The way of cultivating tea: Hand book on tea.* Tea Research Institute: Thalawakele.

Wenzlhuemer, R. (2008). *From Coffee to Tea Cultivation in Ceylon, 1880-1900: An Economic and Social History.* Brill Academic Pub: Boston.

Yang, C., Shropshire, G. J., & Peterson, C. L. (1997). Measurement of ground slope and aspect using two inclinometers and GPS. *Transactions of the ASAE, 40*, 1769-1776.

Ziyad-Mohamed, M. T., &Zoyas, A. K. N. (2006). Current status and future research focus of tea in Sri Lanka. *The Journal of Agricultural Sciences, 2*(2), 32-42.

11

Does Riparian Filtration Reduce Nutrient Movement in Sandy Agricultural Catchments?

Robert Summers[1], David Weaver[2], Nardia Keipert[3] & Jesse Steele[4]

[1] Department of Agriculture and Food, Western Australia, WAROONA, Western Australia, Australia

[2] Department of Agriculture and Food, Western Australia, ALBANY, Western Australia, Australia

[3] Peabody Energy Australia, 14/259 Queen Street Brisbane 4000, Australia

[4] Newmont Mining Corporation Elko, Nevada 89801, Australia

Correspondence: Robert Summers, Department of Agriculture and Food, Western Australia, 120 South Western Highway, WAROONA, Western Australia. E-mail: robert.summers@agric.wa.gov.au

Abstract

Fencing and re-vegetating the banks of agricultural drains is a widely used practice thought to improve drainage water quality. Monitoring of small nested headwater catchments draining sections with fencing and vegetation (Veg), and without fencing (UF) showed that after scraping the base of the drain in the Veg catchment, the phosphorus (P) and sediment concentration was lower than the UF section. When the site was revisited 9 years later this study showed the Veg drain lost a third of the sediment load of the UF drain but the Total Phosphorus (TP) per unit area load from the Veg drain was approximately 3 times higher than the UF drain. The Veg drain also had a higher TP and Filterable Reactive Phosphorus (FRP) concentration than the UF drain. The impact on nitrogen was variable but both the Total Nitrogen (TN) and nitrate (NO_3) concentration were higher from the Veg drain than from the UF drain. This suggests that the absence of fencing and the presence of livestock in the UF section allowed streamflow to mobilise and importantly, expose sediment, which adsorbed soluble forms of P that was retained in the drain. The vegetation within the fenced area appeared to have little impact on P or sediment loss. Other techniques that favour chemical retention by P sorption rather than physical filtration would need to be employed to reduce P loads in these sandy catchments where soluble P forms dominate.

The P concentration in the Veg section dropped relative to the UF section after later treatment with a 30 mm layer of cracked lateritic gravel but it was less effective than the earlier excavation of the bed of the Veg section of the drain, exposing the clay subsoil. The increase in P retention from the gravel is likely to be short lived because the P sorption capacity of the gravel was exhausted in 12 months.

Keywords: riparian vegetation, water quality, nutrients, phosphorus, nitrogen, sediment

1. Introduction

The water quality of the Peel Inlet and Harvey Estuarine system has declined since the 1950's. The estuarine system has been recognized by the Western Australian Environmental Protection agency as being in need of urgent catchment-wide waterway protection and enhancement for the purpose of water quality protection (Environmental Protection Agency, 2008). The eutrophication of the estuary through excess phosphorus (P) inputs has been well reported in several studies pointing to the need to control the losses from the surrounding catchment (McComb & Davis, 1993; Hodgkin, Birch, Black, & Humphries, 1980).

The use of riparian buffers to improve water quality and reduce the impacts of surrounding agriculture has been attributed to the process of retaining nutrient laden sediment from surface runoff water through vegetative filtering (Osborne & Kovacic, 1993, Hoffmann et al., 2009) and by direct stabilisation of the drainage system (Thorne, 1990).

Riparian buffers have been investigated under similar conditions of sandy soils and in a similar environment south of the Peel-Harvey near Albany in Western Australia (McKergow, Weaver, Prosser, Grayson & Reed, 2003; McKergow, Prosser, Weaver, Grayson, & Reed, 2006ab). The riparian vegetation had a dramatic effect on suspended sediment but had little impact on Total P (TP) concentration or load. McKergow et al. (2003) found there was a change in the form of P in streamflow from 50% particulate P (PP) prior to riparian management to 25% PP after riparian management that had improved riparian vegetation. This aspect of overestimating the impact of riparian buffers has been noted in reviews which emphasize the impact of the dissolved forms of N and P bypassing riparian buffers in overland and subsurface flow (Drewry, Newham, Greene, Jakeman, & Croke, 2006, Hoffmann, Kjaergaard, Uusi-Kämppäc, Hansend, & Kronvang, 2009).

A study of water quality over one winter in 1997 in the Peel-Harvey reported 30 to 60% retention of P, which was attributed to stream riparian management (Cronin, 1998). To understand the mechanisms that may have led to these large reported differences in water quality response for catchments with sandy soils, the study site of Cronin (1998) was revisited and re-assessed. This paper reports on the reassessment of the impact of stream riparian management on water quality over 4 years using automated monitoring equipment. Following this the drain was treated with cracked lateritic gravel to simulate the exposure of P retentive sediments using a different medium and without the removal of existing sediments.

2. Materials and Methods

A small ephemeral headwater ditch that drained an area of 171 ha was monitored at 2 points equipped with automatic water samplers. The 2 points allowed the measurement of water quality differences arising from an upstream fenced and vegetated section of drain (Veg), and a downsteam unfenced and unvegetated section of drain (UF). The upstream part of the drainage system (Veg) had 720 m of drain that had been fenced 10 m either side in 1993 and vegetated with native shrubs and trees on one side. The other side of the Veg drain was used as an access track if drain maintenance was required, and this access section was vegetated with volunteer grasses. The Veg drain catchment was 80 ha, of which 39 ha was actively-draining (Figure 1). Minor revegetation of the steep stream bank and to a lesser extent in the stream bed occurred through volunteer weeds and pasture species as a result of stock exclusion. It should be noted that although there was considerably more vegetation in the drain bed of the Veg section than the UF section, no attempt to introduce wetland or riparian plant species was made. Upstream of the Veg drain was an area of predominantly native vegetation (41 ha out of the 80 ha upstream catchment). A drain had been constructed through this area of native vegetation to drain a small area of farmland upstream, however due to the relatively low rainfall and high water use of the native vegetation, this section of the catchment did not flow at any point in the monitoring period and contributed no water to the upstream monitoring point. Below the upstream monitoring point, the drain then flowed into another 860 m section of drain that was unfenced (UF) and grazing cattle had full access to the drain throughout the year. This UF drain had a catchment of 91 ha when the Veg drain catchment was excluded. The ephemeral drainage system was monitored over 6 winters from 2006 to 2011. During summer there was no flow and the drainage system dried up. The landuse surrounding the drain was cattle grazing. Different landholders managed each catchment and an estimate of the fertilizer application was made for the catchment.

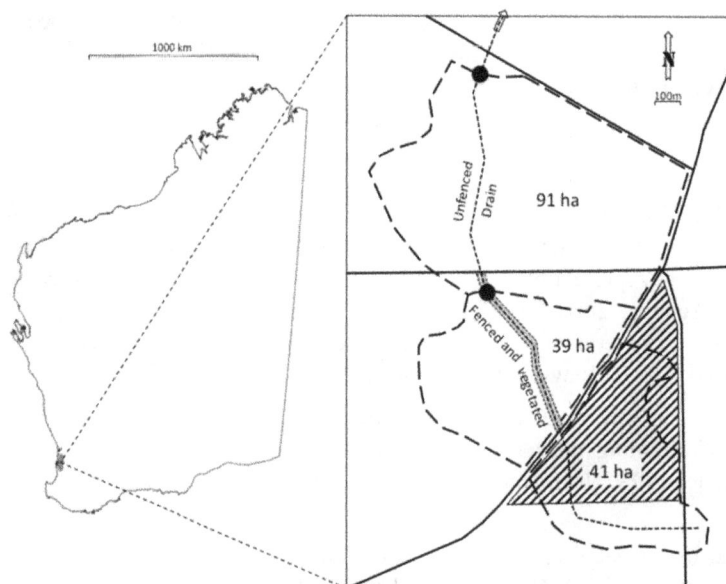

Figure 1. Location and layout of the site showing the gauging stations as filled circles on the drain. Areas of natural vegetation areas are shaded

In 2010 cracked lateritic pea gravel was applied to the base of entire section of the Veg drain to a depth of 30 mm. The cracked gravel was screened to between 3 and 5 mm and a total of 20 tonnes was applied. The base of the drain was approximately 1 metre wide and 745 m long. The pea gravel was analysed for P sorption and compared with results from whole gravel separated from agricultural soils (Weaver et al., 1992). The P sorption was markedly greater in the cracked gravel which showed at 50 mg P L^{-1} solution the P sorbed was 488 mg kg^{-1} and 60 mg kg^{-1} for the cracked gravel and the natural gravel respectively. The P sorption isotherm for the cracked pea gravel indicated a P sorption of 369 mg P kg^{-1} of gravel at an equilibrium solution concentration of 1 mg P L^{-1} (similar to the P concentration in the drain water).

Most of the surface flow from the paddocks into the Veg drain occurred through small feeder drains that concentrated the flow from the paddock, thereby limiting wide-scale interaction of surface flow with the vegetated strip. The vegetated strip was slightly elevated by spoil from drain maintenance, although much of the spoil had been removed during construction and maintenance prior to planting the vegetation. Most interaction between the drainage water and vegetation was through the limited amount of volunteer grasses and weeds that had colonized the banks and to a lesser extent the drain bed. The spoil from construction and maintenance of the UF section of drain had not been removed at any time and resulted in a low, eroded bank, from intermittently dispersed spoil on either side of the shallow drain. The upstream Veg section of drain was more deeply incised than the lower UF section of drain. Both sections of drain were incised through the sandy upper layer and down to a clay layer of the duplex soil below, i.e. the upstream Veg section of drain had approximately 300 to 500 mm more sandy soil profile than the lower UF section which had an upper sandy layer of 500mm.

In the catchment of the upstream Veg section of drain, P was surface applied annually at 17 kg P ha^{-1} as superphosphate with 12 kg P ha^{-1} applied in May and 5 kg P ha^{-1} in August. Nitrogen was applied in August at 25 to 35 kg N ha^{-1}. In the UF section of drain, the adjacent paddocks were not fertilized in 2006 and 2007 and from 2008 to 2011 they applied 17 kg P ha^{-1} of in June.

Records from 1984 and 1985 show plant available or bicarbonate extractable P (Colwell, 1965) were 42 and 36 mg kg^{-1} on average at the UF and Veg sites respectively. Ammonium oxalate extractable iron (Tamm, 1922) was 765 and 341 mg kg^{-1} on average at the UF and Veg sites, respectively. Both catchments soils can be described as having high P concentrations, and contained almost twice as much plant available P as was required for pastures to achieve 90% of maximum pasture production (Bolland et al., 2010).

The water quality of the sites had previously been monitored for one winter in 1997 (Cronin, 1998) for the same parameters monitored here but using manual sampling on a weekly basis. Indeed this site was chosen because of the previous study and to more closely monitor the impact of the stream fencing and riparian establishment.

Automatic water samplers and water level recorders (Teledyne, ISCO 6712, Lincoln NE USA) were installed at the upstream and downstream sites. Samplers were programmed to measure water height (stage) every 15 minutes using a flow bubbler module. The channel cross-section was surveyed at each site to determine the area and discharge at each stage height measurement, and discharge was calculated by the product of velocity and cross-sectional area. Flow and stage height was manually measured at each visit to provide calibration and datum information and for the development of stage discharge relationships. The samplers were programmed to initiate the sampling program when the in-channel stage height exceeded the 'cease to flow' level. The 'cease to flow' stream level was determined through multiple field observations through the flow season. Once initiated the sampler was programmed to collect a sample every 3 hours. Discharge and loads were calculated using the Hydstra version 9 time-series analysis software (Hydstra/TS, 2007) when water quality and stage height time-series data were combined. In 2011 repeated failure of the height monitoring equipment made it impossible to determine the annual load.

All N and P analyses were performed with standard colorimetric methods (Rice, Bridgewater, APHA & AWWA, 2012; Rayment & Higginson, 1992) on a segmented flow auto-analyser (OI Analytical, College Station, Texas USA). Total N and P contents were determined from analysis of samples that had been digested to convert all N forms to nitrate, and all P forms to ortho-phosphate. Digestion was achieved after autoclaving 5 mL samples at 120°C for 30 minutes in a 1:1 mix of sample and 4% alkaline persulphate solution (Ebina, Tsutsui & Shiriu,1983). Nitrate N and Filterable Reactive P (FRP) were determined on filtered (<0.45µm) and undigested samples. Total Suspended Solids (TSS) was determined using a gravimetric filter paper method outlined by Grace Analytical Laboratory (1994). In 2011 only turbidity (Nephelometric Turbidity Unit, NTU) was analysed and this was converted to TSS (mg L^{-1}) using the relationship derived by regression from earlier data (log_{10} TSS $= -0.298 + 0.953 * log_{10}$ turbidity, R^2 $= 0.84$).

The P sorption of the fresh gravel and gravel that had resided in the drain for 12 months was determined at 0 mg L^{-1} and 0.7 mg L^{-1} P solution concentration (the median concentration of the drainage water in the Veg section). The concentration of the P in solution was determined after gravel was left standing in the above solutions for 2 hours, 1 day, 2 days, 5 days and 12 days in a 20:1 solution to soil ratio (3 replicates). Gravel was left standing rather than shaken to avoid unrealistic P sorption measurements due to potential abrasion of gravel and exposure of fresh surfaces.

Exploratory data analysis was carried out using Data Desk (Velleman, 1997). Assessment of the data indicated that they were non-normally distributed, and therefore the data was often presented on log scales using notched box and whisker plots (Velleman & Hoaglin, 1981; McGill, Tukey, & Larsen, 1978), where the box depicts the central half of the data between the 25th and 75th percentiles, and the line across the box displays the median value. The whiskers extend from the 5th to the 95th percentiles. Values beyond the 5th and 95th percentiles (outliers) are plotted as a circle. Notches on the box and whisker plots approximately define the 95% confidence interval of the median (McGill et al., 1978). Visual assessment of notches on box and whisker displays for various treatments was used to assess the likelihood that population medians were different in preference to formal statistical treatments. Non-overlapping notches for different treatments could be regarded as being significantly different at the 95% level (Velleman & Hoaglin, 1981), whilst formal statistical procedures such as ANOVA are beset by problems from pseudo replication, and serial correlation. Non-parametric measures such as median and 95% confidence interval of median are reported and discussed because of the skewed nature of the concentration data. The 95% confidence interval of the median was calculated using the half width method described by McGill et al. (1978).

The loads and flows measured from the upper catchment were subtracted from those from the lower catchment to reflect the losses only from the lower section of the drain. Although it is expected that there will be some interaction between the incoming P and the drain, the P concentration throughout the drainage system is relatively similar and the impact on the equilibrium will remain constant. The impact on loads would be mainly due to the characteristics of the drainage system measured here.

Along each drain section, composite samples of soil were taken at 0-5 cm below the drain bed surface, in the adjacent paddocks, and also within the vegetated area of the Veg drain. This shallow depth was used to measure stratified P in the drain bed material that can interact readily with streamflow, and allow direct comparison with soil in adjacent paddocks. Phosphorus Retention Index (PRI), a single point measure of the P sorption characteristics of a soil, was determined using the method described in Allen and Jeffery (1990). Organic carbon was measured using the Walkley and Black (1934) procedure, plant available P was measured using the Colwell (1965) method and TP analysis was determined on samples digested with sulfuric acid-potassium-copper sulphate and the P concentration measured colorimetrically at 880 nm as described by Allen and Jeffery (1990). Ammonium oxalate extractable iron (AmOxFe) was also determined (Tamm, 1922). Whilst PRI estimates the amount of P sorption remaining in the soil, AmOxFe is a surrogate for the P sorption capacity that existed in the virgin soil prior to the application of P. An estimate of P saturation was determined and expressed as the ratio of Colwell P to AmOxFe. This ratio indicates the proportion of binding sites (AmOxFe) that are occupied with Colwell P and is a simplification of that described by Kleinman, Bryant and Reid (1999).

3. Results and Discussion

The number of water samples collected at each site in any year varied from 40 to 310 (Table 1) depending upon the flow in the drainage system as influenced by the seasonal rainfall pattern. As the automatic samplers were triggered to pick up flow events, and the flow was higher in the downstream UF catchment, slightly more samples were collected there. The flow from the Veg catchment in 2006 was too low to be accurately calculated. The flow from both catchments was too low to be calculated in 2010.

Table 1. Number of samples collected in 1997 to 2011 for assessment of water quality parameters (*Cronin, 1998)

Year	Veg	UF
1997*	11	11
2006	83	101
2007	244	310
2008	55	70
2009	48	40
2011	210	192

3.1 Contaminant Concentration in Drainage Water

3.1.1 P Concentration

In 1997, Cronin (1998) found the Veg drain discharging P with a lower TP than the UF section (Figure 2).

In contrast, the Veg section of drain discharged water with higher TP and FRP concentrations than the UF section over the monitoring period 2006-08 with no apparent difference in TP during 2009 (Figure 2). The median TP concentration (2006-2009) was 1.5 ± 0.1 mg L^{-1} (median $\pm$ 95% confidence interval) and 1.4 ± 0.1 mg L^{-1} in the Veg and UF sites respectively. Similarly the median FRP concentration (2006-2009) was higher in the Veg section (1.1 ± 0.1 mg L^{-1}) than it was in the UF section (0.9 ± 0.1 mg L^{-1}).

In 2011 after the addition of the cracked lateritic pea gravel to the Veg section of drain, TP and FRP) was lower than the UF section (Figure 2).

The ratio of FRP/TP concentrations from 2006 to 2009 appeared higher in the Veg section ($71\pm2\%$) than in the UF ($63\pm2\%$). The FRP/TP concentration ratio from the 1997 measurements of Cronin (1998) showed $28\pm0.1\%$ and $26\pm0.2\%$ in the Veg and UF sections respectively. After gravel addition in 2011 the FRP/TP concentration ratio was $78\pm0.1\%$ and $80\pm0.1\%$ respectively in the Veg and UF sections.

Cronin (1998) interpreted the effect of reduced P flow from the Veg section of drain to be due to the vegetation but had noted that the Veg section of the drain had the bottom and sides of the drain scraped clean of vegetation and deposited sandy sediment by the drainage manager early on in the sampling period, exposing the clay subsoil.

It is now hypothesized that excavation of sediment in the Veg drain in 1997 exposed and disturbed fresh clay subsoil that could retain P and produced a less stable bed and bank. This increased suspended sediment and reduced the proportion of P attached to particulates, resulting in a much lower FRP/TP ratio. This would have the overall effect of reducing the TP concentration in the Veg section through both sorption to the drain bed and banks and re-sedimentation of sorbed P derived from the freshly destabilized drain relative to the UF section.

Similar results were found when dredged drainage ditches were compared with undredged drain sediments by Shigaki, Kleinman, Schmidt, Sharpley and Allen (2008) who found that added dissolved P would be removed from stream flow by freshly dredged drains. They found biological processes accounted for only 30% of the P removal in the dredged sediment. High residence times and sediment interaction can reduce nutrient losses in some drainage systems (Jarvie et al., 2008) and these conditions may occur further downstream. Keipert, Weaver, Summers, Clarke and Neville (2008) estimated that in the catchment of the Peel Inlet and Harvey Estuary, the loss of P from the paddock was substantially more than that which reaches the receiving water bodies. The sediment from the base of the drainage system may also become a source of nutrients (Whithers & Sharpley, 2008).

Phosphorus retention on clay in the drain bed would be accentuated during low flows and less likely in higher flows during the sampling of Cronin (1998) because of shorter contact times. This was found to be the case where samples collected at higher flows had similar P concentrations at both sites.

After 9 years without maintenance (2006) the Veg drain bed and banks had stabilized and were covered with volunteer grasses, and had accumulated P. The proportion of P attached to suspended sediment reduced (approx. 70% PP in 1997, approx. 30% PP in the period 2006-2009), whilst the proportion of FRP increased (approx. 30% FRP in 1997, approx. 70% FRP in the period 2006-2009). The high TP and FRP/TP concentration ratio in the Veg section of drain measured from 2006 to 2009, is most likely due to several factors including desorption of P that had accumulated in the drain, and reduced sorption of P discharged from adjacent paddocks. In contrast, the UF section was continuing to generate sediment due to livestock disturbing the drain banks and bed. This sediment in the UF section of drain was functioning to remove P through P sorption and re-sedimentation. The presence of vegetation appeared to have little effect on the overall retention of P. Rather, in this case, a positive impact of livestock to the UF section of drain appeared to be a modest reduction of the large FRP load. It is unlikely that the vegetation would have much opportunity to physically filter water discharged from paddocks because of feeder drains and low PP in surface runoff due to low slopes and sandy textured soils, which would favour leaching. The use of a similarly relatively small grassed waterway elsewhere had little effect on Dissolved Reactive P (DRP) (Fiener & Auerswald 2009) and a review of the efficiency of buffers by Hoffman, *et al.* (2009) found highly variable effects of riparian buffers on DRP, noting DRP retention through to DRP release.

In 2011 the gravel provided a fresh sorbing surface for internal sources of FRP that could be released from accumulated P in the Veg drain, and from the FRP from the paddock. This had the effect of reducing FRP, and suspended sediment from the drain bed as the sediment was covered by the gravel. The overall reduction in P concentration from the use of gravel was less effective than excavating (exposing fresh subsoil and removing bed sediments) of the drain seen in 1997 by Cronin (1998).

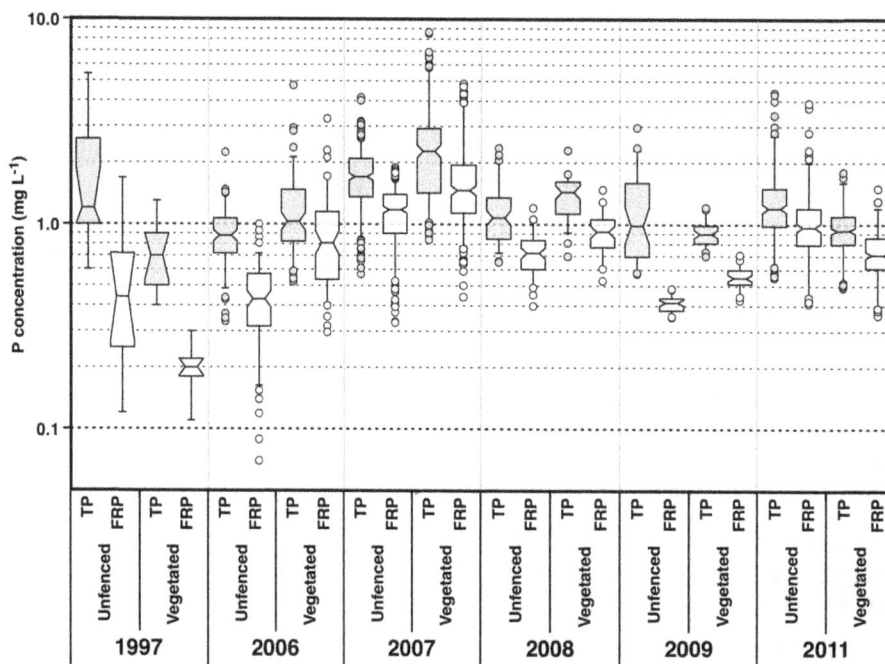

Figure 2. Notched box and whisker plots showing temporal variation in concentration (mg L^{-1}) of TP (grey) and FRP (white) from the UF and Veg drains. Boxes show median, 25th and 75th percentiles. Whiskers range from 5th to 95th percentiles. Notches show 95% confidence interval of the median. Circles show outliers

3.1.2 N Concentration

The impact of vegetating the drain, clearing the drain out or adding gravel had little consistent impact on the N concentrations or forms in the drain.

In 1997 (Cronin, 1998) the median TN concentration was lower in the Veg section (4.2±1 mg L^{-1}) than in the UF section (6.0±0.8 mg L^{-1}). The median NO$_3$ concentration was also lower in the Veg section (1.1±0.8 mg L^{-1}) than in the UF section (1.8±0.9 mg L^{-1}) (Figure 3).

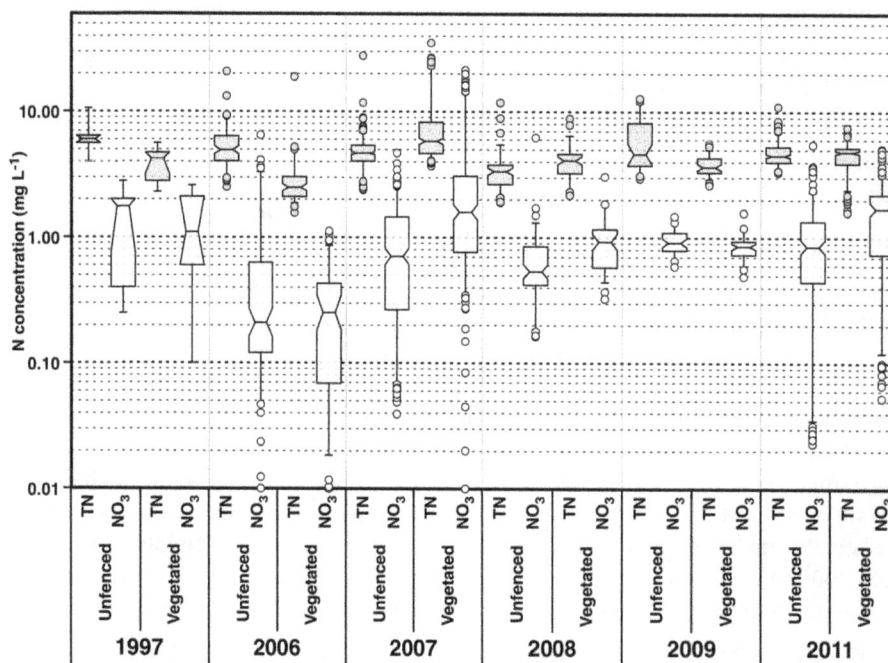

Figure 3. Notched box and whisker plots showing temporal variation in the concentration (mg L^{-1}) of TN (grey) and NO$_3$ (white) from the UF and Veg drains. Boxes show median, 25th and 75th percentiles. Whiskers range from 5th to 95th percentiles. Notches show 95% confidence interval of the median. Circles show outliers

Years of higher rainfall (2007 and 2008) resulted in increased TN and NO_3 from the Veg site. In 2006 and 2009 there appeared to be a slight reduction in TN and NO_3 when the rainfall was extremely low (Figure 3). From 2006 to 2009, the Veg section of drain discharged water with more TN and NO_3 than the UF section.

The median TN concentration (2006 to 2009) was 3% higher in the Veg (4.6±0.2 mg L^{-1}) than the UF site (4.5±0.1 mg L^{-1}). The median NO_3 concentration (2006 to 2009) was 34% higher in the Veg section (0.9±0.1 mg L^{-1}) than it was in the UF section (0.6±0.1 mg L^{-1}).

In 2011, after the addition of the lateritic gravel to the Veg section of drain, the Veg section was discharging a slightly higher median TN (Veg: 4.8±0.1, UF: 4.5±0.2 mg L^{-1}) and NO_3 (Veg: 1.7±0.2, UF: 0.85±0.1 mg L^{-1}) than the UF section. This was similar to the N concentrations found in the period 2006 to 2009.

3.1.3 TSS Concentration

In 1997 the median TSS of the Veg and UF sections of the drain was 6±1 and 75±37 mg L^{-1} respectively, or 92% less in the Veg drain (Cronin, 1998).

The excavation of the base of the Veg drain in 1997 still generated less sediment than the UF section. This suggests that the presence of cattle in the drain contributed to the measured sediment. The coarse textured sandy soils and low slopes in the catchment also supports the idea that very little sediment is contributed from paddock sources and the main source of sediment is therefore likely to be derived from exposed subsoils in the drain, and re-suspension of previously transported sediment in the drain. Fencing livestock out of the drain appeared to have more impact on sediment than the vegetation in the fenced area.

From 2006 to 2009 the UF section of the drain had TSS concentrations that were 65% greater than the Veg section of drain, with a median of 38±4 and 23±2 mg L^{-1} respectively. The difference between the two sites was slightly less in the higher rainfall years of 2007 and 2008 (Figure 4).

In 2011, after the addition of the lateritic gravel to the Veg section of drain in 2010, the Veg section was discharging a lower TSS (1.4±0.3) than the UF (2.8±0.3 mg L^{-1}) section. The low TSS reflected the low rainfall during the winter of 2011, which also had a larger number of smaller rainfall events.

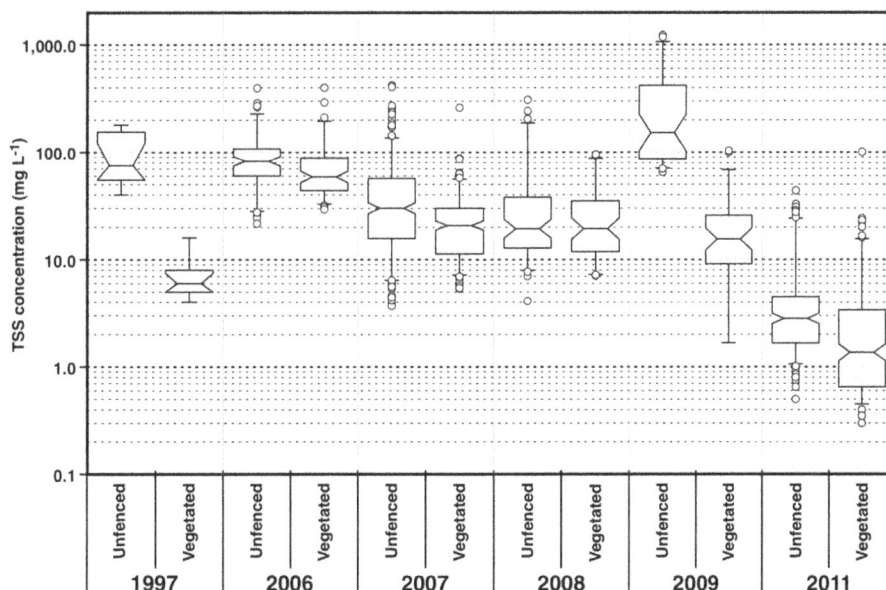

Figure 4. Notched box and whisker plots showing temporal variation in the TSS concentration (mg L^{-1}) from the Unfenced and Vegetated drains. Boxes show median, 25th and 75th percentiles. Whiskers range from 5th to 95th percentiles. Notches show 95% confidence interval of the median. Circles show outliers

3.2 Flow and Runoff

There was slightly more runoff (mm) from the upper Veg section of catchment compared with the UF catchment (Table 2). This may be due to a slightly greater depth of the Veg drain increasing runoff and flow speed. The shallower UF drain was wider and there may have been slightly more opportunity to pool the water and retain it in the soil of the catchment and drain.

Table 2. Cumulative and individual sub catchment area, flow and runoff for 2006 – 2011 (no flow in 2010)

		Area (ha)	Flow (ML)					Runoff (mm)				
Cumulative	Veg	39	0	79	197	69	61	0	203	505	177	156
	UF	130	10	208	522	108	73	8	160	402	83	56
Sub catchment	Veg	39	0	79	197	69	61	0	203	505	177	156
	UF	91	10	129	325	39	12	11	142	357	43	31

Runoff was characterised by two distinctly different seasons (Figure 5). The year 2006 was very dry with intermittent runoff lasting just over 1 month and the only site to receive measurable flow was the UF section of drain. In 2007 there was a more typical winter with close to average rainfall and runoff recorded at both sites between June and October. In 2008 the rainfall was 910 mm (only 97 mm below the average annual rainfall) and had the most runoff in the study. By 2009, the rainfall was 240 mm below the average annual rainfall and again did not produce as much runoff. In 2010 there was no runoff due to the extremely low rainfall. The annual rainfall in 2011 was closest to the long-term average. In 1997 when the previous study of the site was made by Cronin (1998) the rainfall was 792 mm, which was similar to the rainfall of 2009 (Table 3).

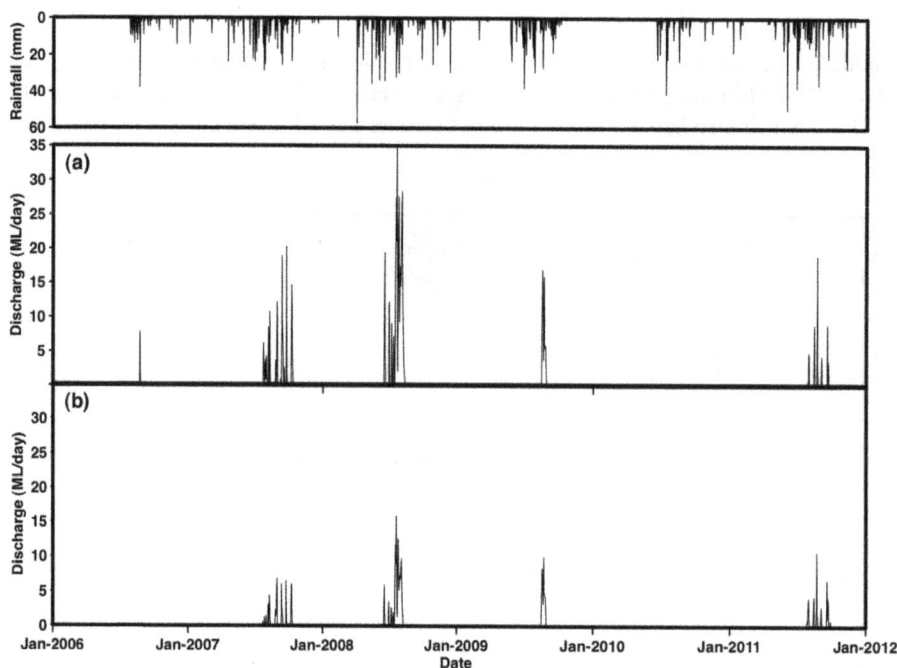

Figure 5. Rainfall (top), flow at the monitoring point below the UF section of drain (middle), flow at the monitoring point below the Veg section of drain (bottom)

Table 3. Annual rainfall during monitoring compared with the long-term average

Year	Annual rainfall	Annual average rainfall	Relative to average
1997	792		-215
2006	568		-439
2007	882		-125
2008	910	1007	-97
2009	767		-240
2010	485		-522
2011	946		-63

During high rainfall periods in mid winter when the soil profile is saturated, P enters the drain over the surface mainly through saturation excess, overland flow. Surface runoff carries desorbed P from P saturated soil, mineralized P from organic matter, low molecular weight organic P and P dissolved from current and previous fertilizer residues shown by the high proportion of FRP in the runoff. During low rainfall periods when the soil profile is not saturated, P may enter the drain via leaching and subsurface pathways. However under these conditions it is likely that less P enters the drain since interaction with the P retentive subsoil would reduce P transport. During these periods the drain sediments themselves may become sources of P as they have the same TP content as those in the paddock (Table 7), and much more Colwell P. Alternatively, P sourced from the topsoil may bypass the subsoil through macro-pore flow and directly enter the drain. All of these mechanisms preclude substantial interaction with vegetation on the banks of the drainage system. In 1997 the sediment of the Veg section of drain was freshly exposed during excavation and was likely to have very low levels of P. However, by 2006, the P in the sediment in the drain bed of this Veg section of drain was found to be similar to that of the surrounding paddock. This potential accumulation of P in the drain bed may explain the earlier observations of Cronin (1998) where P was retained in the Veg drain. Additionally it is clear from the FRP/TP concentration ratios (approx. 30% in 1997, approx. 70% in 2006 to 2009, approx. 80% in 2011) that in-drain sediment supply due to management can alter the form of P measured in the drain. In these sandy catchments it is likely that soluble forms of P are supplied from adjacent agricultural soils, which is then transformed to particulate P dependent on sediment supply (Weaver & Summers, 2014).

3.3 Nutrient Loads

The unit area TP and FRP discharge (kg ha^{-1}) of the Veg section of drain was more than double the UF section (Table 4).

The form of P appeared to be transformed slightly from FRP to PP during transport from the Veg to the UF area. Over the entire monitoring period the FRP/TP load ratio from the Veg area averaged 68% compared to 61% in the UF area.

Table 4. Phosphorus loads for individual sub-catchments for 2006 to 2011 (39 and 91 ha Veg and UF respectively)

Year	TP (kg)		TP (kg ha^{-1})		FRP (kg)		FRP (kg ha^{-1})		PP* (kg)		PP* (kg ha^{-1})		FRP/TP (%)	
	Veg	UF	Veg	UF	Veg	UF	Veg	UF	Veg	UF	Veg	UF	Veg	UF
2006	0	12	0.00	0.13	0	8	0.00	0.09	0	4	0.00	0.04	-	69
2007	218	158	5.59	1.74	150	91	3.85	1.00	68	67	1.74	0.74	69	57
2008	242	224	6.21	2.46	157	168	4.03	1.85	85	56	2.18	0.62	65	75
2009	48	71	1.23	0.78	28	7	0.72	0.08	20	64	0.51	0.70	59	10
2011	62	30	1.58	0.33	48	28	1.25	0.31	14	2	0.36	0.02	79	94
Ave	114	99	2.92	1.09	77	60	1.97	0.66	37	39	0.96	0.42	68	61

* Particulate P (PP) was derived by subtraction of Filterable Reactive P (FRP) from Total P (TP).

The TN and NO$_3$ load per unit area (kg ha^{-1}) from the Veg section of drain was more than double that measured in the UF section (Table 5). The proportion of unit area TN (kg ha^{-1}) measured as nitrate was 30% from the Veg area and 19% from the UF area.

Table 5. Total N loads for the individual catchments from 2006 to 2011 (39 and 91 ha Veg and UF respectively)

Year	TN (kg)		TN (kg ha^{-1})		NO$_3$ N (kg)		NO$_3$ N (kg ha^{-1})	
	Veg	UF	Veg	UF	Veg	UF	Veg	UF
2006	0	41	0.00	0.45	0	1.7	0.00	0.02
2007	666	361	17.1	3.97	274	71	7.03	0.78
2008	697	824	17.9	9.05	184	159	4.72	1.75
2009	190	344	4.87	3.78	37	40	0.95	0.44
2011	273	102	7	1.12	51	51	1.31	0.56
Ave	365	334	9.37	3.67	109	65	2.80	0.71

3.4 Sediment

When catchments are considered separately, the UF drain produced 48% higher TSS unit area loads (kg ha^{-1}). The cumulative loads show the scale of the deterioration in the UF section of drain where nearly 3.5 times as much sediment was discharged from 2006 to 2011 compared to the Veg section. In the 820 m of UF drain TSS increased by over 17000 kg in 2009 alone (Table 6).

This increase in sediment in the water column and the resulting exposure of fresh material in the drain bed may have allowed greater binding and retention of P.

Table 6. Suspended sediment loads for the individual catchments from 2006 to 2011 (39 and 91 ha Veg and UF respectively)

Year	TSS (kg)		TSS (kg ha^{-1})	
	Veg (39 ha)	UF (91 ha)	Veg (39 ha)	UF (91 ha)
2006	0	818	0	9
2007	2171	11 361	56	125
2008	6796	6673	174	73
2009	1831	18 845	47	207
2011	165	30	4.2	0.32
Ave	2192	7545	56	83

3.5 Soil, Drain-Bed Sediment and Lateritic Gravel

The TP in paddock soil and drain sediment was the same for both the Veg and UF sections of drain, whilst there was an increase in Colwell P in sediment for both sections of the drain relative to the paddock indicating that there had been a build-up of P sorbed in drain sediment (Table 7). The soil within the riparian buffer of the Veg section was lower in TP and Colwell P than all of the other sites sampled (Table 7). This suggests that little P is accumulating in this zone due to either deposition of soil particulates, or sorption of FRP from paddock runoff.

The PRI of the sediment in both the Veg and the UF section of drain was the same and also positive indicating that the sediment may adsorb P if P enriched water was in contact with the sediment for sufficient time. The initial potential for P sorption of the soils before fertilizing is represented by AmOxFe and shows that the UF paddocks were higher than Veg paddocks. The PRI is also higher suggesting that the paddock around the UF drain is less likely to discharge dissolved P than that around the Veg drain.

The soil in the paddock surrounding the Veg section of the drain had higher P saturation, lower PRI and lower AmOxFe than the paddock around the UF drain. This may have contributed to lower P in runoff into the UF drain due to the higher P sorption if drainage water travels across or through the surface layer.

The PRI of the fresh gravel was 1.9 and the PRI after being exposed to drainage water for 12 months was < 0.1. Although this was lower than in the sediment in the drain (PRI 12), the sediment was saturated with P shown by the Colwell P/AmOxFe ratio (Table 7). Covering the sediment with a 30 mm layer of sorbing gravel would restrict circulation with the drain water reducing P desorption from the sediment in the drain bed.

The lateritic gravel appears to be only of value in retaining nutrients in the short term. When the gravel was exposed to the median P concentration of the Veg section of drain of 0.7 mg L^{-1}, fresh gravel strongly sorbed P, but gravel that had been in the drain after only one winter slightly desorbed P showing its P sorbing capacity was exhausted (Figure 6). When this used and exhausted gravel was exposed to water without P it became strongly desorbing indicating that if catchment actions were successful in halting P entering the drain, then the drain sediment would become a source of P as the equilibrium was reversed (Figure 6).

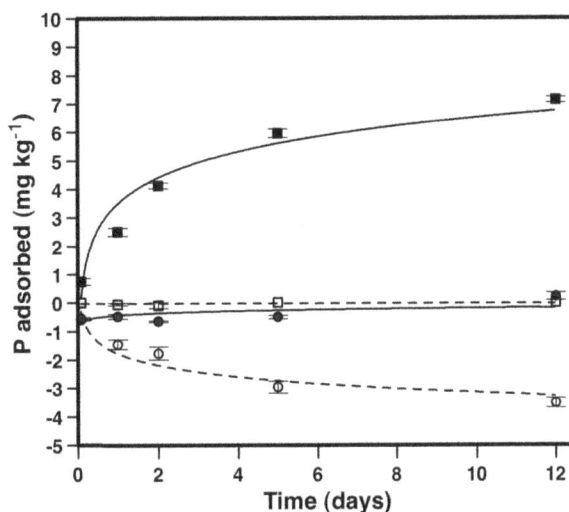

Figure 6. Phosphorus sorption over time for lateritic gravel taken from the drain (dashed lines and open symbols) and fresh unused lateritic gravel (solid lines and closed symbols). Squares are at an initial solution concentration 0.7 mg P L^{-1} and the circles are at an initial solution concentration of 0 mg P L^{-1} (error bars are standard errors of the means)

The runoff from farmed paddocks has carried with it P from paddocks into the drainage system where it travels downstream and accumulates in the drain-bed sediment at concentrations and degrees of saturation similar to that found in the paddock (Table 7).

The soil under the vegetation had a much lower capacity to retain P than the paddocks or the drain-bed sediments shown by the low PRI and AmOxFe. This area may not be a sink for P because the capacity of the soil to retain P is very low or the interaction between the surface water flow was limited by it being on a slightly raised spoil bank.

Table 7. Median soil characteristics from the paddock surrounding the drain, the sediment in the drain bed and the soil inside the fenced and vegetated area

	Paddock soil (0-50 mm)		Drain bed (0-50 mm)		Within the vegetated area (0-50 mm)
	Veg	UF	Veg	UF	Veg
TP (mg kg^{-1})	387 [a]	435 [a]	506 [a]	357 [a]	91 [b]
Colwell P (mg kg^{-1})	70 [a]	53 [a]	156 [b]	129 [b]	15 [c]
PRI	1.9 [bcd]	42 [a]	12 [acd]	7.4 [ad]	1.2 [b]
AmOxFe (mg kg^{-1})	1402 [b]	4490 [c]	861 [bc]	1557 [b]	127 [a]
Colwell P/AmOxFe ratio(%)	7.1 [a]	1.6 [b]	12.9 [a]	7.7 [a]	9.5 [a]
Organic carbon (%)	3.5 [a]	3.6 [a]	3.6 [a]	3.5 [a]	2.9 [b]

Within a row, treatments with different letters are different when means of log transformed values are compared $P<0.05$.

The catchment surrounding the drain had more P (as single superphosphate) applied than the section. This may have contributed to more P being lost from the catchment

The artificial drainage system investigated here is slightly more incised than that of natural streams occurring in the region, hence there may be differences in the effectiveness of riparian management to control nutrients where the banks of natural streams have a lower slope. Additionally the site was only vegetated on one side, making comparison with riparian management with vegetation on both sides difficult. These aspects were controlled by McKergow et al. (2003) in a sandy catchment near Albany Western Australia where they found a similar impact of riparian management that significantly reduced sediment transport, but allowed the passage of the majority of the P in a soluble form. Similarly they also found that during a period where the stream was unfenced and unvegetated a higher proportion of P was attached to sediment. McKergow et al. (2003) also showed no difference in TP loads before and after riparian management was implemented.

In these flat sandy soil catchments, water discharges mainly over P saturated surface soils during winter, and also travels through superficial subsurface pathways, interacting with soil that may be saturated with P. The dominant form of P that reaches the riparian zone is soluble or bound to extremely small particulates ($<0.45\mu m$), which are not physically filtered by vegetation or retained through sedimentation. Although the sediment in the Veg drain had some remaining P retention as shown by PRI, the residence time, degree of contact, and P sorption capacity appears to be insufficient to retain all the soluble P in the drain base sediment. The main mechanism remaining for the riparian zone is to take up P during transpiration, which is clearly inadequate as shown by the presence of runoff water. This mechanism of uptake is also limited due to the low P requirement of mature native vegetation. Less than 1 hectare of riparian vegetation is present in the Veg catchment contributing to approximately 3 kg ha^{-1} yr^{-1} of P uptake (Attiwill, 1980), or 6 kg ha^{-1} yr^{-1} if both sides were vegetated. The retention of only 3 - 6 kg ha^{-1} yr^{-1} in plant growth is also inadequate considering the annual average FRP runoff of 77 kg, hence the impact due to P uptake from riparian vegetation is unlikely to be detected. The relatively low concentration of P in the soil of the vegetated area of the drain banks is another indication of the lack of retention of P in this area and the lack of interaction of this area with surface flows of water entering the drain. The area which is selected to be vegetated needs to be carefully considered because the water appears to be by-passing this region at many points of concentrated flow rather than as sheet flow around the drain sides. This effect may be reduced in natural drainage systems that are less incised and where the banks flood more, but the form of P in the water entering the drain must also be considered.

A number of factors such as differing fertilization or slightly different drain incision may have contributed to differences, or lack of them reported here.

Although this was not an ideally controlled trial, the apparent reductions in P load from a catchment with riparian vegetation could not be replicated, nor could the anticipated or documented reductions in P load found elsewhere in the literature.

4. Conclusion

This fenced drain with well established riparian vegetation increased channel stability and the capture of sediment in runoff, reducing sediment load by more than 27 kg ha^{-1}. However, there was little measurable impact on the P concentration and appeared to have little impact in retaining P in soluble forms.

P retention was reduced most by mechanical exposure of the drain bed, followed by livestock exposing the drain bed and vegetation appeared to have little impact. Fencing appeared to have a greater impact on sediment retention than the vegetation even when the drain bed was scraped during maintenance. The vegetation appeared to have little impact on P or sediment retention.

The results suggest that the absence of fencing allowed livestock to mobilise sediment and expose fresh sorbing drain-bed surfaces, which interacted with and retained soluble P entering drains from adjacent paddocks. There was also a small amount of conversion of the soluble P to particulate P forms in the drain, and subsequent retention over the length of the UF reach of drain.

In these flat sandy catchments, fencing and re-establishment of riparian vegetation appears useful in stream stability and particulate retention but of limited or minor value for nutrient retention where P is not in a particulate form or where vegetation has little interaction with drainage water. Other techniques and mechanisms would need to be employed to reduce nutrient loads. These results also suggest that regularly cleaning out fenced sections of drains performs the dual function of retaining P and reducing the generation of sediment. Similarly, alternate sections of unfenced and vegetated drains may be partially successful in generating nutrient sorbing drain-bed sediment followed by fenced drains designed to trap the generated suspended sediment. Additionally, cracked lateritic pea gravel can retain some soluble P when placed in the stream, but this appears to only be a short-term solution.

Alternatively, the use of fencing with or without revegetation needs to accompany the use of inorganic nutrient retentive materials either on the paddock or in the drainage system to retain the soluble or low molecular weight organic forms of P on-site. This would enable other potential environmental benefits of revegetation to be combined with maximum improvement in water quality.

The impact of this assessment of a first order drain must also be considered in the context of the entire drainage system. This needs to be understood and taken into account and targeted intervention in the drainage system may be needed to maximize the impact of the drainage system on improving the water quality.

Acknowledgements

The role of the landholders of the Pitter and Nancarrow families of Coolup are gratefully acknowledged for their tireless assistance and cooperation over the years of the study. Also, of importance is the assistance of Mark Rivers, Martin Clark, Tony Allen, and Fred Robinson for establishing and operating site equipment. Financial and in-kind assistance from the Peel-Harvey Catchment Council, the Department of Agriculture and Food, WA and the Department of Water through with support from the Australian Government and the WA Government.

References

Allen, D. G., & Jeffrey, R. C. (1990). Methods for analysis of phosphorus. *Western Australian Soil Report of Investigation No.37 Chemistry Centre WA, p. 37*, cited in Burkitt L. L., Moody P. W., Gourley C. J. P., Hannah M. C. (2002) A simple phosphorus buffering index for Australian soils. *Australian Journal of Soil Research, 40*, 497–513. http://dx.doi.org/10.1071/SR01050

Attiwill, P. M. (1980). Nutrient cycling in a Eucalyptus Obliqua (L'herit) forest [in Victoria] iv Nutrient uptake and nutrient return. *Australian Journal of Botany, 28*, 199-222. http://dx.doi.org/10.1071/BT9800199

Bolland, M. D. A., Russell, B., & Weaver, D. M. (2010). Phosphorus for high rainfall pastures. *Bulletin 4808. Department of Agriculture and Food Western Australia.* Retrieved from http://www.agric.wa.gov.au/objtwr /imported_assets/content/past/bn_phosphorus_high_rainfall_pastures.pdf. Accessed 28 August 2014

Colwell, J. D. (1965). An automated procedure for determination of phosphorus in sodium hydrogen carbonate extracts of soils. *Chem Ind, 22*(May), 893-895.

Cronin, D. (1998). *The Effectiveness of 'Streamlining' in Improving the Water Quality of Agricultural Drains in the Peel-Harvey Catchment, Western Australia.* School of Biological and Environmental Science, MSc thesis. Murdoch University, Perth Western Australia.

Drewry, J. J, Newham, L. T. H., Greene, R. S. B., Jakeman, A. J., & Croke, B. F. W. (2006). A review of nitrogen and phosphorus export to waterways: context for catchment modelling. *Marine and Freshwater Research, 57*, 757-774. http://dx.doi.org/10.1071/MF05166

Ebina, J., Tsutsui, T., & Shiriu, T. (1983). Simultaneous determination of total nitrogen and total phosphorus in water using peroxidisulphate oxidation. *Water Research, 17*, 1721-1726. http://dx.doi.org/10.1016/0043-1354(83)90192-6

Environmental Protection Agency. (2008). *Water Quality Improvement Plan for the Rivers and Estuary of the Peel-Harvey System, Environmental Protection Authority, Perth Western Australia.* http://www.epa.wa.gov.au/Policies_guidelines/other/Pages/phwqip.aspx. Accessed 22 September 2014.

Fiener, P., & Auerswald, K. (2009). Effects of Hydrodynamically Rough Grassed Waterways on Dissolved Reactive Phosphorus Loads Coming from Agricultural Watersheds. *Journal of Environmental Quality, 38*, 548-559. http://dx.doi.org/10.1016/j.ecoleng.2006.02.005

Grace Analytical Laboratory. (1994). *Standard Operating Procedure for the Sampling and Analysis of Total Suspended Solids in Great Lakes Waters.* Chicago, IL 60605, August 2, 1994, viewed 30 September 2008. Retrieved August 28, 2014, from http://www.epa.gov/glnpo/lmmb/methods/tss2.pdf

Hodgkin, E. P., Birch, P. B., Black, R. E., & Humphries, R. B. (1980). *The Peel-Harvey Estuarine System Study, 1976-80.* Department of Environment and Conservation, Perth, Report No. 9.

Hoffmann, C. C., Kjaergaard, C., Uusi-Kämppäc, J., Hansend, H. C. B., & Kronvang, B. (2009). Phosphorus Retention in Riparian Buffers: Review of Their Efficiency. *Journal of Environmental Quality, 38*, 1942-1955. http://dx.doi.org/10.2134/jeq2008.0087

Hydstra/TS. (2007). Hydstra Time-Series Data Management Software Package, Kisters Pty Ltd.

Jarvie, H. P., Withers, P. J., Hodgkinson, R., Bates, A., Neal, M., Wickham, H. D., ... Armstrong, L. (2008). Influence of Rural Land Use on Streamwater Nutrients and Their Ecological Significance. *Journal of Hydrology, 350*, 166-186. http://dx.doi.org/10.1016/j.jhydrol.2007.10.042

Keipert, N., Weaver D. M., Summers, R. N., Clarke, M., & Neville S. (2008). Guiding BMP adoption to improve water quality in various estuarine ecosystems in Western Australia. *Water Science and Technology, 57,* 1749-1756. http://dx.doi.org/10.2166/wst.2008.276

Kleinman, P. J. A., Bryant, R. B., & Reid, W. S. (1999). Development of pedotransfer functions to quantify phosphorus saturation of agricultural soils. *Journal of Environmental Quality, 28,* 2026-2030. http://dx.doi.org/10.2134/jeq1999.00472425002800060044x

McComb, A. J., & Davis, D. A. (1993). Eutrophic Waters of Southwestern Australia. *Fertilizer Research, 36,* 105-114. http://dx.doi.org/10.1007/BF00747580

McGill, R., Tukey, J. W., & Larsen, W. A. (1978). Variations of Box Plots. *The American Statistician, 32,* 12-16. http://dx.doi.org/10.2307/2683468

McKergow, L. A., Prosser, I. P., Weaver, D. M., Grayson, R. B., & Reed, A. E. G. (2006a). Performance of grass and eucalyptus riparian buffers in a pasture catchment, Western Australia, part 1: riparian hydrology. *Hydrological Processes, 20,* 2309-2326. http://dx.doi.org/10.1002/hyp.6053

McKergow, L. A., Prosser, I. P., Weaver, D. M., Grayson, R. B., & Reed, A. E. G. (2006b). Performance of grass and eucalyptus riparian buffers in a pasture catchment, Western Australia, part 2: water quality. *Hydrological Processes, 20,* 2327-2346. http://dx.doi.org/10.1002/hyp.6054

McKergow, L. A., Weaver, D. M., Prosser, I. P., Grayson, R. B., & Reed, A. E. G. (2003). Before and after riparian management: sediment and nutrient exports from a small agricultural catchment, Western Australia. *Journal of Hydrology, 270,* 253-272. http://dx.doi.org/10.1016/S0022-1694(02)00286-X

Osborne, L. L., & Kovacic, D. A. (1993). Riparian vegetated buffer strips in water quality restoration and stream management. *Freshwater Biology, 29,* 243-258. http://dx.doi.org/10.1111/j.1365-2427.1993.tb00761.x

Rayment, G. E., & Higginson, F. R. (1992). *Phosphorus Australian Laboratory Handbook of Soil and Water Chemical Methods* (Volume 3, pp. 68-70). Melbourne: Australian soil and land survey handbook Inkata Press. http://nla.gov.au/nla.cat-vn685915

Rice, E. W., Bridgewater, L., APHA, & AWWA (2012). *Standard methods for the examination of water and wastewater* (19th ed.). Greenberg A. E. (Ed.). American Public Health Association, American Water Works Association and Water Environment Federation. Washington D.C. USA. ISBN 9780875530130.

Shigaki, F., Kleinman, P. J. A., Schmidt, J. P., Sharpley, A. N., & Allen, A. L. (2008). Impact of Dredging on Phosphorus Transport in Agricultural Drainage Ditches of the Atlantic Coastal Plain. *Journal of the American Water Resources Association, 44,* 1500-1511. http://dx.doi.org/10.1111/j.1752-1688.2008.00254.x

Tamm, O. (1922). Eine Methode zur Bestimmung de der anorganischen Komponente des Bodens. *Meddelanden fran Statens skogsforsoksanstalt Stockholm, 19,* 387-404.

Thorne, C. R. (1990). *Effects of vegetation on riverbank erosion and stability.* In: Vegetation and Erosion: Processes and Environments. New York: Wiley.

Velleman, P. F. (1997). *Data Desk Version 6.0.* Statistics Guide Vol. 3, Data Description Inc, PO Box 4555 Ithaca, NY. Retrieved from http://www.datadesk.com accessed 22 Sept 2014

Velleman, P. F., & Hoaglin, D. C. (1981). *Applications, Basics and Computing of Exploratory Data Analysis.* Boston: Duxbury Press.

Weaver, D. M., & Summers, R. N. (2014). Fit-for-purpose phosphorus management: do riparian buffers qualify in catchments with sandy soils? *Environmental Monitoring and Assessment, 186,* 2867-2884. http://dx.doi.org/10.1007/s10661-013-3586-4

Weaver, D. M., Ritchie, G. S. P., & Gilkes, R. J. (1992). Phosphorus sorption by gravels in lateritic soils. *Australian Journal of Soil Research, 30,* 319-30. http://dx.doi.org/10.1071/SR9920319

Withers, P. A., & Sharpley, A. N. (2008). Characterization and Apportionment of Nutrient and Sediment Sources in Catchments. *Journal of Hydrology, 350,* 127-130. http://dx.doi.org/10.1016/j.jhydrol.2007.10.054

12

The Social Value of Environmental Improvements in the Tarim Basin – toward a Comprehensive Assessment in a Heterogeneous Setting

Michael Ahlheim[1], Oliver Frör[2], Jing Luo[3], Sonna Pelz[1], Tong Jiang[4,5] and Yiliminuer[6,7]

[1] University of Hohenheim, Institute of Economics, Stuttgart 70593, Germany

[2] University of Koblenz-Landau, Institute for Environmental Sciences, Landau 76829, Germany

[3] Research Center for China's Borderland History and Geography, Chinese Academy of Social Sciences, Beijing 100732, China

[4] National Climate Center, China Meteorological Administration, Beijing 100081, China

[5] School of Geography and Remote Sensing / Collaborative Innovation Center on Forecast and Evaluation of Meteorological Disasters, Nanjing University of Information Science & Technology, Nanjing 210044, China

[6] China Academy of Forestry Sciences in Xinjiang, Urumqi 830011, China

[7] Institute of Applied Physical Geography, Catholic University of Eichstätt-Ingoldstadt, 85072 Eichstätt, Germany

Correspondence: Oliver Frör, University of Koblenz-Landau, Institute for Environmental Sciences, Landau 76829, Germany. E-mail: froer@uni-landau.de

Abstract

The benefits of environmental restoration projects are frequently underestimated because decision makers tend to ignore the non-use values of such projects. Using data from a representative contingent valuation survey conducted in Beijing in 2013 and of information gathered during five focus group workshops in Xinjiang, we show that environmental improvements in the Tarim Basin in Northwest China would not only enhance the wellbeing of the local population but also of people living in other parts of the country. In both study sites we find similar preferences for various ecosystem services and the mitigation of environmental problems. Furthermore, respondents from Xinjiang and Beijing were willing to contribute approximately equal shares of their income to a prospective environmental project aimed at the restoration of the Tarim Basin's natural ecosystems. We conclude that government representatives should consider the preferences of people from all parts of China when deciding on future land and water management strategies in Northwest China.

Keywords: environmental protection, Tarim River Basin, cost-benefit analysis, non-use values, contingent valuation method

1. Introduction

The Tarim Basin owes its name to Central Asia's longest inland river, which is the Tarim River. Because of the extremely arid climate, little precipitation and high evaporation humans and nature in the Tarim Basin strongly depend on water provision by the Tarim River (Thevs, 2011). The Tarim Basin is located in the southern part of Xinjiang Uyghur Autonomous Region in Northwest China. In spite of its enormous size of almost one million km^2, the basin is scarcely populated. This is because a large part of the Tarim Basin is covered by the Taklimakan Desert or mountains and therefore uninhabitable. About half of the eight million inhabitants live in the basin's oasis cities along the Tarim River (Huang et al., 2010).

Unique but highly vulnerable dryland ecosystems, which are of great significance for local people's livelihoods, can be found in the Tarim Basin. Due to an increasingly intensive land use in that area the quantity and quality of natural ecosystems has been decreasing. The still existing riparian poplar forests, reed beds, grasslands and shrub vegetation provide essential ecosystem services (ESS) to the local population. Among other things, local farmers use the riparian forest and the grasslands as pasture for hay-making to feed sheep and goat. Some native plants are traditionally used as natural medicines. The riparian forests and shrubs protect the oasis cities from sandstorms and dust. The flooded areas of the riparian forests and reed beds facilitate the recharge of groundwater with fresh river water. Since groundwater is a main freshwater resource, local households directly benefit from this service of the riparian ecosystems. Furthermore, the poplar forests along the Tarim River are an attraction for Chinese and international tourists. Most importantly, these forests build a 'green corridor' which separates the Taklimakan Desert from the southern part of the Gobi Desert (Halik et al., 2005).

However, the importance of the natural ecosystem for local people's livelihoods has been recognized only lately by the Chinese authorities. For a long time, environmental protection played a minor role for water and land management decisions in the region. Since water was mainly allocated to industry, households and man-made ecosystems like cotton fields, the natural ecosystems continuously deteriorated. Without doubt, one of the reasons for overlooking the benefits from natural ecosystems is that nature provides them for free so that none of the usual indicators of value, like e.g. market prices, exist to inform decision makers about the actual value and the increasing scarcity of natural ecosystems and the services they provide to society. For Chinese policy makers managing the ecosystem services in the Tarim region it is, however, essential to have a valid measure of the social value of such services at hand.

In the present study we make use of the Contingent Valuation Method (CVM) – one of several nonmarket valuation techniques – to assess the social value of more sustainable water- and land-management strategies in the Tarim Basin which would lead to a restoration and long-term protection of the natural vegetation along the Tarim River. Unlike previous studies, which made use of market-price based approaches to assess the use value of natural ESS (e.g. Xu et al., 2014), the present paper aims to derive a more comprehensive value assessment, including both use and non-use values of the natural environment along the Tarim River. Following Ahlheim et al. (2013), it is argued that not only the benefits accruing to the local population but also those accruing to people living in other parts of China should be assessed. In the present study we combine the results of a nonmarket valuation study carried out in Xinjiang with those of an exemplary comprehensive CVM survey conducted in Beijing. We show that, for the case of ecosystem services of nationwide importance, both regional and long-distance values need to be thoroughly accounted for in order to obtain a comprehensive value estimate of such services. An important contribution of this paper is the joint analysis of environmental values based on methodologically different valuation studies adapted to their specific survey contexts in a heterogeneous country.

2. Background and Purpose of the Present Study

2.1 Degradation of Natural Ecosystems in the Tarim Basin

Until the 1950s mainly Uighur farmers lived in the Tarim Basin and the region was hardly developed economically. The economic development of Xinjiang has been successfully driven forwary by the Chinese government since the 1950s. More and more Han Chinese workers settled down in the Tarim Basin. One of the main drivers of Xinjiang's economic development and population growth is the cotton industry. Today one sixth of China's total cotton output is produced in the Tarim Basin (Hirji & Davis, 2009). Along with the population growth, GDP and income levels have been increasing in the region. The oasis cities along the Tarim River have been growing and local firms as well as households have benefited from public investments into the region's economy and infrastructure. At the same time, the growth of the population, intensive agricultural activities, especially the water-consuming cotton production, and industrial development have contributed to permanent water shortage in the lower reaches of the Tarim River thereby causing severe environmental deterioration (Zhang et al., 2010). The environmental consequences of the progressing deterioration of the natural riparian ecosystems caused by the reduced runoff of the Tarim River include, to list but the most severe problems, more sandstorms than in the past, an increasing number of dust days, desertification of the landscape and dramatic loss in biodiversity (Thevs, 2011). Obviously, the economic development of the region has been realized at the cost of a degrading environment. Natural degradation is likely to become even more serious under the impact of global warming, at least in the long run. Chen et al. (2013) and other climate experts predict increasing temperatures, changes in seasonal precipitation and melting of glaciers. Under these circumstances the Tarim River may desiccate completely. This would imply the merger of the Taklimakan Desert and the southern part of the Gobi Desert. In other words, the Tarim Basin would not be habitable anymore and unique dryland ecosystems would be lost forever.

Since the 1990s the degradation of the natural ecosystems in the Tarim Basin has gained increasing attention, both at the Chinese and the international level. The impact of land and water management decisions on the state of the environment has been readily understood and China's Central Government has undertaken a number of measures to restore and protect the natural vegetation along the Tarim River. With the support of the World Bank institutions responsible for the water allocation in the upper, middle and lower reaches of the Tarim River have been created. According to Lu et al. (2010) the Chinese government has invested approximately 11 billion RMB into water management projects aiming at a restoration of the natural ecosystems in the Tarim Basin. Furthermore, scientists have contributed to a better understanding of how water and land use affect environmental conditions in the region (cf. e.g. Xu et al., 2008, Zhang et al., 2010, Huang et al., 2010). In spite of an increasing awareness of the water-related environmental issues, the integration of natural ecosystems into land and water management decisions appears to be difficult and insufficient – the deterioration of the natural environment in the lower reaches of the Tarim River continues.

2.2 Objectives of the Present Study

As highlighted above, the natural ecosystems along the Tarim River contribute to the wellbeing of the local population in multiple ways. Furthermore, natural scientists have developed sophisticated methods and models to predict and analyze the impact of changes in the water allocation on the state and quantity of the natural ecosystems (e.g. Xu et al., 2007, Huang et al., 2010). Moreover, several researchers have made concrete proposals to the responsible authorities concerning the technical and political measures needed to restore, preserve and protect the natural vegetation along the Tarim River in the long term (cf. e.g. Zhang et al., 2010, Chen et al., 2013). Based on this information the costs of different water distribution and land management schemes can be determined rather straightforwardly on the basis of market prices like wages, capital costs and material costs. However, to weigh these costs against the benefits, the value of the restoration of the natural ecosystems through more sustainable land and water management strategies needs to be quantified and monetized as well. Obviously, there are no market prices available for environmental goods such as wildlife, landscape beauty, improved air quality, etc. However, such data is indispensable for a comprehensive cost-benefit analysis of alternative water- and land-management policy options. Before implementing a particular environmental project, decision makers should verify whether society would be better off subsequent to the implementation of this project. In other words, they should ensure that the social benefits accruing from the environmental project in question outweigh its costs.

Huang et al. (2010) made a first attempt to estimate the change in value of the ecosystems in the lower reaches of the Tarim resulting from changes in water allocation. The authors computed unit prices for the cropland, forest, grassland, wetland and desert ecosystems for the period 1970-2005. Market prices of the outputs of these ecosystems served as the basis for the valuation approach. The unit price for croplands, for example, is based on natural grain output of croplands and market prices for crops. Huang et al. (2010) showed that the value of forest, grassland and wetland had been gradually shrinking due to decreases in the total area covered by these ecosystems and diminishing capacities in ESS supply. The results of Huang et al.'s (2010) market-price based evaluation are a suitable reference for the evaluation of environmental projects implemented in the Tarim Basin in the past. Nevertheless, these results are hardly helpful for cost-benefit assessments of prospective environmental projects. In addition to that, market price based methods generally underestimate the true social value of ecosystems. This is because so-called non-use values accruing from the restoration and preservation of an ecosystem, which are not reflected by market prices, are likely to alter its social value (cf. Nunes, 2002).

In the present study a different approach for determining the monetary value of the natural ecosystems along the Tarim River will be introduced. We make use of the contingent valuation method (CVM) to assess the social benefit of a prospective environmental restoration project in the Tarim Basin. This survey-based technique aims at determining the value of an environmental improvement by eliciting households' maximum willingness to pay (WTP) for a hypothetical environmental restoration project. In economics, the overall benefit of an environmental project is commonly computed as the sum of the WTP of all individuals affected by that project (cf. section 3.1). The results of a comprehensive CVM survey implemented in different Chinese cities in 2013 regarding common people's perceptions of the degradation of the natural ecosystems along the Tarim River as well as their WTP for an environmental restoration project will be presented and analyzed in the following.

Until now hardly any comparable CVM study has been published. Firstly, the assessment of 'long-distance benefits' of environmental projects by means of the CVM has attracted the attention of only a few researchers (cf. e.g. Jørgensen et al., 2013; Ahlheim et al., 2013). Secondly, survey studies from Xinjiang in general, and CVM studies in particular, are extremely rare. The political and social circumstances in Northwest China are likely to be one of the reasons for this lack of research. Ethnic unrest, increased security measures and mistrust towards social scientists complicate the conduct of interviews with Xinjiang's population. However, a few researchers successfully implemented surveys on environmental topics in this region. Deng et al. (2011) and Deng et al. (2012), for example, investigated how a number of climate change adaptation measures, such as public water-saving programs, were perceived by residents in the Urumqi River Basin and the Aksu River Basin. However, they did not aim to assess the social value of a particular ecosystem or environmental restoration project. Xu et al. (2014) conducted a CVM survey with Han Chinese farmers working in the State Farms at the upper, middle and lower reaches of the Tarim River to assess the social value of the protection of the local poplar forests. Most interviewed farmers expressed a favorable opinion concerning the protection of poplar forests, but relatively few were willing to make a financial contribution to it. In contrast to Xu et al. (2014), who exclusively focused on the preferences of a particular group of local residents (Han-Chinese farmers employed by the state farms), in a pilot study Ahlheim et al. (2013) addressed the question how the degradation of the natural ecosystems along the Tarim River was perceived by people living far away from the Tarim Basin. Based on an intercept CVM survey in the city of Beijing, they showed that also people who are only indirectly affected by the environmental problems in the Tarim Basin were willing to make financial contributions to the realization of an environmental restoration project in this region. They stressed that decision makers should not only consider the preferences of people living at a particular environmental site but also of people living in other parts of the country when determining the value of environmental projects of national importance, like the Tarim project. In this pilot study, of course, it was not possible to conduct a comprehensive assessment of use and non-use values to be expected from such an environmental restoration project. With only 300 individuals interviewed it was necessary to employ a second-best elicitation question format like the payment card instead of the incentive compatible but less efficient dichotomous choice format. Further, no suitable validity tests could be carried out with this small sample size. Finally, the pilot study exclusively focused on Beijing citizens' preferences but did not include the question as to how the local population in Xinjiang perceives the environmental deterioration of the Tarim Basin. To overcome these limitations of the pilot study, a more comprehensive and sophisticated CVM study was carried out in 2013.

In this study, we make use of a refined WTP question and a larger sample to assess the 'long-distance value' of an environmental restoration project in the Tarim area. Using econometric techniques, the validity of the WTP estimate will be tested. Furthermore, the preferences of people indirectly affected by the public project in question will be compared to the preferences of the local population, including Han Chinese residents and also residents with other ethnic backgrounds.

3. Method

3.1 The Contingent Valuation Method

In economics the social value of a particular environmental improvement is measured as the change in wellbeing of all households affected by the improvement in question. The change in wellbeing, or utility, of a single household can be written in terms of the indirect utility function $\upsilon_h(.)$, which describes the maximum utility a household can reach given its income (I_h), market prices (p) and the state of the environment (z).

$$\Delta U_h \quad U_h^1 \quad U_h^0 \quad \upsilon_h(p, z^1, I_h) \quad \upsilon_h(p, z^0, I_h), \tag{1}$$

where z^0 is a vector of environmental parameters that refer to the state of the environment in the initial situation and z^1 a vector of environmental parameters describing the state of the environmental subsequent to the environmental improvement. The indirect utility function is monotonically increasing in environmental quality z. Thus, in the case of an environmental improvement ($z^1 > z^0$), the change in utility ΔU_h is positive (Stephan & Ahlheim, 1996).

The utility change described in equation (1) can also be expressed in terms of the expenditure function $e_h(.)$, which describes the minimum (monetary) expenditures a household has to make to reach a particular utility level. The expenditure function is monotonically increasing in utility, i.e. higher utility levels are reached by an increase in consumption. Taking the new state of the environment (z^1) as reference (assuming that prices stay constant) we obtain the Hicksian Compensating Variation (HCV) for a household h:

$$HCV_h \quad e_h(p, z^1, U_h^1) \quad e_h(p, z, U_h^0) \quad I_h \quad e_h(p, z^1, U_h^0), \tag{2}$$

Assuming that the household spends its entire income on consumption goods, the HCV_h just equals the household's income minus the (fictional) expenditure it would have to make to get back to its initial utility level U_h^0, given constant market prices p and the new state of the environment z^1. In the case of an environmental improvement, the HCV_h is positive because the household's income exceeds the minimum expenditure necessary to reach the initial utility level. The HCV_h can also be included in the indirect utility function, leading to the following expression:

$$\upsilon_h(p, z^1, I_h \quad HCV_h) \quad \upsilon_h(p, z^0, I_h) \quad 0, \tag{3}$$

As can be seen in equation (3), the HCV_h is the maximum amount of money that can be taken away from the household's income without making it worse off than in the situation before the environmental improvement. It is typically interpreted as the household's maximum willingness to pay (WTP_h) to obtain the environmental improvement (Carson & Hanemann, 2005).

The sum of individual HCVs yields a measure of the overall value or social benefit (B^{soc}) of all households H affected by the environmental improvement in question. In practice, it is approximated by summing up the individual WTP_h amounts, i.e.

$$\text{B}^{soc} \quad \sum_{h\,1}^{H} \text{HCV}_h \quad \sum_{h\,1}^{H} \text{WTP}_h . \qquad (4)$$

The obtained value can then be compared to the cost of an environmental project, for example, the public funds needed to realize an environmental restoration project.

Over the past decades several methods for the assessment of people's WTP for environmental improvements have been developed (cf. e.g. Ahlheim & Frör, 2003, Atkinson & Mourato, 2008). The present study makes use of the Contingent Valuation Method. The CVM is an interview-based technique where a market for an environmental good is simulated and the good's social value is inferred from respondents' choices on this contingent market. Randomly selected households are presented an environmental project that is expected to increase their wellbeing. Subsequently, they are asked to directly state whether they would be willing to pay for this environmental project, for example by accepting or rejecting a particular policy that would imply a tax increase and hence higher household expenditures. Ideally, the obtained sample is representative for all households affected by the environmental improvement in question so that the sample's average WTP reflects the average WTP of all affected households. If this is the case the social benefit can be computed by multiplying mean (or median) WTP of the sample by the total number of households affected by the environmental improvement.

Unlike revealed preference methods (travel cost method, hedonic pricing, etc.) not only use values but also non-use values can be assessed by means of the CVM. This makes the CVM particularly attractive in the context of the present study which aims to assess the social benefit of an environmental project that is supposed to generate mainly non-use values (cf. section 3.2). At the same time, as a stated preference technique which relies on people's answers to survey questions rather than drawing inference from actual market behaviour, this method is frequently critisised. Critics point to the possibility that WTP statements are systematically biased and doubt that CVM surveys are a suitable tool for the assessement of environmental values (cf. Hausman, 1993; Hausman, 2012). The list of potential errors and biases inherent to WTP estimates is long and includes issues like the hypothetical bias (i.e. the divergence between actual and stated WTP), strategic behaviour (i.e. respondents' tendency to over- or understate their true WTP for strategic reasons), embedding effects (i.e. the insensisivity of WTP to the scope or scale of environmental improvements) as well as information effects (i.e. the systematic influence of the information provided by a survey on WTP). The latter and additional factors that threaten the validity of the results of CVM studies have been comprehensively reviewed elsewhere (c.f. e.g. Venkatachalam, 2004; Carson, 2005). Even though the usefulness of CVM for the assessment of environmental values can be questioned, the shortcomings of this method have been extensively studied and are today relatively well understood. Hence, like many other CVM practioners, we believe that a carefully designed CVM survey and an attentive analysis of the gathered data, combined with validity testing, is a useful tool to assess the benefits of propspective environmental projects. Conducting CVM surveys in People's Republic of China goes along with the challenge of adapting the CVM to the particular cultural and social context of this country. For example, Chinese people's distrust of government may be an obstacle for obtaining valid WTP statements for public investement projects in China (Chen & Hua, 2015). However, CVM is increasingly used by Chinese scholars to inform decision makers of the the the value of environmental goods (cf. Zhang & Zhou, 2012).

3.2 The Social Value of Environmental Projects

A crucial step in any environmental valuation study consists of identifying the population actually affected by the environmental project in question. This is because the magnitude of the social value of an environmental project critically depends on the number of households whose WTP is taken into account. Since the social benefit is approximated as the sum of individual WTP values (cf. equation 4), the chances that an environmental project passes the cost-benefit test increase with the magnitude of individual WTP measured and also with the number of individuals considered when aggregating individual WTP values.

Most obviously, households living close to the area where a prospective public project ought to be implemented are affected by the environmental consequences of this project. Taking the environmental restoration project in the Tarim Basin as an example, people living in the oasis cities along the Tarim River are the direct beneficiaries of the resulting improved environmental conditions. These people would be better off because of future water supply security, less frequently occurring sandstorms, a more enjoyable climate, etc. Since these benefits arise from the utilization of water, ambient air, local climate and other local environmental goods, they are summarized as the project's *use values* in the following. Furthermore, local people might also be happy about several more indirect effects of the environmental restoration project such as the protection of native plants and animals and enhanced living conditions for their descendants. Following Krutilla (1967), this second group of benefits shall be summarized as *non-use values*, because people can enjoy an environmental good without actively using it. In addition to bequest values (benefits for future generations) and existence values (benefits accruing from the pure knowledge about an environmental resource), option values accruing from preserving the option of actively using the environmental good in the future and altruistic values arising from the satisfaction of knowing that other people can enjoy an environmental improvement fall into the category of non-use values (cf. Ahlheim et al., 2013, Nunes, 2002). Since people can enjoy non-use values without actively using an environmental good it is speculated that not only local people's utility would change but that also the wellbeing of people living far away would be affected when a particular environmental good's quality or quantity changes. Accordingly, also people living at a distance might be willing to give up part of their income for the realization of public projects like a restoration of the natural ecosystems along the Tarim River. Not accounting for this 'long-distance value' would lead, as exemplarily shown by Ahlheim et al. (2013), to a substantial underestimation of the benefits of an environmental project for society as a whole.

The above argument implies that CVM surveys on environmental projects should be conducted nation-wide whenever the project in question is expected to generate considerable non-use values. The environmental restoration project in the Tarim Basin is a good example for such a project because, apart from its rather immediate features like water supply security, most environmental effects accruing from this project will affect the living conditions of future generations. Naturally, conducting interviews with a representative sample of all Chinese households was out of the scope of the present study. Furthermore, it can also be questioned that conducting a nation-wide survey would be reasonable. In view of China's large population, researchers would have to conduct interviews with an impressively high number of individuals in order to generate a representative sample. Furthermore, it is expected that a large share of China's population would have a zero WTP, for example because of the low education level of the rural population which may impede the perception of the environmental project's non-use level; or, even more obviously, because many Chinese households are too poor to contribute any money to an environmental project.

Given our limited budget, we opted for implementing a CVM survey at two exemplary sites, namely in several cities in Xinjiang to assess the perceptions of people directly affected by the environmental restoration project and in the city of Beijing to scrutinize the opinions of people indirectly affected. Beijing serves as an example for a Chinese megacity; we expect that the comparatively affluent and well-educated inhabitants of megacities are particularly likely to perceive the non-use values of environmental projects.

While conducting CVM interviews with Beijing residents went relatively smoothly, realizing a large-scale survey in Xinjiang turned out to be highly sensitive due to the political and social situation in this region. Due to these issues, a large and representative sample from Beijing but only a considerably smaller sample from Xinjiang was obtained. In the following we explore and compare local people's preferences for the environmental project in question to the preferences of people living at large distance from the project site.

3.3 The Survey

In 2013 standardized CVM interviews were conducted in Beijing and Xinjiang. The questionnaires employed in both study sites were broadly identical and consisted of the following parts: 1) an introductory section where the survey's topic and purpose were briefly presented; 2) several warm-up questions concerning a respondent's characteristics (age, education, marital status, etc.); 3) a presentation of the Tarim River and its environment, including maps and pictures, followed by several questions referring to a respondent's acquaintance with the project area, his or her perception of several ESS and environmental problems in the Tarim area; 4) the project scenario, i.e. a text passage introducing the 'Tarim Environmental Preservation Plan', a hypothetical environmental project; 5) a payment scenario describing the kind of contributions individuals would have to make to finance the environmental restoration project followed by the WTP elicitation question; 6) debriefing questions to scrutinize a respondent's motivation for (not) paying; 7) follow-up questions concerning a respondent's household (disposable income, number of household members, etc.) and attitudes towards several aspects of life.

Prior to the WTP question we informed the respondents that the 'Tarim Environmental Preservation Plan' could not be financed out of existing funds alone. Therefore, the Central Government would have to collect additional resources through a tax increase. However, the project would only be realized if the majority of Chinese households tolerated an increase in their monthly expenditures. In Beijing, respondents obtained the information that the monthly expenditures of an average household in Beijing would increase by a particular amount. Afterwards, we asked them whether or not they would support the environmental project, given that their monthly expenditures would increase. In the economic literature, this kind of WTP question is discussed as the dichotomous choice or referendum format. The participants of dichotomous choice surveys are randomly assigned to different amounts of money (so-called bids) and asked whether or not they would be ready to pay this amount of money to finance a particular environmental project. In the present study six predefined bids, reaching from 10 to 200 RMB, were employed. The bid design was based on the results of 200 pretest-interviews and followed a distribution that is recommended in the relevant literature (following Kanninen (1995), we made sure that the minimum bid of 10 RMB was accepted by approximately 85% and the maximum bid of 200 RMB was rejected by 85% of the pre-test participants).

Respondents of CVM surveys generally perceive dichotomous choice questions as simpler than alternative elicitation formats, such as open-ended WTP questions ('What would be the maximum amount you would tolerate to pay (...)?'). This is because the choice task resembles everyday purchase situations, i.e. thinking of paying a given price for a desired commodity. During pre-test interviews we also observed that respondents who were confronted with the dichotomous choice question completed the choice task much faster than those who had to answer open-ended WTP questions. Furthermore, as highlighted by Carson & Groves (2007), single-bounded dichotomous choice questions do, in contrast to other elicitation formats, give the respondents an incentive to give a truthful answer rather than under- or overstate their actual WTP. One of the main drawbacks of the dichotomous choice format is, however, that rather large samples are needed. This is because the estimation of average WTP involves econometric techniques which only yield valid and reliable results when the sample size is sufficiently large (a sample size of at least 1 000 observations is recommended in literature, cf. Arrow & Solow, 1993). Since only a much smaller number of interviews could be conducted in Xinjiang due to restrictions regarding social-science research in that area an alternative elicitation format had to be employed. The version of the questionnaire used in Xinjiang contained a two-step elicitation question. In a first step respondents were asked whether or not they would support the environmental restoration project and the related payment in general. Those who answered 'yes' had to indicate their maximum WTP on a payment card ranging from 5 RMB to 100 RMB.

To draw inference from the sample a CVM survey should be representative for the population in question. Representativeness is typically ensured by random sampling. Furthermore, face-to-face interviews at people's homes are the preferable interview mode regarding the overall response rate, item non-response rate as well as the quality of the collected data (Bateman et al., 2002). Unfortunately, we had to deviate from these guidelines in both study sites. In Xinjiang safety reasons were a main obstacle. Social instability, separatism and terrorism are alleged to be major issues in Xijiang. Under these circomstances it has become next to impossible to obtain a research permission, even for the experienced Chinese scholars involved in our research team. Since we could not obtain an official permission for conducting household interviews, we organized five workshops in different cities, namely in Ürümqi, Lop Nor, Korla and Kuche. Participants were recruited by our local research partners who informed their local network about the workshops. Naturally, it was impossible to obtain a large representative sample like this; however, we managed to recruit people with diverse socio-demographic characteristics, including young and elderly people, Han Chinese and ethnic minorities, farmers and scholars, etc. At the beginning of each workshop the CVM questionnaire was read out question by question in Mandarin or Uighur and the participants filled it in by themselves. The completed questionnaires were collected immediately and debates about water-related environmental problems in the Tarim Basin were opened up only when all participants had handed in their questionnaires. Compared to standard CVM interviews, the workshop-method has several shortcomings including sample selection bias, lack of anonymity of the participants and the possibility that a respondent's answers to several survey questions are influenced by other participants' presence and/or comments. Nevertheless, conducting interviews during workshops was the only possibility for getting WTP data from the local population at all. In the literature the idea of using a workshop setting for environmental valuation has featured quite prominently (e.g. Lienhoop et al. 2007, Macmillan et al., 2002). While our workshop approach, however, explicitly excluded group discussions prior to completing the questionnaires in order to ensure a comparable situation with the Beijing survey, we consider the joint presentation of the scenario information in the group and the presence of other respondents busy at completing the questionnaires as helpful for the motivation of respondents to take the valuation task seriously and complete the questionnaire thoroughly. Using the workshop valuation method described above, we consulted a total of 70 residents from Xinjiang during five workshops in July 2013. 61 valid questionnaires were gathered and all of them contained a valid WTP statement.

In Beijing things went more smoothly, however, a random household sample could not be drawn either because household registration lists turned out to be inaccessible. In addition to that, local experts experienced in market and opinion research advised against household interviews due to the difficulty of entering people's apartments. In the urban areas of Beijing people mostly live in large compounds with guarded entrances. Furthermore, even if an interviewer manages to pass the security guards it is unlikely that the suspicious residents let him enter their apartments. In view of these issues and after consultation with Chinese experts in this field we decided to conduct intercept interviews in the six urban districts of Beijing. A quota sampling approach ensured the representativeness of the intercept survey, thereby controlling for gender, age and education when approaching the survey participants. 50 students from the Minzu University conducted personal interviews. All of them had been trained in conducting standardized CVM interviews prior to the main survey. Interviewers were assigned to multiple locations in different parts of the city in order to represent the population of the six urban districts proportionally to their actual sizes. The interviewers were told to stop people randomly, but had to take account of a number of selection criteria, namely that the persons who agreed to participate in the interview had been living for at least five years in Beijing, that she was currently living in one of the six urban districts and that she fulfilled the required quota. A total of 2472 interviews were completed in August and September 2013. 34 questionnaires had to be discarded because they lacked relevant information, like a respondent's age, home district or because the respondent did not fulfill one or several of the selection criteria. All but 7 of the 2438 valid questionnaires contained an answer to the referendum question, yielding an impressively high response rate for the elicitation question of 99.7%. Also, the overall survey response rate (relation of the number of people willing to be interviewed to the number of people contacted) of 40% is quite acceptable for an intercept survey in a busy megacity.

4. Results and Analysis

4.1 Sample Characteristics

55.7% of the respondents from Xinjiang are male. They are on average 40 years old and 75.4% of them have minority nationalities (mainly Uighur). 73.8% hold a university degree and their average disposable household income amounts to 4 721 RMB per month. Although the sample is quite diverse in terms of the socio-demographic characteristics of the respondents, it is not fully representative of the population of Xinjiang. Men, ethnic minorities and people with higher education are overrepresented. However, the average household income of the survey participants comes close to the official figure for urban households in Xinjiang reported in the Statistical Yearbook (Statistics Bureau of Xinjiang Autonomous Region, 2012). 37.7% of all respondents in the sample live in Lop Nor, a city located at the lower reaches of the Tarim River with approximately 100 000 inhabitants; 24.9% live in Kucha or Korla (middle reaches) and 8.2% are from Aksu (upper reaches). 26.2% of the respondents are from Ürümqi, Xinjiang's capital situated at approximately 400km north of the Tarim Basin. Table 1 and Figure 1 summarize these figures.

Table 1. Demographic characteristics of the survey samples

Variable	Xinjiang N=61	Beijing N=2438
	Mean (std. deviation)	
Gender	0.557	0.504
(1=male, 0=female)	(0.500)	(0.500)
Age	39.7	40.2
(years)	(8.9)	(15.4)
Han	0.350	0.919
(1=yes, 0=otherwise)	(0.481)	(0.273)
Native	0.754	0.366
(1=yes, 0=no)	(0.434)	(0.482)
Higher education	0.738	0.382
(1=yes, 0=no)	(0.044)	(0.486)
Disposable monthly income (in 1000 RMB)	4.721	8.485
	(3.700)	(8.485)

Regarding the Beijing sample, 50.4% of the respondents are male, on average 40 years old and 38.2% hold a university degree. These figures are very similar to the official numbers in the Statistical Yearbook (Beijing Municipal Bureau of Statistics, 2013). Merely 36.6% of the respondents are originally from Beijing, but this figure exactly corresponds to the official share of non-native residents. Han Chinese residents are slightly underrepresented in the sample (-4%). This might be due to both the survey's topic (environmental restoration in a region which is mainly populated by ethnic minorities) and also the ethnic background of the interviewers (28.8% were ethnic minorities). Furthermore, the average income of the sample is significantly higher than the average income in urban Beijing of 7 640 RMB as measured in the 2011 household census (ibid). However, this difference may simply reflect a general raise in income since 2011. The representation of the six urban districts in terms of their population is displayed in the second histogram in Figure 1. Very similar to the official figures, most respondents live in the two largest urban districts, Chaoyang and Haidian. Residents from Dongcheng and Xicheng are slightly overrepresented (+4% and +2% respectively), while residents from Fengtai and Shijingshan are underrepresented (-2% each). Taken together, the quota-based sampling procedure has yielded a sample that reflects the characteristics of Beijing's urban population very closely.

With the exception of average age the two samples differ significantly in all demographic characteristics listed in Table 1. These dissimilarities have to be kept in mind when comparing and analyzing the results from both survey sites. The most striking differences include the respondents' levels of education and household income. The income of respondents from Beijing is on average twice as high as the income of respondents from Xinjiang. At the same time, the likelihood of holding a university degree is much greater in the case of the local sample as compared to the long-distance sample. The latter issue demonstrates a biased selection of respondents; since the participants of the Xinjiang survey were contacted by word-of-mouth, people with academic backgrounds had a higher probability to enter the sample.

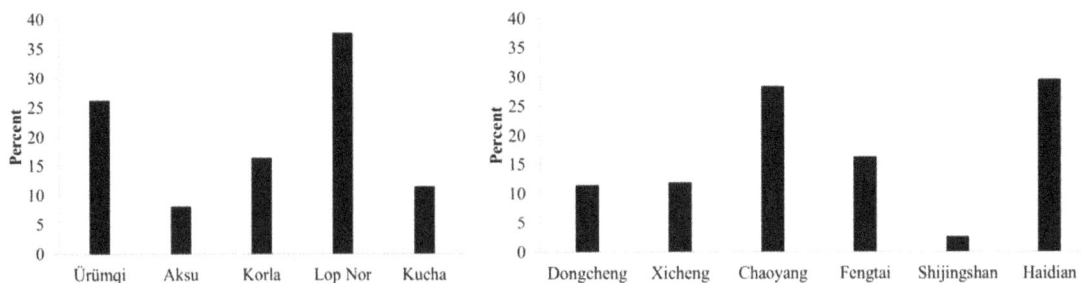

Figure 1. Respondents' residence (town or district)

4.2 Acquaintance with the Project Area and Environmental Perceptions

The second part of the questionnaire contained a number of questions regarding respondents' familiarity with the area along the Tarim River. 95.1% of the respondents from Xinjiang had already been to the Tarim River; merely three respondents from Ürümqi had never been there. Most of them had been there as tourists, many went there for work and some had visited relatives living in the area along the Tarim River. In Beijing merely 4.4% of all respondents had been to the Tarim River, however, a broad majority of 72.2% said that they had already heard of the river and 42.2% had heard of environmental problems in the region. Like in Xinjiang, those who had already travelled to the Tarim River mostly mentioned tourism or business trips as reasons.

All respondents obtained detailed information about the key ESS provided by the natural vegetation along the Tarim River. Afterwards, we asked them to express their preferences concerning these ESS by selecting the two ESS they considered most important. Based on this choice task we derived ESS-rankings for both samples. As displayed by the first chart of Figure 2, the rankings are very similar for both samples. Both groups of respondents considered 'mitigation of dust and sandstorms' as most important. However, while almost all respondents from Xinjiang listed this ESS as 'most important' only two thirds of the respondents from Beijing shared this opinion. The largest difference regarding the perceptions of local people, on the one hand, and people living far away on the other, can be observed for 'stabilization of soils'. While approximately half of the respondents from Beijing selected this ESS as 'most important', only a third of the respondents from Xinjiang did so. Similarly, 'beauty of landscape' and 'provision of useful herbs' were more frequently perceived as important in Beijing than in Xinjiang.

Subsequent to the ranking task, respondents were informed about the reasons and consequences of the degradation and destruction of the natural ecosystems in the Tarim area. Afterwards, we asked them which environmental problem they considered most serious. Again, the results from Xinjiang resemble the results from Beijing; 'desertification of the landscape' ranks first, followed by sandstorms and dust (cf. right chart in Figure

2). Only a small number of respondents said that the extinction of animals and plants was the most serious issue. However, comparatively more people from Beijing (18.5%) than from Xinjiang (10.0%) shared this opinion. Little surprisingly, sandstorms and dust, i.e. environmental problems immediately affecting the living conditions in the cities along the Tarim River, were considered as more serious by locals (42.0%) than by people living in Beijing (29.1%). Regarding the ranking of ESS and environmental problems we conclude that environmental perceptions of the population directly affected and the population indirectly affected are quite similar. However, environmental impacts which directly and immediately affect the livelihood of local people, namely sandstorms and dust and the associated ESS, matter more to people from Xinjiang than to people in Beijing.

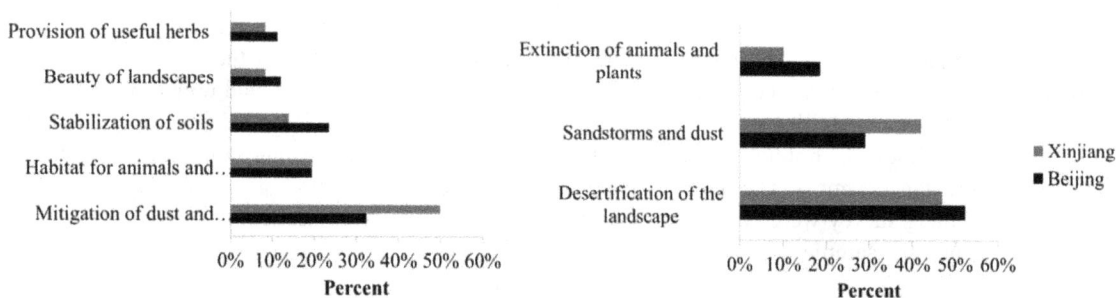

Figure 2. Ranking of ecosystem services (l.) and environmental problems (r.) (Rumbaur et al. 2015)

Table 2. Responses to debriefing questions

Item wording	Xinjiang	Beijing	p-value derived from t-test of equal means
	Agreement		
Environmental conditions in the Tarim area will improve through the TEPP.	93.0%	74.7%	0.000
Not households, but central government alone should pay for the TEPP.	75.0%	76.2%	0.162
Did you consider your chances of spending time at the Tarim area some day in the future? (Option value)	28.3%	38.6%	0.000
Did you consider the TEPP's positive effects for other people's livelihood? (Altruistic value)	41.7%	46.2%	0.000
Did you consider the TEPP's positive effects for future generations? (Bequest value)	90.0%	61.3%	0.000
Did you consider the TEPP's positive effects for plants and animals? (Existence value)	60.0%	52.7%	0.000

After having listened to the presentation of the project scenario and answered the WTP question respondents were presented a number of debriefing questions. These questions covered reasons for supporting or not supporting the environmental restoration project financially. Particular attention was paid to the issue whether or not respondents had taken into account the non-use values of the environmental restoration project before answering the WTP question. Some selected results are displayed in Table 2. We find that an impressively high share of respondents reported that they believed in the positive effects of the 'Tarim Environmental Preservation Plan'. Xinjiang residents' answers reveal that the population directly affected by the prospective environmental improvement was even more optimistic than the population indirectly affected. The t-test of equal means indicates that respondents from Xijiang were more likely to agree with the statement 'Environmental conditions in the Tarim area will improve through the TEPP' with a significance level of < 0.001%. Given this result, we conclude that most respondents perceived the environmental project to be valued as realistic and expect that they answered the related payment question seriously as well. At the same time, approximately two quarters of the respondents in both samples opposed the idea that households should bear the costs of the project in question.

This result is somewhat surprising in view of the results presented in the next sections regarding respondents' WTP. However, overall agreement with this protest statement may simply reflect people's mostly sceptical opinion towards government representatives' ability to manage existing funds efficiently rather than respondents' rejection of the payment scenario. Furthermore, non-use values were perceived differently in both study sites with a significance level of 0.0%. The share of respondents who took into account the option value of the natural vegetation along the Tarim River is relatively low in both samples, compared to other value components. The non-use value which respondents considered most frequently is the project's benefit for future generations. The answers to the associated question reveal that the bequest value mattered comparatively more for the WTP statements of people interviewed in Xinjiang than for respondents from Beijing.

4.3 WTP for the Restoration of Natural Ecosystems

In a next step respondents' answers to the WTP question shall be analyzed and compared. As displayed in Table 3, respondents in the Xinjiang sample are willing to pay on average 57 RMB[1] per month. This corresponds to approximately 1.2% of their household's disposable income. It is to be noted that the maximum amount displayed on the payment card employed in Xinjiang was 100 RMB. Given the fact that average income in Xinjiang is about half as high as in Beijing, this overall cap seemed to make sense when designing the questionnaire. Surprisingly, a considerable number of respondents, namely 29.6%, selected the 100 RMB bid from the payment card when asked about their WTP. It is speculated that some of these people had a WTP of even more than 100 RMB but could not express their preferences because of the limited range of the payment card. Thus, the estimate of 57 RMB is likely to underrate the actual WTP of the respondents from Xinjiang. In Beijing average WTP amounts to 108 RMB[2] which corresponds to 1.3% of an average household's disposable monthly income (cf. second column of Table 3). The t-test indicates that mean WTP is higher in Beijing that in Xinjiang with a significance level of 0.0%. However, when taking into account the income differences of the two samples and the methodological weaknesses of the payment card employed in Xinjiang, the WTP of the population directly affected by the environmental restoration project is very similar to the WTP of the population indirectly affected. Furthermore, the share of respondents who agreed with the highest bid – 100 RMB in the case of the Xinjiang survey and 200 RMB in Beijing – is equal with a significance level of 98.9%.

Table 3. WTP results

	Xinjiang	Beijing	p-value derived from t-test of equal means
Mean WTP			
- absolute value (in RMB per month and household)	57.361	107.61	0.000
- % of income	1.2%	1.3%	
Agreement with highest bid (100 RMB in Xinjiang; 200 RMB in Beijing)	29.6%	33.3%	0.989

As explained above, we employed different elicitation formats in the two study sites. Due to the individual features of the dichotomous choice format on the one hand and the payment card format on the other, the two estimates for average WTP are not strictly comparable. However, some more insights concerning the WTP of the population directly affected and the population living far away from the project site can be gained when looking at the share of respondents who agreed with the different bids and who selected a particular amount from the payment card, respectively. For example, if a respondent from Xinjiang selected 'max. 50' RMB on the payment card it is supposed that this respondent would also have agreed with the dichotomous choice question whether he or she would support the environmental project, although it would increase his or her household's expenditures by 50 RMB monthly. Hence, it makes sense to compare the share of respondents who agreed with each bid in the Beijing sample to the share of respondents who indicated a WTP of the same amount or higher on the payment card in the Xinjiang sample. As displayed in Figure 3, the proportion of respondents with a WTP of at least 10 RMB is approximately 8% higher in Xinjiang than in Beijing. The proportions of respondents with a WTP of at least 25 RMB and 50 RMB do not significantly differ in the two samples. Only in the case of 100 RMB, the proportion of respondents from Xinjiang who would have been willing to pay this amount is significantly lower than in Beijing. The latter result is very plausible, given that respondents from Xinjiang dispose of an income which is about half of an average Beijing resident's income. Hence, the 'Tarim Environmental Preservation Plan' seems to be greatly welcome at both research sites.

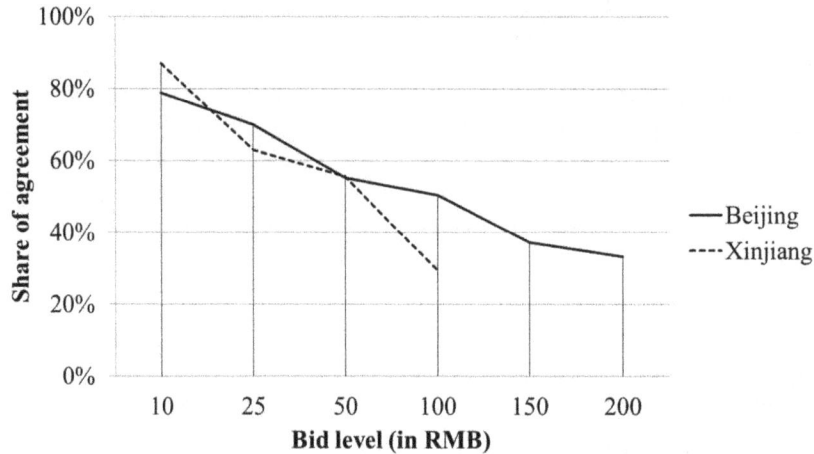

Figure 3. Agreement with bids

4.4 Validity of 'Long-Distance' WTP

As shown in the previous paragraph not only the population directly affected is willing to make considerable financial contributions to the environmental restoration program but also people living far away from the Tarim River would tolerate an increase in their monthly expenditures in exchange for implementing this project. The question arises how plausible these WTP estimates actually are. A major criticism against assessing the benefits of public projects by means of survey questions is that respondents merely give an answer to a hypothetical question but do not effectuate an actual payment. Hence, stated WTP is necessarily higher than actual WTP, according to the critics. Since people's actual WTP for prospective environmental improvements is unknown, the plausibility of the obtained measure is commonly tested by checking whether answers to the WTP elicitation question systematically vary with a number of variables which are expected to affect a respondent's WTP. For example, the probability of answering 'yes' to the dichotomous choice question should be negatively related to the bid level; it should increase with a household's income; and it should be the higher the more favorable a respondent's attitudes towards the project and the related payment are. In line with most CVM practitioners we make use of multiple regression models to identify the determinants of WTP. Due to the small number of interviews conducted in Xinjiang, the analysis of the main determinants of WTP is limited to the Beijing sample.

The results of a probit model, where the dichotomous choice variable is regressed against the bid and several explanatory variables, are shown in Table 3. Coefficient estimates, standard errors as well as the marginal effects of each control variable are displayed. A description of all variables can be found in the appendix. Model 1 merely includes bid, dummies for experimental treatments[3] and some standard demographic variables. As expected, the magnitude of the bid negatively affects the likelihood of answering 'yes' to the referendum question. Furthermore, holding other factor fix, men are more likely to agree with the referendum question (i.e. have a higher WTP) than women. WTP increases with a respondent's level of education and with the respondent's disposable household income. These effects remain robust when additional explanatory variables are added to the regression model (cf. Model 2 in Table 3). Similar effects were also identified in other CVM studies conducted in China, pointing to the validity of our WTP data. A positive relationship between WTP and income was reported in several studies (c.f. e.g. Wang et al., 2013, Ahlheim et al., 2015, Day & Mourato, 2002). In line with the results of the present study, Ahlheim et al. (2015) identified being male and a respondent's education level as predictor variables for WTP. The results of the second regression model provide further evidence for the WTP estimate's plausibility. In Model 2 a dummy variable indicating whether a respondent has already been to the area along the Tarim River (BEEN_TARIM), two attitudinal variables and a variable accounting for the number of non-use value components (i.e. option, bequest, existence and altruistic values) considered before answering the WTP question are included. Plausibly, respondents who have already been to the Tarim River, i.e. those who experienced the harsh environmental conditions in this area, have a higher WTP. In addition to that, those who believe that environmental conditions will improve through the environmental project in question (NODOUBTS) state a higher WTP, while those who think that government alone should pay for this project (GOVPAYS) state a lower WTP. The latter effect adds to the relevance of an argument made in a recent paper by Chen & Hua (2015): Distrust of government in terms of manageing tax revenue efficiently and appropriately affects Chinese people's WTP for environmental goods substantially. Finally, the consideration of several non-use values alters the probability of agreement. The marginal effect of NONUSE is also comparatively strong. If one more non-use value component is considered, the likelihood of stating 'yes' increases by 16.2%, ceteris paribus on other factors. This finding indicates that people's WTP for environmental projects that provide mainly non-use values to them depends on a person's awareness of such non-use values. Taken together, the effects of the four additional variables are fully plausible, which is an indicator for the validity of the WTP data collected in urban Beijing.

Table 4. Probit regression models displaying determinant of supporting the environmental project

Dependent variable: WTP	Model 1			Model 2		
	Coef.	s.e.	dy/dx	Coef.	s.e.	dy/dx
CONSTANT	0.171	0.155		-0.058	0.234	
BID	-0.006***	0.000	-0.002	-0.006***	0.000	-0.002
TREATMENT	x			x		
MALE	0.127**	0.055	0.045	0.103*	0.058	0.034
AGE	-0.001	0.002	0.000	-0.002	0.002	-0.001
EDUCATION	0.055**	0.023	0.020	0.051**	0.024	0.017
CHILD	0.040	0.059	0.014	0.096	0.061	0.031
INCOME	0.008**	0.004	0.003	0.009**	0.004	0.003
BEEN_TARIM				0.288*	0.150	0.094
NODOUBTS				0.286***	0.031	0.093
GOVPAYS				-0.267***	0.028	-0.087
NONUSE				0.496***	0.084	0.162
Observations	2244			2244		
Log likelihood	-1411			-1289		
Pseudo R^2	0.089			0.168		

Note: *** means significant at 1%-level, ** 5% level, *10% level.

5. Conclusions and Recommendations

This study adds to the evidence that many environmental projects are not only beneficial for people living at the project site but also for those living far away. We found that residents from Xinjiang as well as Beijing residents, which serve as an example of the population living at a distance, have a positive WTP for a prospective public project that would enhance the environmental conditions in the Tarim Basin. Based on a small sample gathered in Xinjiang, local people's WTP for the environmental project in question amounts to 57 RMB per month. The results of a representative CVM survey implemented in Beijing reveal that the residents living in this megacity are willing to contribute approximately 108 RMB per month to the same environmental project. In absolute terms, the WTP of Beijing residents is substantially higher than the WTP of the local population. This result is somewhat surprising because people in Xinjiang would experience both use and non-use-values while people in Beijing would merely benefit from the non-use values accruing from the environmental project in question. However, the validity of the 'long-distance WTP' estimate is supported by its plausible determinants identified econometrically. Furthermore, when accounting for income differences in both survey sites, relative WTP is approximately the same in both samples. We also find that people at both survey sites perceive the importance of several ecosystem services and the seriousness of environmental problems caused by the deterioration of the natural ecosystems along the Tarim River in a similar way. Finally, based on our analysis of the non-use values of a more sustainable water and land management in the Tarim Basin, we conclude that most Chinese people are concerned about the consequences of environmental mismanagement for future generations.

Naturally, the present study has a number of limitations. The standards of best practice for the conduct of CVM surveys could not be followed in Beijing and even less in the politically sensitive region of Xinjiang. None of our two samples is random, i.e. some degree of sample selection is present in both samples, so that the results cannot easily be generalized. Furthermore, the CVM survey implemented in Beijing can merely serve as an example to demonstrate that also people living far away from the Tarim Basin would enjoy the consequences of a more sustainable policy in this region. The validity and reliability of our results should be further explored by assessing people's WTP in other Chinese megacities. This would allow a generalization of non-use values to a broader scope of the Chinese population affected by the project. We suppose that researchers would find similar results when assessing the social benefit of other environmental projects of national importance in China or in other countries. It would be interesting, for example, to explore whether Chinese people's WTP for other environmental projects significantly differs from the figures assessed in the present study or whether WTP is just an expression of some positive environmental attitudes that does not vary with the kind, scope and scale of different environmental improvements.

In spite of these limitations, this study demonstrates that it is possible to obtain a valid measure of 'long-distance WTP' by means of the CVM in the context of China while the scientific assessment of such values in a region like Xinjiang remains very difficult in practice. An environmental valuation study focused on the benefits of local people only would clearly underestimate the social value of a more sustainable land and water management in the Tarim Basin. The present study conducted both in Xinjiang, a region where hardly any data on the preferences of the local population for preserving ecosystem services exist, and in Beijing proves to be of special scientific and practical interest since it can serve as the starting point for a rigorous comparison of environmental values of the broad and heterogeneous Chinese population for environmental goods of national importance. For this reason, on the basis of our study we recommend that decision makers interested in the social benefit of a more sustainable land and water management in the Tarim Basin should initiate an environmental valuation study that covers both the preferences of Xinjiang's population and of the inhabitants of metropolis such as Beijing, Shanghai, Guangzhou, etc. Ideally, a random household sample should be drawn so that the overall social value of a more sustainable land and water management in the Tarim Basin can be assessed based on the sample's average WTP. While it has turned out impossible to obtain such a completely representative sample for the CVM study by scientific researchers, government or administrative institutions in China possess the necessary means to conduct such high quality surveys. Of course, the latter recommendation does not only apply to the particular case of the Tarim Basin considered here. The social benefit of other environmental projects - in China or elsewhere - of similar scope, scale and national importance, should be comprehensively assessed as well.

Acknowledgments

This research was part of the Sino-German joint research project SuMaRiO which is funded by the German Ministry of Education and Research and approved by the National Basic Research Program of China (973 Program, 2012CB955903). We also want to thank Prof. Dai Tang-Ping and his students from Minzu University of China for supporting the survey in Beijing.

References

Ahlheim, M., Börger, T., & Frör, O. (2015). Replacing rubber plantations by rain forest in Southwest China—who would gain and how much? *Environmental monitoring and assessment, 187*(2), 1-20. http://dx.doi.org/10.1007/s10661-014-4088-8

Ahlheim, M., & Frör, O. (2003). Valuing the non-market production of agriculture. *Agrarwirtschaft, 52*(8), 356-369. http://dx.doi.org/10.1093/erae/19.3.351

Ahlheim, M., Frör, O., Luo, J., Pelz, S., & Jiang, T. (2013). How do Beijing Residents Value Environmental Improvements in Remote Parts of China. *Advances in Climate Change Research, 4*(3), 190-200. http://dx.doi.org/10.3724/SP.J.1248.2013.190

Arrow, K., & Solow, R. (1993). Report of the NOAA Panel on Contingent Valuation. Washington, DC: Resources of the Future. Retrieved from http://www.darrp.noaa.gov/library/pdf/cvblue.pdf

Atkinson, G., & Mourato, S. (2008). Environmental cost-benefit analysis. *Annual review of environment and resources, 33*, 317-344. http://dx.doi.org/10.1146/annurev.environ.33.020107.112927

Bateman, I., Carson, R. T., Day, B., Hanemann, M., Hanley, N., Hett, T., ... Swanson, J. (2002). Economic Valuation with Stated Preference Techniques : a manual. Cheltenham: Elgar.

Beijing Municipal Bureau of Statistics. (2013). *2012 Beijing Statistical Yearbook*. Beijing: China Statistics Press.

Carson, R. T., & Hanemann, W. M. (2005). Contingent Valuation. In K.-G. Maler, J. R. Vincent (Ed.), *Handbook of Environmental Economics* (Vol 2). Valuing Environmental Changes (pp. 821-936). North-Holland: Elsevier.

Chen, Y., Xu, C., Chen, Y., Liu, Y., & Li, W. (2013). Progress, Challenges and Prospects of Eco-Hydrological Studies in the Tarim River Basin of Xinjiang, China. *Environmental Management, 51*(1), 138-153. http://dx.doi.org/10.1007/s00267-012-9823-8

Chen, W. Y., & Hua, J. (2015). Citizens' distrust of government and their protest responses in a contingent valuation study of urban heritage trees in Guangzhou, China. *Journal of Environmental Management, 155*(0), 40-48. http://dx.doi.org/10.1016/j.jenvman.2015.03.002

Day, B., & Mourato, S. (2002). Valuing river water quality in China. In D. Pearce (Ed.), *Valuing the enviroment in developing countries*. Cheltenham: Elgar.

Deng, M., Qin, D., & Zhang, H. (2011). Public Perceptions of Cryosphere Change and the Selection of Adaptation Measures in the Ürümqi River Basin. *Advances in Climate Change Research, 2*(3), 149-158. http://dx.doi.org/10.3724/SP.J.1248.2011.00149

Deng, M., Qin, D., & Zhang, H. (2012). Public perceptions of climate and cryosphere change in typical arid inland river areas of China: Facts, impacts and selections of adaptation measures. *Quaternary International, 282*, 48-57. http://dx.doi.org/10.1016/j.quaint.2012.04.033

Halik, Ü., Küchler, J., & Kleinschmit, B. (2005). Bevor die Erde zur Wüste wird. *TU International, 57*, 34-37. Retrieved from https://www.alumni.tu-berlin.de/fileadmin/Redaktion/ABZ/PDF/TUI/57/halik.pdf

Hanemann, W. M. (1989). Welfare evaluations in contingent valuation experiments with discrete response data: reply. *American journal of agricultural economics, 71*(4), 1057-1061. Retrivied from http://www.jstor.org/stable/1242685

Hausman, J. A. (Ed.). (1993). Contingent valuation : a critical assessment. Amsterdam: North-Holland.

Hausman, J. (2012). Contingent Valuation: From Dubious to Hopeless. *The Journal of Economic Perspectives, 26*(4), 43-56. http://dx.doi.org/10.1257/jep.26.4.43

Hirji, R., & Davis, R. (2009). Environmental flows in water resources policies, plans, and projects: case studies Environment Department Papers. *Natural Resource Management Series* (Vol. P. 117). Washington, D.C.: The World Bank Environment Department.

Huang, X., Chen, Y., Ma, J., & Chen, Y. (2010). Study on change in value of ecosystem service function of Tarim River. *Acta Ecologica Sinica, 30*(2), 67-75. http://dx.doi.org/10.1016/j.chnaes.2010.03.004

Jørgensen, S. L., Olsen, S. B., Ladenburg, J., Martinsen, L., Svenningsen, S. R., & Hasler, B. (2013). Spatially induced disparities in users' and non-users' WTP for water quality improvements—Testing the effect of multiple substitutes and distance decay. *Ecological Economics, 92*(0), 58-66. http://dx.doi.org/10.1016/j.ecolecon.2012.07.015

Kanninen, B. J. (1995). Bias in Discrete Response Contingent Valuation. *Journal of Environmental Economics and Management, 28*(1), 114-125. http://dx.doi.org/10.1006/jeem.1995.1008

Krutilla, J. (1967). Conservation reconsidered. *American Economic Review, 56*, 777-786. Retrieved from http://www.jstor.org/stable/1815368

Lienhoop, N., MacMillan, D.C. (2007). Contingent Valuation: Comparing Participant Performance in Group-Based Approaches and Personal Interviews. *Environmental Values, 16*(2), 209-232. http://dx.doi.org/10.3197/096327107780474500

Lu, Z., Zhao, L., & Dai, J. (2010). A study of water resource management in the Tarim Basin, Xinjiang. *International Journal of Environmental Studies, 67*(2), 245-255. http://dx.doi.org/10.1080/00207231003693274

Macmillan, D. C., Philip, L., Hanley, N., & Alvarez-Farizo, B. (2002). Valuing the non-market benefits of wild goose conservation: a comparison of interview and group based approaches. *Ecological Economics, 43*(1), 49-59. http://dx.doi.org/10.1016/S0921-8009(02)00182-9

Nunes, P. A. L. D. (2002). The contingent valuation of natural parks : assessing the warmglow propensity factor. Cheltenham: Elgar.

Rumbaur, C., Thevs, N., Disse, M., Ahlheim, M., Brieden, A., Cyffka, B., ... & Zhao, C. (2015). Sustainable Management of River Oases along the Tarim River in North-Western China under Conditions of Climate

Change, Special Issue: Climate Change and Environmental Pressure: Adaptation and Resilience of Local Communities in the Hindu-Kush-Himalaya. *Earth Syst. Dynam., 6*, 83-107. http://dx.doi.org/10.5194/esd-6-83-2015

Statistics Bureau of Xinjiang Autonomous Region. (2012). 2011 Xinjiang Statistical Yearbook. Urumqi: China Statistics Press.

Stephan, G., & Ahlheim, M. (1996). Ökonomische Ökologie. Berlin: Springer.

Thevs, N. (2011). Water Scarcity and Allocation in the Tarim Basin: Decision Structures and Adaptations on the Local Level. *Journal of Current Chinese Affaires, 3*, 113-137. Retrivied from http://hup.sub.uni-hamburg.de/giga/jcca/article/view/456

Venkatachalam, L. (2004). The contingent valuation method: a review. *Environmental Impact Assessment Review, 24*(1), 89-124. http://dx.doi.org/10.1016/S0195-9255(03)00138-0

Wang, H., He, J., Kim, Y., & Kamata, T. (2013). Willingness-to-pay for water quality improvements in Chinese rivers: An empirical test on the ordering effects of multiple-bounded discrete choices. *Journal of Environmental Management, 131*(0), 256-269.

Xu, C. Z., Li, Q., & Wang, W. X. (2014). Contingent Valuation Method Assessment of the Ecological Benefits of Populus Euphratica Forest Based on Consumer Perspective-Taking the Typical Area in Tarim River Basin as an Example. Advanced Materials Research, 962-965, 703-709. http://dx.doi.org/10.4028/www.scientific.net/AMR.962-965.703

Xu, H.-l., Ye, M., & Li, J.-m. (2007). Changes in groundwater levels and the response of natural vegetation to transfer of water to the lower reaches of the Tarim River. *Journal of Environmental Sciences, 19*(10), 1199-1207. http://dx.doi.org/10.1016/S1001-0742(07)60196-X

Xu, H., Ye, M., & Li, J. (2008). The water transfer effects on agricultural development in the lower Tarim River, Xinjiang of China. *Agricultural water management, 95*(1), 59-68. http://dx.doi.org/10.1016/j.agwat.2007.09.004

Zhang, J., Wu, G., Wang, Q., & Li, X. (2010). Restoring environmental flows and improving riparian ecosystem of Tarim River. *Journal of Arid Land, 2*(1), 43-50. http://dx.doi.org/10.3724/SP.J.1227.2010.00043

Zhang, X., & Zhou, L. (2012). Development of contingent valuation method in evaluating non-market values of resources and environment in China. *Sciences in Cold and Arid Regions, 4*(6), 536-543. http://dx.doi.org/10.3724/SP.J.1226.2012.00536

Notes

Note 1: The answers of respondents who said that they would not support the program are coded as a $WTP_n=0$. Respondents who answered "don't know" when asked to state their maximum WTP are excluded from the analysis (12% of the sample). For all others, WTP_n corresponds to the midpoint bid of the interval between the value a respondent ticked on the payment card and the next higher one. In the case of the highest value on the payment card (100 RMB), WTP_n is set to 125 RMB. Average WTP is computed as the mean of all valid WTP statements, i.e.

$$\overline{WTP}_{sample} \ (\ WTP_1 \quad WTP_2 \quad ... \quad WTP_N\ )/N\ .$$

Note 2: As shown by Hanemann (1989) dichotomous choice data can be modelled by the following linear utility model: $\blacksquare\blacksquare_h \quad \blacksquare \quad \blacksquare BID_h$, where α is a constant integrating all observable and unobservable household characteristics, BID_h is the randomly assigned bid and β the corresponding parameter. A respondent is expected to agree with the dichotomous choice question whenever the change in utility is greater or equal to zero. The probability to answer 'yes' can be modelled by a probit model and maximum WTP can be estimated based on the corresponding parameter estimates, namely as $\overline{WTP}_h \quad \blacksquare/\blacksquare$. In the present work the parameters of a probit model with the binary choice variable as dependent and the bid as the only independent variable have been estimated with the econometric software package STATA, resulting into Prob (yes)=0.642-0.006BID.

Note 3: We employed three treatments to test the effect of monetary incentives for participating in the survey and different WTP answer formats on WTP. Several of these treatments have significant effects on the probability of agreeing with the WTP questions. However, the analysis and interpretation of these effects is outside the scope of this summary paper.

Appendix

Description of variables used in the regression models

Table A1. Description of variables used in the regression models

Variable	Description	Mean	Std. dev.	Min.	Max.
WTP	"Considering that your monthly household expenditures would increase by approximately [BID] RMB through the program would you personally be willing to support it?" (1=yes, 0=no)	0.537	0.499	0	1
BID	Bid amount	89.730	68.248	10	200
MALE	Gender of the respondent (1=male, 0=female)	0.509	0.500	0	1
AGE	Age of the respondent	39.890	15.315	18	84
EDUCATION	Level of education of the respondent (1=did not graduate from primary school, 7=master degree or higher)	4.332	1.321	1	7
CHILD	There are children living in the respondent's household (1=yes, 0=no)	0.346	0.476	0	1
INCOME	Monthly disposable household income in 1000 RMB	8.557	7.686	1	50
BEEN_TARIM	The respondent has already been to the Tarim area (1=yes, 0=no)	0.045	0.208	0	1
NODOUBTS	Level of agreement with the statement "Environmental conditions in the Tarim area will improve through the TEPP" (1=strongly disagree, 5=strongly agree)	4.134	0.967	1	5
GOVPAYS	Level of agreement with the statement "Not households, but central government should pay for the TEPP" (1=strongly disagree, 5=strongly agree)	4.105	1.081	1	5
NONUSE	Number of non-use aspects considered (0=none, 0.25=one out of four aspects, (…) 1=all four aspects)	0.497	0.355	0	1

13

Using Bioretention Retrofits to Meet Virginia's New Stormwater Management Regulations: A Case Study

Brett A. Buckland[1], Randel L. Dymond[1] & Clayton C. Hodges[1]

[1] Via Department of Civil and Environmental Engineering, Virginia Tech, Blacksburg, Virginia, United States

Correspondence: Brett A. Buckland, Via Department of Civil and Environmental Engineering, Virginia Tech, Blacksburg, Virginia, United States. E-mail: bbrett08@vt.edu

Abstract

Virginia's new stormwater regulations involve the use of the Runoff Reduction Method (RRM), a methodology to estimate a volume reduction in predicted runoff. Regulations require that for downstream erosion control, the product of the peak flow rate and runoff volume (Q*RV) from one-year storm events in the post-development condition be reduced to less than pre-development Q*RV. This study models different bioretention sizing scenarios in a developed watershed in Blacksburg, Virginia to determine the performance at both the sub-watershed and watershed levels. In addition, models of "optimal" bioretention cells, sized to meet the RRM for each sub-watershed, are evaluated. A direct relationship is observed between the size of the cell required to meet the RRM and the sub-watershed's developed Natural Resources Conservation Service (NRCS) curve number, and a sizing analysis is provided. Modeling shows that the required size of "optimal" cells for many sub-watersheds exceeds conventional bioretention designs. Upon applying the RRM for all sub-watersheds, the resulting hydrograph at the watershed outlet more closely resembles the pre-development hydrograph than existing development.

Keywords: bioretention, stormwater management, Virginia, low impact development

1. Introduction

Low Impact Development (LID) is a design methodology that seeks to restore a developed site's hydrologic response to a storm to its pre-development condition (Prince George's County, 1999). Bioretention, a common LID practice, accepts runoff, allows the water to pond on top of it, and then lets water percolate through its engineered soil media to either the underlying soil or an underdrain. In Virginia, bioretention cells with an underdrain are referred to as "bioretention filters", and those without underdrains are called "bioretention basins" (DCR, 2011). Bioretention cells often utilize an outlet structure or overflow weir to allow any water in excess of the intended treatment volume that enters the cell to be routed in an efficient manner to a desired location downstream. Retrofitting urbanized areas with LID and Best Management Practice (BMP) technologies is an effective way of reducing runoff in a watershed (Damodaram et al., 2010). Although many BMPs exist that could be used in urban stormwater infrastructure retrofits, bioretention is a practice that has increasingly become attractive to designers. This study strictly focuses on the volume reduction benefits of bioretention; however, other reasons that make it a widely used practice are its high removal efficiency of nutrients and pollutants and creation of canopy and wildlife habitat for small species in urban settings.

In September 2011, the Commonwealth of Virginia's (VA) Department of Conservation and Recreation (DCR) made substantial revisions to the Virginia stormwater management regulations. Since then, authority has been transferred to the Virginia Department of Environmental Quality (DEQ), which has taken the lead in development and implementation of the regulations. These regulations were divided into two main categories: quantity and quality of stormwater runoff (DEQ, 2013). Although improvements to storm water quality and its ultimate effect on the Chesapeake Bay were a huge driving force in development of these regulations, this study deals exclusively with the stormwater quantity aspect of the regulations; specifically, those dealing with channel erosion.

The stormwater quantity regulations have changed significantly with the recent revisions, which previously required the peak developed flow rates from the 2- and 10-year storm events to be returned to the pre-development flow rates (DCR, 1999). In addressing stormwater runoff quantity, the new regulations consider channel protection and flood protection as the two primary components of interest. Discharge requirements are based on the type and condition of the receiving channel. When discharging into a natural conveyance system, for instance, a primary channel protection criterion requires comparison of the 1-year, 24-hour storm event's peak runoff rate and total runoff volume for both pre- and post-development conditions (VA, 2011).

To minimize erosion, the new channel protection requirements use a method that is unique to Virginia. Equation 1 (VA, 2011) is used for channel protection calculations when discharging to a natural channel. The purpose of the equation is to calculate the maximum allowable peak flow rate for the developed condition during the 1-year storm event (Q_{Dev}). Rearranging the equation by multiplying both sides of Equation 1 by the developed runoff volume (RV_{Dev}), yields Equation 2, where the peak flow rates (Q) are multiplied by the volumes (RV) of flow for the 1-year storm event for both the pre- and post-development conditions. This product is used as the basis for analysis in the rest of this study and referred to as Q*RV. Note that in Equation 2, the developed Q*RV must be less than or equal to 80% (*I.F.*) of the pre-development Q*RV for sites greater than 0.4 hectares, which constitutes all of the sites in this study.

$$Q_{Dev} \leq I.F.* Q_{Pre} * \frac{RV_{Pre}}{RV_{Dev}} \tag{1}$$

$$Q_{Dev} * RV_{Dev} \leq I.F.* Q_{Pre} * RV_{Pre} \tag{2}$$

Where: *I.F. (Improvement Factor) = 0.8 for sites > 0.4 hectares or 0.9 for sites ≤ 0.4 hectares*

   Q_{Dev} = *peak flow rate for the developed condition 1-year storm (m^3/s)*

   Q_{Pre} = *peak flow rate for the pre-developed condition 1-year storm (m^3/s)*

   RV_{Dev} = *volume of runoff - developed condition using RRM for the 1-year storm (cm)*

   RV_{Pre} = *volume of runoff - pre-developed condition for the 1-year storm (cm)*

The Q*RV value seems to be unique to Virginia regulations, as none of the surrounding states have regulations that incorporate this value. However, several large municipalities in Northern Virginia, including Fairfax and Prince William Counties, have begun using the similar channel and flood protection criteria to the new Virginia regulations that include the Q*RV value (Fairfax County, 2014 and Prince William County, 2014). Rolband and Graziano (2012), who describe this method as the "Energy Balance" method, aided in the method's development with VA's DCR. For flood protection, calculation of the 10-year, 24-hour runoff is required to ensure that the post-development peak flow rate is at, or below, the pre-development peak (VA, 2011).

In an attempt to estimate stormwater volume reductions through BMPs, the new regulations use a tool called the Virginia Runoff Reduction Method (RRM). The RRM is used to adjust post-development curve numbers downstream of proposed BMPs. Although the use of a particular hydrologic method for calculations is not explicitly required in the regulations, the integrated computation of the curve number adjustment practically forces design engineers in Virginia to use NRCS methodology for site design, without regard to the size of the contributing drainage area. The strategies for this method were developed for Virginia by the Center for Watershed Protection (CWP) and the Chesapeake Stormwater Network (CSN) in an attempt to better emulate pre-development hydrologic conditions on the developed site (CWP & CSN, 2008) and estimate the effects of BMPs in series. A review of the Virginia Runoff Reduction Spreadsheet shows that it incorporates a number of BMPs with varying runoff reduction and pollutant removal capabilities. One of the most efficient BMPs is bioretention. Brown and Hunt (2010) stated that bioretention improves both water quality and quantity aspects. Due to both water quality and quantity requirements in Virginia's regulations and bioretention's benefits in both of these aspects, it is likely to become more frequently implemented.

Many studies have demonstrated that bioretention is an effective means of stormwater management for both quantity and quality, especially at the site scale. Bioretention is especially effective for less intense, more frequent storm events (Davis, 2008; James and Dymond, 2012). The results of the installation of two bioretention cells in a Maryland parking lot suggest that bioretention can greatly reduce the volume of runoff, lower peak flow rates, and increase lag times (Davis, 2008). Bioretention retrofits are becoming more popular due to their hydrologic benefits. Winston et al. (2013) found a substantial reduction in runoff volume can be achieved in a developed watershed through the addition of bioretention cells along the roads, permeable pavement parking stalls, and a tree filter device. A retrofit bioretention cell installed in the Stroubles Creek watershed in Blacksburg, Virginia was shown to reduce the average peak and volume of runoff by over 90%, even though its surface area is only 2% of the drainage area, which is below the recommended and commonly used percentages (DeBusk and Wynn, 2011). However, there were very few large, intense storms studied due to the timing of the monitoring.

The location of bioretention in a watershed is critical for maximizing its efficiency. James and Dymond (2012) found that bioretention is more efficient when it is treating large impervious areas, than when it is treating areas that have a higher percentage of pervious cover. Gilroy and McCuen (2009) had similar conclusions, and also determined that installing BMPs in series compounds their effects. Proper sizing, maintenance, and construction practices are also critical to the performance of bioretention and, if designed correctly, can result in a practice that reduces both the peaks and volumes of flows leaving a site (Brown and Hunt, 2010). Li et al. (2009) studied four locations with bioretention cells in Maryland and North Carolina and found that cells with larger storage volumes, either through a larger cell area or deeper media depths, more closely replicated pre-development conditions, even for larger storms, by reducing peak flow rates, reducing outflow volumes, and promoting more infiltration. Although these studies provide insight on the functionality of bioretention cells, they do not go as far as determining what size facilities would be required to meet water quantity regulations for their respective jurisdictions.

The purpose of this study is to provide insight into bioretention sizes required to meet peak reductions as required by the channel protection criteria in the new Virginia stormwater regulations. The study is performed on a watershed in the Town of Blacksburg, Virginia for which a calibrated rainfall-runoff model was developed. Using several different modeling scenarios, various sizes of bioretention cells are modeled to simulate their retrofitted installation throughout the watershed. In addition, the "optimal" scenario is found for each sub-watershed within the watershed, so that it can meet the channel protection criteria of the RRM. Furthermore, effects of the RRM is studied at the watershed outlet for the scenario when all sub-watersheds within the watershed are meeting the requirements outlined for channel protection.

2. Method

The watershed modeled in this study is the "North Stroubles" watershed in Blacksburg. It is a 192-hectare watershed consisting of residential, commercial, industrial, institutional, and open space land uses in the headwaters of Stroubles Creek, a tributary of the New River. There is a flow sensor within the stream at the outlet of the watershed, near Webb Street, that is owned and monitored by the Town of Blacksburg. The flow sensor used is an Acoustic Doppler velocimeter, which measures the velocity of the flow and the flow area (as a function of stage). Upstream of the flow sensor, the watershed has been delineated into 41 sub-watersheds, or catchments, ranging from just over 0.4 hectares up to approximately 14 hectares, as shown in Figure 1. The catchments were delineated based on key points of interest, such as ponds or intersections of major conveyances. In addition to the flow sensor, there is a rain gauge less than a mile outside of the boundary of the watershed. Flow and rainfall data measured by these devices were used for model calibration.

Virginia regulations require the volumetric sizing of bioretention facilities to be based on a composite weighted runoff coefficient which incorporates impervious, turf, and forested components of runoff. However, Virginia regulations do not require that bioretention facilities be designed to fully meet the requirements of the downstream erosion protection requirement of the regulations since it is realized that multiple BMPs may be required on site to achieve the various quantity and quality improvement goals. Therefore, strict adherence to the Virginia sizing methodology will not ensure that bioretention sized by the Virginia method will meet the downstream erosion protection goals examined in this study. Because of this, a simplified sizing methodology using a percentage of contributing drainage area has been used throughout. This technique is similar to previous Virginia methods which have correlated sizing of bioretention surface area with the percentage of upstream impervious area. This study examines a number of modeling scenarios which are summarized here for clarification and further described below.

- The existing conditions scenario is used as a base for other modeled scenarios

- The pre-development scenario is used for comparison at the watershed scale and to determine target values for design of each catchment's "optimal design" bioretention cell.

- There are four different "performance" scenarios that evaluate the performance of bioretention cells with surface areas sized as a percentage (3%, 5%, 7%, and 10%) of the contributing drainage areas. As mentioned previously, the bioretention basins are sized strictly based on upstream drainage area and are not strictly related to the impervious percentage in the sub-watershed. Because this study is focused on volumetric improvement and not water quality removal efficiency, it is believed by the authors that varying the percentage of the entire contributing drainage area may be more appropriate than varying the percentage of a single land cover (upstream impervious area).

- There are also two "optimal design" scenarios where the area percentage is adjusted such that the flow leaving the bioretention cell exactly meets the channel protection requirements of the RRM.

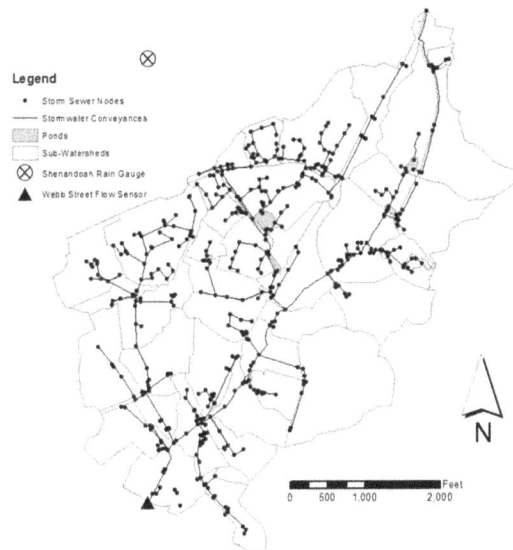

Figure 1. Study Watershed

2.1 Existing Conditions Model

In order to provide a baseline for comparison, the existing condition of the North Stroubles watershed was modeled and then calibrated to observed flows. One existing bioretention cell and 11 existing detention ponds are present in the watershed and were included in the model along with other existing stormwater infrastructure, such as manholes, catch basins, pond outlet structures, pipes, and open channels.

The stormwater infrastructure included in this base model is part of an ongoing partnership between the Town of Blacksburg and researchers at Virginia Tech. This infrastructure information was field collected and defined with the aid of GPS and aerial imagery. Attributes of the stormwater nodes and conveyances were entered into a Geographic Information System (GIS). Using LiDAR data, elevation and slope attributes were determined for the nodes and conveyances. The physical attributes of the infrastructure were then transferred into the SewerGEMS V8i modeling environment (Bentley, 2013), which was utilized for the hydrologic and hydraulic modeling in this study.

Data collected from the field, aerial imagery, and LiDAR data were used to delineate the sub-watersheds based on the drainage areas of the existing ponds and the intersection of major confluences within the stormwater network. Following delineation of the sub-watersheds, the modeled flows contributed by each sub-watershed were introduced at their respective downstream node located in each sub-watershed. The flows were calculated for each sub-watershed with the TR-20 (SCS, 1992) unit hydrograph methodology. Use of an NRCS methodology is required by the RRM in Virginia. After the sub-watershed flows were input, the modeling software implemented an implicit solver to route the flows downstream (Jin, 2002).

Using detailed aerial photos to assign land cover classes and the Soil Survey Geographic Database (SSURGO) (USDA, 2009), an area-weighted NRCS curve number (CN) for each sub-watershed was produced. The high-resolution aerial imagery was digitized into land cover classes, such as buildings, asphalt, concrete, meadow, light forest, or dense forest. This information was combined with the hydrologic soil group information in the SSURGO data to produce the NRCS CN. For each sub-watershed that was delineated, a time of concentration was also determined by commonly used flow equations from the Virginia Department of Transportation (VDOT) Drainage Manual and was based on the slopes developed from the LiDAR data and the land cover data. For each sub-watershed, the most hydraulically-remote point was estimated by checking the resulting time of concentration of several possible locations along the border of each sub-watershed and selecting the longest. This process, along with other aspects of model development, is discussed in Hixon (2009) and Aguilar and Dymond (2013).

The existing conditions model was designed to mimic existing watershed conditions, so it was calibrated using data from the flow measurement device at the watershed outlet. One of the main parameters altered during the calibration process was the NRCS CN of the sub-watersheds, with calibrated values shown in Figure 2. For calibration, the CN for each sub-watershed was increased uniformly across all sub-watersheds, and the time of concentrations (t_c) for the sub-watersheds were modified. Both changes were performed to produce more runoff, as it seems that the method of area-weighting the CNs resulted in underestimation of the CNs in this urbanized watershed. This was deemed necessary because the flow sensor's flow readings were much higher than those initially calculated by the model with initial estimates of CN. With the CN and t_c adjustments, the model was calibrated using four, single-peak storm events in 2009. These storm events were selected because they more closely replicate the single-peak nature of NRCS design storms used in the remainder of the analysis. One of the primary parameters used to calibrate the model was the difference between the measured and modeled Q*RV. This was considered a critical parameter due to a desire to meet the RRM's standards. For the selected four storm events, the differences in Q*RV between the modeled and measured values ranged from -40% to +21%, as shown in Table 1. The other primary calibration parameter used was the Nash-Sutcliffe model efficiency coefficient (Nash and Sutcliffe, 1970), which ranged from 0.64 to 0.81. The Nash-Sutcliffe coefficient measures the relative difference between a measured value and the modeled value compared to the measured value and the average of all of the measured values in a time series, with an ideal value of 1.0. Possible sources of discrepancies between the model and actual watershed response include errors in data collection, the effects of aggregating the parts of the sub-watersheds, or errors caused by the assumption of uniform rainfall across the watershed using precipitation data from a rain gage that is located just outside of the watershed boundary. Figure 3 shows the measured flow and the calibrated model's flow for the May 14, 2009 storm event.

Table 1. Calibration Summary

Storm Event	5/8/2009	5/14/2009	6/15/2009	7/17/2009
Duration of Runoff (hr)	15.5	8.0	4.5	13.5
Precipitation Depth (cm)	1.9	2.2	1.3	2.0
Time Step (hr)	0.25	0.25	0.25	0.25
Nash-Sutcliffe R^2 (1970)	0.78	0.81	0.66	0.64
Measured Runoff Depth (cm)	0.56	0.67	0.46	0.78
Modeled Runoff Depth (cm)	0.75	0.90	0.35	0.81
Measured Peak Flow (m^3/s)	2.72	3.54	2.40	3.17
Modeled Peak Flow (m^3/s)	2.49	3.12	1.79	2.49
Deviation in Volume	32%	34%	-24%	4%
Deviation in Peak Flow	-8%	-12%	-21%	-21%
Deviation in Q*RV	21%	18%	-40%	-18%
Peak Time Shift (hr)	0	-0.25	0	0
Model Continuity Error	0.0%	0.2%	0.0%	0.2%

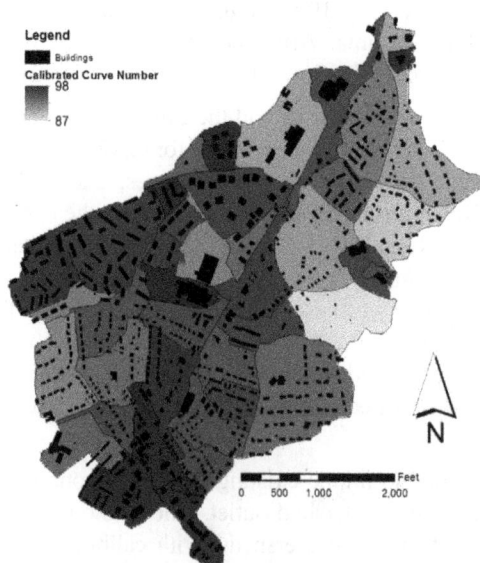

Figure 2. Calibrated Curve Numbers of Sub-Watersheds

5/14/2009

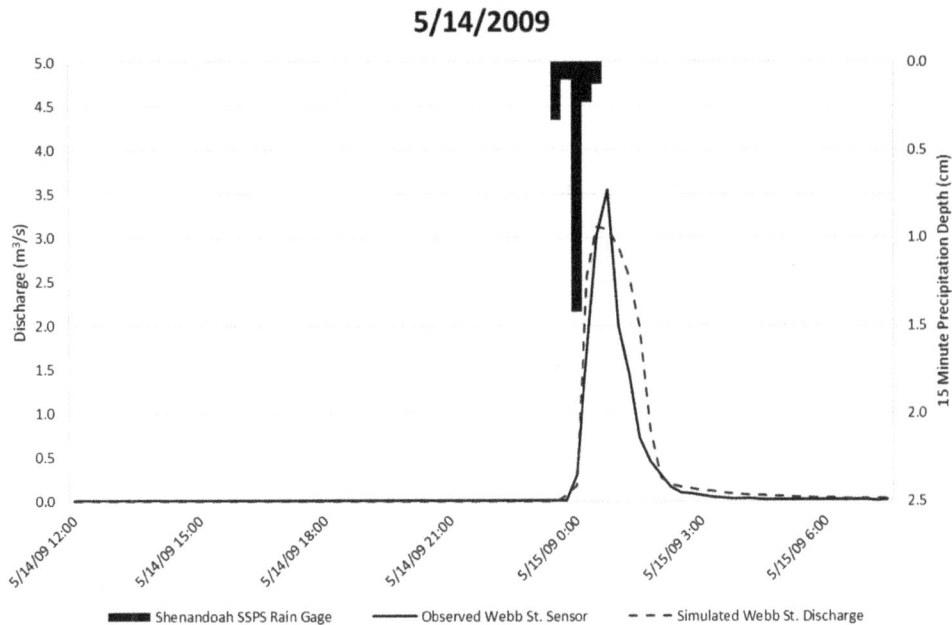

Figure 3. May 14, 2009 Storm Event

2.2 Pre-Development Model

In order to satisfy the requirements of the channel protection criteria, the runoff from the watershed in its developed condition must be compared to the pre-development condition. Similar to the approach in Hixon (2009), the pre-development condition was assumed as meadow in good condition with soils of Hydrologic Soil Group (HSG) C, which is considered a typical pre-development condition in the study area. This represents a CN of 71, which was distributed uniformly across the study watershed. For comparison at the watershed outlet, the entire watershed was combined into a single catchment with no man-made infrastructure, and the path used for the time of concentration assumed that the elevation of the land within the watershed remained approximately the same. For most of the analyses performed in this study, similar pre-development conditions were assumed for each of the sub-watersheds when comparisons were being made at the sub-watershed level. When analyzing the flow at the watershed outlet for the whole watershed scale, the single catchment pre-development scenario was used, as discussed above.

2.3 Design of Bioretention Cells in Model

The "performance" scenarios modeled the sub-watersheds with the bioretention cells sized as a percentage of the drainage area and located at the local outfall. All runoff from each sub-watershed was directed to its bioretention cell. The areas of each cell differed, but the vertical structure of the cells remained the same. Primarily, the cells were designed with 91 cm of engineered soil media with a porosity of 25% and then 15 cm of surface storage (100% porosity), per the Level 2 Design in the Virginia Bioretention Design Specifications (DCR, 2011). The surface storage can be increased from 15 cm to 30 cm, but 15 cm was used in this model as it is the recommended value to ensure the long term viability of the plantings within the cells. At the top of the surface storage, the weir outlet structure was designed to pass the 10-year storm for that sub-watershed. Flow entering the cell inundates the storage volume until the storage capacity is overwhelmed. Uncaptured hydrograph flow beyond the fixed storage volume exits the facility through the weir outlet structure and flows to the most downstream structure within each sub-watershed. None of the captured water in the cell reenters the modeled system, and it is assumed that it would be removed by infiltration and evapotranspiration. This modeling approach was assumed to be valid due to the high infiltration rates of the engineered soil media that is typically installed in bioretention cells and the lack of an underdrain in the design. In Virginia, the bioretention modeled in this study would be considered bioretention basins because they rely on infiltration and not an underdrain. Without an underdrain, the time that it would take the water to infiltrate into the groundwater and then return to streams as base flow would be large enough to make it insignificant to the model. Although it is not expected that infiltration would be possible in all locations, this assumption was made in order to simplify the model. In practice, soil testing would be required to determine whether an underdrain would be required.

Since the cross-section of the cells remained uniform in the model, the variation of cell surface areas provides the means for directly calculating the changes in cell storage volume. Due to the large area of some of the sub-watersheds modeled, the cells associated with these large drainage areas are much larger than typical bioretention cells. However, it is assumed that these large cells can represent a distributed network of cells located throughout each of the sub-watersheds. Elliott et al. (2009) determined that it is acceptable to aggregate a network of bioretention cells for modeling purposes, and Gilroy and McCuen (2009) found that BMPs in series have their effects compounded regardless of the distance between them. Although the bioretention cells are designed as retrofits, they can also be implemented upon the initial development of the land.

By keeping the same vertical structure of each cell and changing the surface area, the volume of each cell is changed in a consistent manner. Since each cell only receives flow from a single sub-watershed, they were sized based on a percentage of the area of their respective sub-watershed. The four consistent percentages used for sizing the surface areas of the cells were 3%, 5%, 7%, and 10% of the sub-watershed's area.

2.4 "Optimal" Models

Along with the sizing scenarios based solely on the percentage of the sub-watershed's drainage areas, two "optimal" scenarios were tested. The design of the "optimal" cells was achieved by adjusting the surface area for each cell until the calculated Q*RV leaving the cell for the 1-year storm event equaled 80% of the related pre-development value. The 80% value was chosen to meet the RRM's channel protection requirements.

The first of these two scenarios maintained the typical cell's cross-section used in all other non-optimal scenarios. In the second optimal scenario, the engineered soil media depth was increased from 91 to 122 cm. This scenario using cells of increased depth would represent an urban area where space is limited and constructing a deeper cell would be desired.

2.5 Flood Protection Analysis

In the RRM, the flood protection requirements call for reducing the peak flow rate from the 10-year storm event in the developed condition back to, or below, the pre-development peak. The model was run with the 10-year storm event for the pre-development scenario, the existing conditions scenario, and the 91-cm optimal scenario. Existing stormwater management ponds are present in the watershed and affected the flood protection analysis at the watershed outlet, but did not affect any analysis at the sub-watershed scale because the bioretention was modeled to be upstream of the pond, if present in the sub-watershed.

3. Results

Model scenarios were run for the 24-hour, 1- and 10-year return frequency NRCS design storm events for Blacksburg, Virginia. Rainfall depth for each storm was obtained from NOAA Atlas 14, Volume 2 (Bonnin et al., 2004) partial duration series. These values were 5.8 cm and 10.4 cm for the 24-hour storm events for return frequencies of 1 and 10 years, respectively.

For each sub-watershed, as well as the watershed as a whole, model results were obtained for each scenario and compared to the pre-development values with respect to the peak flow rate, the volume of flow, and the Q*RV. These values were plotted against the calibrated CN of the sub-watershed's developed condition, as shown in Figures 4 and 5. Generally as the CN of the sub-watershed increased, the peak flow rate and volume of flow increased, and therefore the peak multiplied by the volume, increased as well. Also, as expected, as the area and volume of bioretention installed in each sub-watershed increased, the peak and volume of flow decreased.

When compared to the pre-development peak for the 1-year storm event, the sub-watersheds in the existing condition model (0% bioretention) produced peak flows between 2 and 10 times higher, as shown in Figure 4a. The 3% scenario includes one sub-watershed which had a low CN and was brought below the pre-development peak, and the 5% scenario includes five sub-watersheds achieving that reduction. Almost half of the 41 sub-watersheds in the 7% scenario had peaks at or below the pre-development value, and all of the watersheds in the 10% scenario had peak flows below the pre-development peak.

Meeting the pre-development values for volume was less successful. None of the 3%, only 1 of the 5%, and only 4 of the 7% sub-watersheds met the pre-development threshold (Figure 4b). Only about one-third of the sub-watersheds in the 10% scenario released less total flow than the pre-development scenario. Note that the storage volume in the 10% scenario was so large that it resulted in no flow leaving the bioretention cell for several of the sub-watersheds.

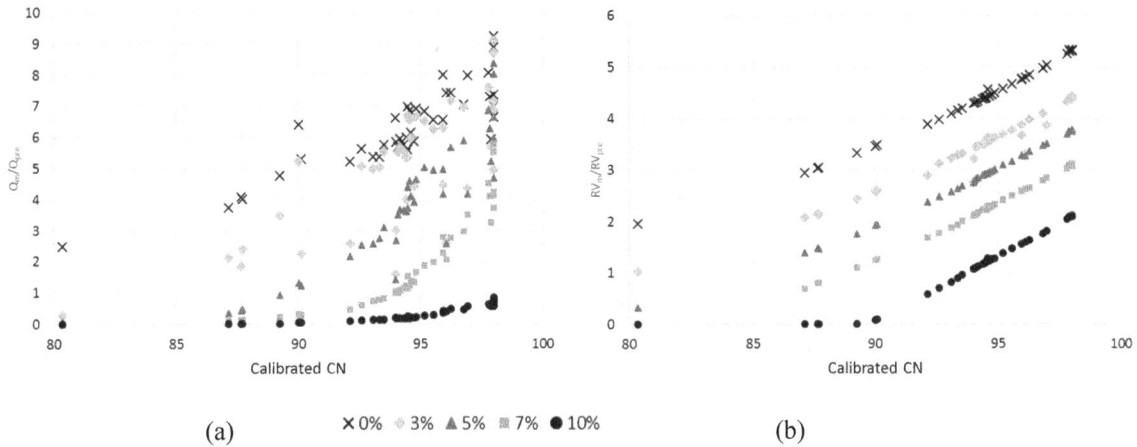

(a)   ✗ 0% ◆ 3% ▲ 5% ▦ 7% ● 10%   (b)

Figure 4. Comparison of Scenarios versus Pre-Development Values of Peak Flow and Volume (1-Year Storm Event)

(a) Ratio of Modeled Peak $[Q_m]$ to Pre-Development Peak $[Q_{pre}]$ vs. CN, (b) Ratio of Modeled Volume $[RV_m]$ to Pre-Development Volume $[RV_{pre}]$ vs. CN

When reviewing the Q*RV in Figure 5, the range of results increased greatly. Several of the sub-watersheds in the existing condition had values almost 50 times greater than those in the pre-development condition. Figure 5a shows the full range of results for the scenarios compared to the pre-development, and Figure 5b shows the same information, but only for those data points below 200% of the pre-development. In the 10% scenario, 76% of the 41 sub-watersheds are below the 80% value (shown by the black line in Figure 5b) needed for the channel protection requirements of the RRM, with far fewer meeting this value in the other scenarios.

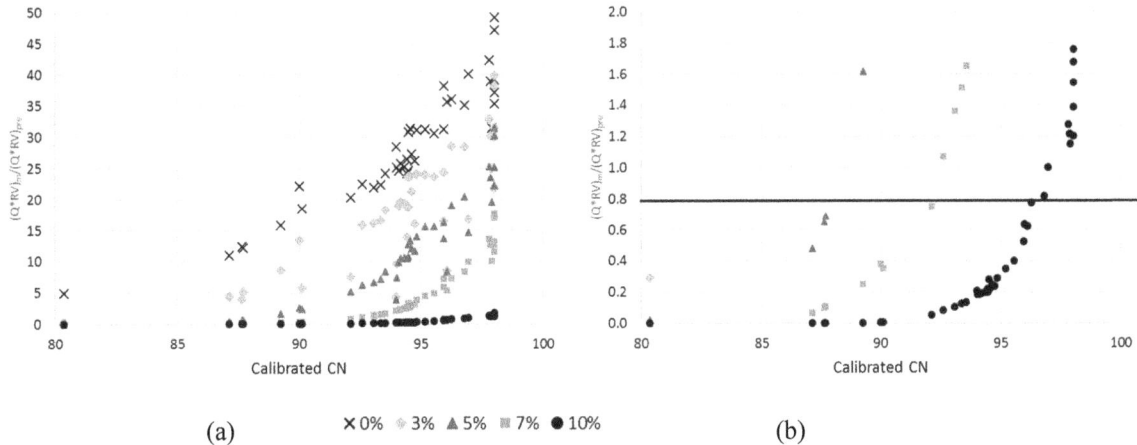

(a)   ✗ 0% ◆ 3% ▲ 5% ▦ 7% ● 10%   (b)

Figure 5. Comparison of Scenarios versus Pre-Development Values of Q*RV (1-Year Storm Event)
(a) Ratio of Modeled Q*RV to Pre-Development Q*RV vs. CN, and (b) Ratio of Modeled Q*RV to Pre-Development Q*RV [under 2.0] vs. CN

The trends shown in Figures 4 and 5 demonstrate the need for a simple method of sizing bioretention based on the CN of the upstream watershed because of the correlation that is seen between the CN and the runoff metrics used in this study. There is no "one-size-fits-all" percentage that meets the requirements in a sensible way. Therefore, for each sub-watershed, the bioretention cell was iteratively sized until it met the 80% value of the Q*RV. The results of this "optimal" design for each sub-watershed can be seen in Figure 6. As expected, more area and volume of bioretention is generally needed to achieve the same results when the CN of the contributing drainage area is higher. Less-intensely developed sub-watersheds required cells in the range of 4-6% of the drainage area, with moderately-developed sub-watersheds needing 7-8% of the drainage area, depending upon the depth of the cell used. Sub-watersheds that are mostly developed required approximately 9 and 11% of their areas, based on the depth of the cells, to be used as bioretention to meet the channel protection criteria. Linear trendlines were fit to the data for both the 91- and 122-cm depths of soil media to demonstrate the approximate linear relationship between the CN of the drainage area and the required size of the bioretention cell. Other forms of regression lines had similar goodness-of-fit measures, but the simpler linear regression line was used to show the basic trend in the data since no other forms of the line have an obvious relationship between CN and volume of runoff. Again, for the target for the conditions in this area in this particular study, the pre-development CN is 71, so using the post-development CN and cell depth, Figure 6 could be used to size the area of a bioretention cell based on the size of the contributing drainage area.

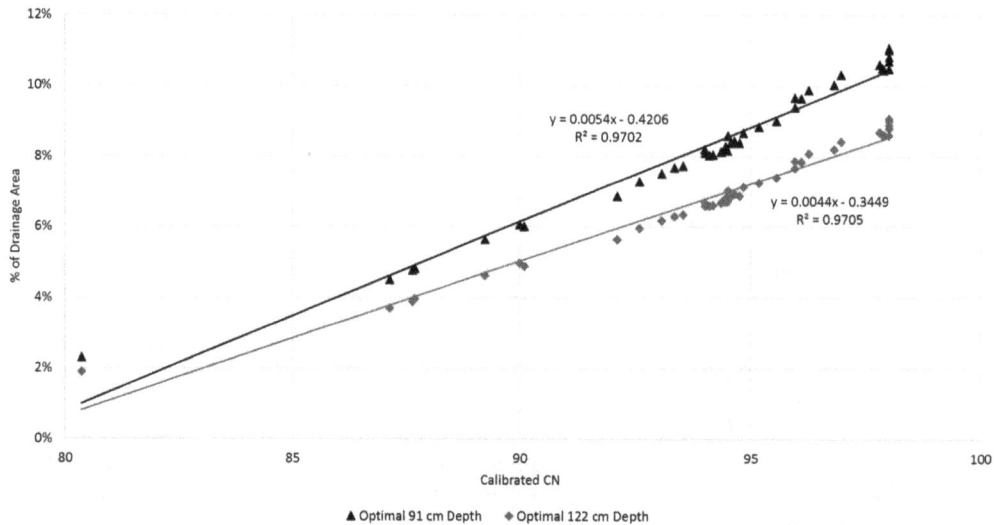

Figure 6. Optimal Bioretention Design

When increasing the depth of soil media from 91 to 122 cm, there is a consistent drop of 18% of surface area needed for the cell for all of the sub-watersheds. Due to the design of the bioretention basins and the porosity of the media, increasing the soil media depth by 30 cm has the same impact on the available storage volume that increasing the surface ponding depth by 7.6 cm would have. Although variations in surface storage depth (typical depths are between 15 cm and 30 cm) would more quickly affect cell storage, at an expected 4:1 ratio compared to variations in media depth, the surface ponding depth was maintained at a constant 15 cm across model runs because that is the preferred depth listed in Virginia DCR Stormwater Design Specification No. 9, Bioretention (2011). By design, all of the data points from both the 91- and 122-cm depths meet the 80% of pre-development Q*RV requirement. However, it is noted that the peak runoff rates and volumes of runoff for the 91- and 122-cm depths do not equal each other for any sub-watershed. The deeper cells typically have a 2-3% higher peak with 2-3% less volume. Also, note that the size of the cell is plotted against the calibrated CN, so consideration should be given to CN values before using the equations for any reference or design practices.

After observing the trends in Figure 6, it was necessary to determine how the optimal cells were achieving the reduction to 80% of the pre-development Q*RV. Figure 7 shows ratios of the peak flow rates and volumes, discharging from the 91- and 122-cm deep optimal cells, to their respective pre-development values versus the CN of each sub-watershed. When analyzing the contributions of the peak flow rates and runoff volumes to the optimization of the cells, it is obvious that the 80% Q*RV was achieved primarily by peak runoff reduction. The peak flows leaving the basins averaged approximately 47% of the pre-development peak flow rates, whereas the volumes averaged approximately 170% of the pre-development volumes. The deeper cells relied slightly less on peak flow rate reduction with a relatively smaller volume increase, and as the CN increased, a larger relative peak flow rate reduction was necessary to meet the target.

To analyze flood protection in the watershed, the 10-year storm event was modeled for the pre-development, existing conditions, and 91-cm optimal scenarios. The presence of the existing stormwater management facilities limits the investigation into the effects of the added bioretention on flood protection because the model results show that the existing facilities already reduce the peak flow rate in the watershed to below the pre-development peak. However, limited analysis suggests that the bioretention, installed for the purpose of channel protection, can decrease the storage required for flood protection at the sub-watershed scale by 15-20%. This brief analysis was performed by comparing the volume of storage that would be required for the 91-cm optimal and the existing conditions scenarios to reduce the peak flow from each sub-watershed to the pre-development peak for the 10-year storm.

A secondary goal of this study was determining the compounding effect on an entire watershed of implementing the RRM's channel protection criteria on many discrete sites within that larger watershed. The percentage scenarios, along with the existing condition, pre-development, and "optimal" scenarios, were modeled for the 1-year storm for the entire North Stroubles watershed, as well as the individual sub-watersheds. The resulting hydrographs at the outlet of the watershed can be seen in Figure 8. Applying the RRM by treating each sub-watershed as a development site (demonstrated by the optimal scenario) results in an outflow hydrograph at the watershed outlet that has a lower peak, but larger volume, than the pre-development hydrograph. Although the peak is lower than the pre-development peak, attenuation results in a descending limb that is higher than the pre-development hydrograph at the same point. Also of interest is that the application of the 10% scenario results in so much storage throughout the watershed that the hydrograph has no true peak and does not resemble a conventional hydrograph as shown in the other scenarios. Scenarios using bioretention are also noted to have shorter rising limbs then the existing condition hydrographs. This was also seen in the individual sub-watersheds and is caused by delayed runoff response caused by the cells filling up with the slower rates of runoff from the lower intensity rainfall at the beginning of the design storm and then the subsequent bypass of the additional runoff at a high rate after complete inundation of the storage volume within the cell. This also seemed to occur near the time of the higher rainfall intensity portions of the design storms used.

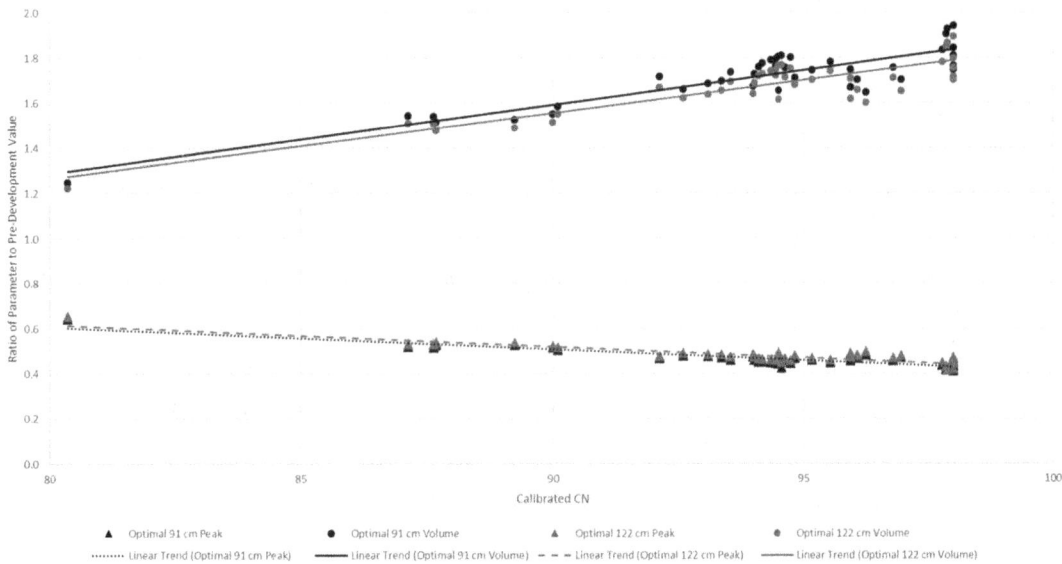

Figure 7. Ratio of Modeled Values from the Optimal Designs to the Pre-Development

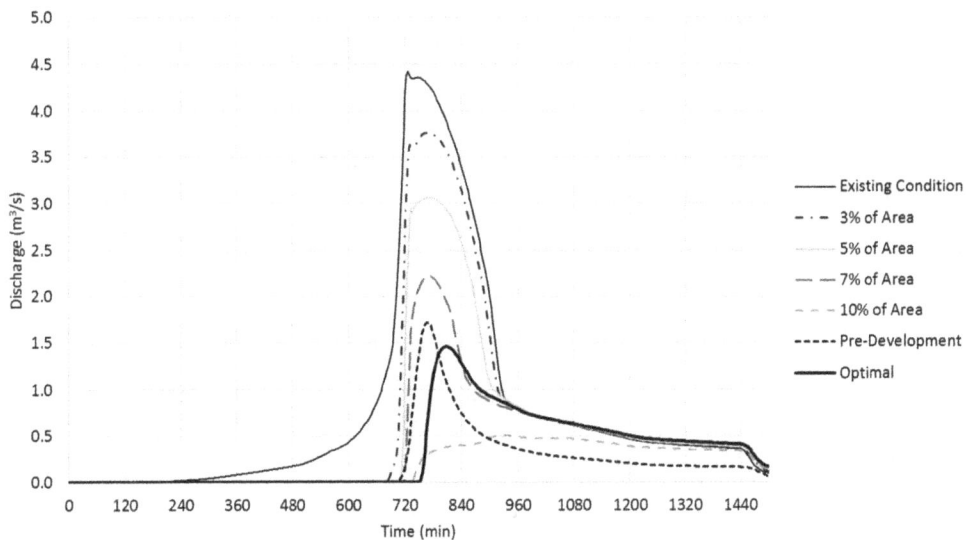

Figure 8. Modeled 1-Year Storm at the Webb Street Station

4. Discussion

This study considered the channel protection criteria of the new Virginia stormwater regulations, how bioretention could be used to meet these new regulations, and the effects that implementing bioretention to meet the new regulations would have on a larger watershed scale. Bioretention cells, located at the outfall of the sub-watersheds that make up a larger, calibrated watershed model, were simulated in variable sizes as discussed previously. Resulting models yielded information regarding the effect of bioretention and overall watershed response as outlined below.

4.1 Bioretention

The installation of bioretention cells can result in the developed hydrology more closely mimicking the pre-development hydrology for both the site- and overall watershed-scales. All sizes of bioretention retrofits that were modeled showed decreased peak flows and volumes of flows from the developed, existing condition sub-watersheds. However, when larger percent area cells (7 and 10%) were modeled, the flows leaving some of the cells were very small, or non-existent, which could have hydrologic ramifications on downstream receiving waters.

4.2 Optimal Sizing

The area, and resulting volume, of bioretention required to meet the RRM is directly related to the difference between the CN of the developed condition and that of the pre-development condition. As this difference increases, larger cells are needed to retain the larger amount of flow volume. The resulting sizes of the bioretention cells needed for the new standard of 80% of the Q*RV are typically larger in area than those traditionally seen in practice. Due to large differences between the developed and pre-development condition, modeling indicates that some sub-watersheds require cell areas to be more than 10% of the drainage area, which is infeasible. However, these large single cells were modeled in this manner to simplify the model and, in reality, depict a distributed network of cells in the sub-watershed. As concluded in Tillinghast et al. (2012), it may be unreasonable to attempt to mimic the pre-development conditions of an intensely-developed watershed through retrofits for several reasons. The sub-watersheds with the most development, and therefore requiring larger bioretention cells, would need to have available land area to accommodate the large cells. However, based on preliminary observation, this amount of open space would not be available in many of the sub-watersheds in the area of study. In addition, the available land may not be in a location that permits the directing of runoff to the cells. Other forms of LID and BMP techniques would likely be required in this situation. In an urbanized area, the combined use of underground sand filters and underground detention can help to achieve the required water quality and quantity metrics, while limiting the impact on the surface area requirements.

4.3 Depth Effects

If available surface area is a major issue, modeling revealed that increasing the engineered soil media depth from 91 cm to 122 cm is a valid option. A consistent 18% reduction in area of the bioretention was shown for this change in design for meeting the regulations. This reduction was consistent across all of the sub-watersheds. Also, increasing the depth of the cells seems to result in slightly more volume attenuation, and slightly less peak attenuation. Therefore, this could be taken into account if regulations apply to either the volume or peak, but not necessarily their product.

4.4 Watershed-Scale Effects

Finally, the retrofit of bioretention in a watershed with the channel protection criteria results in a watershed that has hydrologic characteristics closely approaching the pre-development. When this method is applied at individual sites throughout a watershed, the modeled results at the watershed outlet are shown to result in a peak lower than the pre-development, but with prolonged flows that are somewhat higher than the pre-development levels, as shown in Figure 7. This is a substantial improvement over previous stormwater management methods that have resulted in higher peaks with longer, higher receding limbs. However, implementation of this method requires much more space for the distributed network of smaller facilities.

The sizing of bioretention cells is critical to their performance. If they are sized too small, there is little channel protection benefit in their installation, and if they are too big, the decrease in outflow for lower recurrence interval storms can be so small that it could affect the nature of the receiving waters. When sizing cells for channel protection goals, the required size is directly related to the CN of the contributing drainage area. However, the feasibility of the space requirements for meeting some of these goals, especially in a retrofit environment, is questionable. One possible way to overcome limitations in available area is increasing the depth of the bioretention cell, which corresponds with a decrease in the required surface area. With the RRM applied throughout the study watershed, the hydrograph at the watershed outlet mimicked the pre-development hydrograph much closer than the hydrograph of traditional stormwater management techniques.

Acknowledgments

The authors would like to acknowledge the Town of Blacksburg for their funding of ongoing research for urban stormwater management, as well as other members of the research team for their assistance.

References

Aguilar, M. & Dymond, R. (2013). Innovative technologies for stormwater management programs in small urbanized areas. *Journal of Water Resources Planning Management,* http://dx.doi.org/10.1061/(ASCE)WR. 1943-5452.0000410 (Sep. 27, 2013).

Bonnin, G., Todd, D., Lin, B., Parzybok, T., Yekta, M., & Riley, D. (2004). Precipitation-frequency atlas of the United States. *National Oceanic and Atmospheric Administration (NOAA) Atlas 14*, Vol. 2, NOAA, Silver Spring, MD.

Brown, R. A., & Hunt, W. F. (2010). Impact of Maintenance and (Im)Properly Sizing Bioretention on Hydrologic and Water Quality Performance. *Low Impact Development, 2010*, 82-95. http://dx.doi.org/ 10.1061/41114(371)312.

Center for Watershed Protection and Chesapeak Stormwater Network. (2008). *Technical Memorandum: The Runoff Reduction Method. Ellicott City, Maryland.*

Commonwealth of Virginia. (2011). Virginia Stormwater Management Program (VSMP) Permit Regulations. *Water Quantity*, Virginia Administrative Code, United States.

Damodaram, C., Giacomoni, M. H., Prakash Khedun, C., Holmes, H., Ryan, A., Saour, W., & Zechman, E. M. (2010). Simulation of Combined Best Management Practices and Low Impact Development for Sustainable Stormwater Management. *Journal of the American Water Resources Association, 46*(5), 907-918. http://dx.doi.org/10.1111/j.1752-1688.2010.00462.x.

Davis, A. (2008). Field performance of bioretention: Hydrology impacts. *Journal of Hydrologic Engineering, 13*(2), 90–95. http://dx.doi.org/10.1061/(ASCE)1084-0699(2008)13:2(90).

Department of Conservation and Recreation. (1999). Chapter 5: Engineering Calculations. *Virginia Stormwater Management Handbook.* Department of Conservation and Recreation, Division of Soil and Water Conservation, Richmond, VA.

Department of Conservation and Recreation. (2011). Bioretention. *Virginia DCR Stormwater Design Specification No. 9.* Version 1.8. Retrieved from http://vwrrc.vt.edu/swc/april_22_2010_update/ DCR_BMP_Spec_No_9_BIORETENTION_FinalDraft_v1-8_04132010.htm.

DeBusk, K., & Wynn, T. (2011). Storm-water bioretention for runoff quality and quantity Mitigation. *Journal of Environmental Engineering, 137*(9), 800-808. http://dx.doi.org/10.1061/(ASCE)EE.1943-7870.0000388.

Department of Environmental Quality. (2013). Stormwater Management. Retrieved from http://www.deq. virginia.gov/Programs/Water/ StormwaterManagement.aspx.

Elliott, A., Trowsdale, S., & Wadhwa, S. (2009). Effect of aggregation of on-site stormwater control devices in an urban catchment model. *Journal of Hydrologic Engineering, 14*(9), 975–983. http://dx.doi.org/ 10.1061/(ASCE)HE.1943-5584.0000064.

Fairfax County. (2014). Chapter 124 (Stormwater Management Ordinance). Code of the County of Fairfax, VA.

Gilroy, K., & McCuen, R. (2009). Spatio-temporal effects of low impact development practices. *Journal of Hydrology (Amsterdam), 367*(3–4), 228–236. http://dx.doi.org/10.1016/j.jhydrol.2009.01.008.

Hixon, L. (2009). Making the case for tailored stormwater management. Master of Science thesis, Virginia Polytechnic Institute and State University, Blacksburg, VA.

James, M. B., & Dymond, R. L. (2012). Bioretention hydrologic performance in an urban stormwater network. *Journal of Hydrologic Engineering, 17*(3), 431-436. http://dx.doi.org/10.1061/(ASCE)HE.1943-5584. 0000448.

Jin, M., Coran, S., & Cook, J. (2002). New one-dimensional implicit numerical dynamic sewer and storm model. *Proceedings of the 9th International Conference on Urban Drainage Global Solutions for Urban Drainage,* ASCE, Reston, VA. http://dx.doi.org/10.1061/40644(2002)120

Li, H., Sharkey, L. J., Hunt, W. F., & Davis, A. P. (2009). Mitigation of Impervious Surface Hydrology Using Bioretention in North Carolina and Maryland. *Journal of Hydrologic Engineering, 14*(4), 407-415. http://dx.doi.org/10.1061/(ASCE)1084-0699(2009)14:4(407).

Nash, J. E., & Sutcliffe, J. V. (1970). River Flow Forecasting through Conceptual Models Part I - A Discussion of Principles. *Journal of Hydrology, 10*, 282-290. http://dx.doi.org/10.1016/0022-1694(70)90255-6.

Prince George's County Program and Planning Division. (1999). Low-Impact Development Design Strategies: An Integrated Design Approach. Department of Environmental Resources, Prince George's County, MD.

Prince William County. (2014). Section 700 - Environmental Systems (July 1, 2014). *Design and Construction Standards Manual*, Land Development Division, Prince William County, VA.

Rolband, M. S., & Graziano, F. R. (2012). The 'Energy Balance' method of stormwater management. *Stormwater: The Journal for Surface Water Quality Professionals*, Forester, Santa Barbara, CA, July-August 2012. Retrieved from http://www.stormh2o.com/SW/Editorial/The_Energy_Balance_Method_of_Stormwater_Management_17718.aspx.

SewerGEMS V8i [Computer software]. (2013). Bentley Systems, Inc., Exton, PA.

Soil Conservation Service. (1992). Hydraulics and hydrology tools and models: Other models—TR-20 computer program for project formulation hydrology. Retrieved from http://www.nrcs.usda.gov/wps/portal/nrcs/detailfull/null/?cid=stelprdb1042793.

Tillinghast, E., Hunt, W., Jennings, G., & D'Arconte, P. (2012). Increasing Stream Geomorphic Stability Using Storm Water Control Measures in a Densely Urbanized Watershed. *Journal of Hydrologic Engineering, 17*(12), 1381-1388. http://dx.doi.org/10.1061/(ASCE)HE.1943-5584.0000577.

United States Department of Agriculture. (1986). Urban hydrology for small watersheds, TR-55. *Technical Release 55*, National Resources Conservation Service, Washington, DC.

United States Department of Agriculture. (2009). *Soil Survey Geographic Database*. Retrieved from http://soils.usda.gov

Winston, R., Page, J., & Hunt, W. (2013). Catchment Scale Hydrologic and Water Quality Impacts of Residential Stormwater Street Retrofits in Wilmington, North Carolina. *Green Streets, Highways, and Development 2013*, ASCE, 159-172. http://dx.doi.org/10.1061/9780784413197.014.

14

The Willingness to Pay by Industrial Sectors for Agricultural Water Transfer During Drought Periods in Taiwan

Ya-Wen Chiueh[1] & Cheng Chang Huang[2]

[1] Professor, National Hsinchu University of Education, Taiwan

[2] Associate Researcher, Agricultural Engineering Research Center, Taiwan

Correspondence: Ya-Wen Chiueh, Professor, Department of Environmental and Cultural Resources, National Hsinchu University of Education, Hsinchu City 300, Taiwan. E-mail: yawen.chiueh@gmail.com

Abstract

With continuous climatic change, droughts have begun to occur more frequently. In order for industrial sectors to secure a stable supply of water during the time of droughts, or to maintain the normal functions of industrial production lines, transfer of agricultural water has often been utilized. This will happen more frequently as the climates continue to change. There is a high possibility that continuous climatic change will affect the current water management operations. The IPCC (Intergovernmental Panel on Climate Change) thinks that developing the water market by targeting reasonable distribution of scarce water resources is one of the strategies for water resource management under continuous climatic change (IPCC, 2008). Past studies have often discussed the amount of compensation given in exchange for agricultural water transfer, or the costs lost resulting from water transfer. However, few studies have discussed the industrial sector's willingness to pay for agricultural water transfer from the viewpoint of managing water shortage risks. In general, if the amount of compensation is established based on farmers' WTA information, there might be situations where the amount of compensation established is inclined to become an exploitation of the surplus from transferring agricultural water. Or, the benefits of the agricultural sectors might be sacrificed or exploited. It could also be that the transaction cost would become higher for agricultural water transfers during droughts when the information of the provider and those in need are asymmetrical, which in turn would affect the benefits of the agricultural sectors and the efficiency of water resource distribution. This study uses the Contingent Valuation Method (CVM) to evaluate the amount of money industrial sectors are willing to pay under climatic change to avoid the risk of water shortage in Taiwan. We target the larger industrial areas and science parks as the objects of investigation. Interviews about the amount of willingness to pay (WTP) for transferring agricultural water are conducted in factories in the above mentioned areas, which include the Hsinchu Industrial Park, Chung-Li Industrial Park, Taichung Industrial Park, Lin-Yuan Industrial Park, Hsinchu Science Park, Central Taiwan Science Park, and Tainan Science Park. The results of this study show that the WTP for agricultural water transfer of the abovementioned industrial/science parks are $28NT/ton during drought periods.

Keywords: drought, water transfer, contingent valuation method

I. Introduction

Due to frequent droughts resulting from climate changes, water reserved for agriculture is often transferred into non-agricultural setors in order to ensure a stable water source or maintain normal production in industries. With continuing climate changes, such transferring of water has become increasingly frequent. Climate changes have a high potential of impacting current water management. The Intergovernmental Panel on Climate Change (IPCC) believes that the marketization of water to reasonably distribute scarce water resources is a management strategy for coping with climate changes. Nevertheless, the marketization of water resources or transferring water among different sectors involves the "supply" party's willingness to accept (WTA) and the "demand" party's willingness to pay (WTP). Understanding each party's respective WTA and WTP information can facilitate a fair and effective transferring and compensation mechanism.

Previous literature reviews have mainly focused on the WTA amounts for farmers to transfer out water. For example, Chiueh (2007) estimated the WTA amount for farmers in the Kaohsiung area while Chiueh and Zheng （2007) estimated the WTA amount for farmers in the Taoyuan and Shimen areas. However, very few studies have evaluated the price that non-agricultural sectors are willing to pay to transfer in agricultural water, in view of controlling water shortage risks. Generally, if prices were only established according to WTA, there is concern that such a compensation policy could easily result in the tendency to exploit the agricultural sector of their water surplus, or in other words, undermine their welfare. Moreover, if the supply and demand market information is unbalanced, it is easy for agricultural water transferring to cost more during periods of drought, thereby affecting the welfare of agricultural sectors or the efficiency of water resource distribution. The goal of this research is to design a questionnaire to survey the WTP amount set by non-agricultural sectors to transferr agricultural water in order to eliminate water shortage risks due to climate changes and to ensure a stable water supply for their sectors. Non-agricultural water users from areas that are at higher risk of water shortage due to climate changes were chosen for questionnaire sampling. The contingent valuation method (CVM) was used to evaluate the WTA prices of non-agricultural departments for transferring agricultural water to eliminate water shortage risks due to climate changes and to ensure a stable water supply for their sectors. It is hoped that information can be provided to water markets so that the cost of water allocation can be reduced, thereby facilitating a fair and efficient distribution of water resources.

2. Literature Review

The IPCC believes that the marketization of water to reasonably distribute scarce water resources is a management strategy for coping with climate changes (Huang, 2008; IPCC, 2008). In countries around the world, water resource allocation is often goes from non-agricultural sectors to agricultural sectors. Cortignani and Severini (2009) used Positive Mathematical Programming (PMP) to analyze how deficit irrigation influenced the impact of increased water supply cost, decreased water resources and increased grain prices. It was found that deficit irrigation is advantageous to efficiency and the welfare of farmers, and is helpful for setting appropriate water resource management policies. Taylor and Young (1995) asserted that as long as the benefits of transferring out irrigation water outweighs the benefits and cost of irrigation water use, then irrigation water transferring rights are economically feasible. Willies et al. (1998) used a different irrigation approach and transferring ratio to analysis the cost to temporarily transferring out water for salmon breeding to salmon farmers. Their results indicated that a contract for temporary transferring can provide water for salmon breeding. In his study of water use trends in the western United States, Frederic (2006) found that if the agricultural sectors in the western states can conserve water use by 15%, the future daily and industrial water needs of the western United States can be met. Brewer et al. (2008) collected data on the price and volume of water transferring in the western United States from 1987 to 2005, and found that both the price and frequency of agriculture-to-urban water transferring is higher than agriculture-to-agriculture water transferring. Moreover, short-term leases are gradually shifting toward permanent sales.

The marketization of water or transferring of water among sectors involves the "supply" party's willingness to accept (WTA) and the "demand" party's willingness to pay (WTP). Understanding each party's respective WTA and WTP information can facilitate a fair and effective transferring and compensation mechanism. Garrida (2005) pointed out that prices can be established based on inverse water demand curves of non-agricultural sectors. Rensetti (1992) used the theory of derived demand to collect data on major Canadian manufacturers' intake, treatment, recirculation and discharge, and calculated their demand function. Nauges and Thomas (2000) calculated the water needs for daily living in France, and analyzed the negotiation power between daily water users and water providers. The results showed that the ability of daily water users to negotiate prices and the negotiation power of providers determined the demand function. Chu et al. (2009) used the agent-based theory to analyze daily water use in Beijing, China, and established a pattern of daily water use. The authors pointed out that disclosing water use information and identifying the pattern facilitate the drafting of water management policies in Beijing, China. Danilov-Danilyan and Khranovich (2008) used the production function to analyze how establishing a market mechanism affects dependable water volume, and found that the establishment of a market mechanism provides an incentive for increasing the volume of dependable water resource.

In the past, most studies of agricultural water transferring have focused mainly on the WTA amount that was acceptable to farmers for transferring out agricultural water. For example, Chiueh（2007）evaluated the WTA amount for farmers in the Kaoshiung area while Chiueh and Zheng (2007) evaluated the WTA amount for farmers in the Taoyuan and Shihmen areas. However, few studies have evaluated the WTA amount for non-agricultural sectors based on the perspective of water shortage risk control.

Generally, if prices are only established according to WTA, there is concern that such a compensation policy could easily result in the tendency to exploit the agricultural sector of their water surplus, or in other words, undermine their welfare. Moreover, if the supply and demand market information is unbalanced, it is easy for agricultural water transferring to cost more during periods of drought, thereby affecting the welfare of agricultural sectors or the efficiency of water resource distribution.

This study designed a contingent valuation method questionnaire to evaluate the WTP amount for non-agricultural sectors when transferring in agricultural water in order to eliminate their water shortage risks caused by climate changes and to ensure a stable water supply. Non-agricultural water users from areas at higher risk of water shortage due to climate changes were chosen for questionnaire sampling. Evaluation was conducted using the contingent valuation method (CVM) and econometric analysis. It is hoped that information can be provided to water markets so that the cost of water allocation can be reduced, thereby facilitating a fair and efficient distribution of resources.

This study used the contingent valuation method to analyze the price that non-agricultural users were willing to pay for agricultural water. This method of evaluating non-market goods involved presenting hypothetical scenerios through a questionnaire. The respondents were primarily asked to valuate goods and services that do not involve transactional actions. The main feature of the contingent valuation method is ex ante judgment, that is, predictive assessment. The concepts of non-dry season, dry season and the price of transferring rights for agricultural water have yet to be implemented, and, therefore, would be very suitable for establishing a pre-assessment of pricing mechanism for transferring agricultural water.

The difference between a CVM price inquiry and general direct questionnaires is that CVM emphasizes integrating survey methods and theory. CVM became widespread in the 70's. Following the United Kingdoms Forest Law and the United States' Presidential Decree NO. 12291, using the CVM to evaluate economic benefit became even more common. In the Exxon Valdez oil spill in 1989, the amount of compensation ordered by the United States Federal Courts was based on CVM, further increasing the credibility of CVM. In 1993, due to its common use, the US government determined its natural resource policies according to CVM. The NOAA issued CVM operation guidelines, using CVM as the norm. Research has shown that in the absence of direct or indirect market price, CVM can provide a reasonable valuation of public goods or environmental goods (Smith, 1993). Mitchell and Carson (1989) and Hutchinson et al. (1995) also pointed out that if designed properly, CVM is a reliable valuation method.

3. Methods

3.1 Contingent Valuation Method Theory and Empirical Model

As described below, this study adopted the contingent valuation method to analyze the amounts that the non-agricultural industrial and science park users are willing to pay for agricultural water:

The main feature of the contingent valuation method is its ex ante evaluation, that is, predictive assessment. In this study, the market mechanism concepts for transferring agricultural water have yet to be implemented, and are therefore appropriate for pre-evaluating market mechanisms for transferring agricultural water given climate changes. As such, the WTP prices of industrial and science parks for agricultural water to ensure stable water supply and eliminate risks of water shortage due to climate changes can be further understood.

CVM can be conducted by adopting Hanemann's (1984) random utility model or Cameron's (1988) disbursement function. However, Cameron (1988) pointed out that the dichotomous selection of information in Hanemann's (1984) random utility model was not only ordered, but in the questionnaire, the base prices were also observable. Therefore, Hanemann assumed a non-sequential discrete choice model, and apparently did not adequately utilize the provided base price information. Hence, Cameron adopted a censored dichotomous choice model to directly estimate the parameters for the disbursement function and directly obtain the WTP amount. Economic theory proves that because a dual relationship exists between the indirect utility function and disbursement parameters, they are therefore representative of consumer preference. To prevent excessive bias so that all the gathered information can be fully used, this study adopted a closed dichotomous choice method questionnaire design to gather information, and also used Cameron's (1988) and Cameron and James' (1987) disbursement function to estimate the price parameters for transferring agricultural water.

In terms of manufacturer benefits, in order to use the questionnaire to determine WTP prices, a hypothetical agricultural water transferring mechanism must first be presented to the manufacturers. The questionnaire price inquiry and manufacturers' WTP or acceptable prices were then compared using Cameron's (1988) disbursement function. The estimation process is shown in Fig.1, and the empirical model is as follows:

$$Y(Q_0, Q_1, U_0, S) = E(Q_0, U_0, S) - E(Q_1, U_0, S) \qquad (1)$$

Let $Y(Q_0, Q_1, U_0, S)$ be the function for price offered by industries to transfer in agricultural water; and $E(Q_0, U_0, S)$ and $E(Q_1, U_0, S)$ be the disbursement function, where S represents market goods vectors and individual socioeconomic vectors:

$$S = S\ (P_W, P_X, S_O) \tag{2}$$

and S_O represents its individual socioeconomic vector. If the price offered in the CVM survey is \$T, then when

$$Y\ (Q_0, Q_1, U_0,\ S)\ \geq T \tag{3}$$

the probability that the interviewee will select this offered price can be expressed by Equation (4):

$$Pr = Pr[Y^*(Q_0, Q_1, U_0, S) - T > u] \tag{4}$$

where Y* denotes the observable component and u denotes an observable random component, as expressed in Equation (5):

$$Y\ (Q_0, Q_1, U_0, S) = Y^*\ (Q_0, Q_1, U_0, S) + u \tag{5}$$

The bid function can be estimated using the probit model (Cameron & James, 1987):

$$I_i = 1 \text{ if } Y_i > T_i$$
$$= 0 \quad \text{otherwise}$$
$$Pr(I_i = 1) = Pr(Y_i > T_i) = Pr(u_i > T_I - X_i'B)$$
$$= Pr(u_i/\sigma > (T_i - X_i'B)/\sigma)$$
$$= 1 - \phi((T_i - X_i'B)/\sigma) \tag{6}$$

where $X_i'B$ denotes explainable variables and their coefficients, and φ denotes the cumulative probability density function. The estimated bids of the interviewees can be expressed in Equation (7):

$$Y_i = X_i'B + u_i \tag{7}$$

However, the standard binary probit model is:

$$I_i = 1 \quad \text{if } Y_i > 0$$
$$= 0 \quad \text{otherwise}$$
$$Pr(I_i = 1) = Pr(Y_i > 0_i) = Pr(u_i > -w_i'\delta)$$
$$= Pr(z_i > -w_i'\delta\ /\ v)$$
$$= 1 - \phi(-w_i'\delta/\ v)$$
$$\text{Here,}$$
$$Y_i = w_i'\delta + u_i$$

Through the following transformation,

$$-(T_i, X_i') \begin{bmatrix} -1/\sigma \\ B/\sigma \end{bmatrix} = -w_i'\delta$$

$$\delta^* = (\alpha, \gamma) = (-1/\sigma, B/\sigma)$$

the following are obtained:

$$B = -\gamma/\alpha$$
$$\sigma = -1/\alpha$$
$$Y_i^* = X_i'B \tag{8}$$

Conforming to the probit model, Y_i^* represents a manufacturer's price appraisal of agricultural water, which can be reasonably calculated using this equation.

Supposing u is the logistic distribution, then based on the logistic model, the empirical results can be obtained (Cameron, 1988), as expressed in Equation (9):

$$P(Y) = [1 + e - [Y_i - T_i]] - 1$$

As with the probit method, the following is obtained:

$$Y_i^* = X_i'B \tag{9}$$

Conforming to the probit model, Y_i^* represents a manufacturer's price appraisal of agricultural water, which can be reasonably calculated using this equation.

In the same way, supposing u is the probit distribution, then conforming to the probit model, Y_i^* represents a manufacturer's price appraisal of agricultural water.

3.2 Questionnaire Design

For industrial water users, water supply stability is a critical factor in production. For example, Item 1 of the questionnaire pertained to how manufacturers resolve drought problems. Then Items 2-7 described hypothetical scenarios of varying severity of water scarcity and levels of ensuring water supply. The hypothetical scenarios assumed that the government promised agricultural water of equivalent quality, "water quality guarantee and compensation for loss due to water shortage," "exclusive pipelines for transferring agricultural water with free pipeline connection" and use of agricultural water for non-agricultural sectors during drought induced water shortage. In the hypothetical scenarios, the price lists were based on actual current domestic price for transferring agricultural water. In addition, water prices were taken into consideration in the questionnaire inquiry price. Furthermore, based on the requirements of the theoretical framework of the contingent valuation method, one or more prices ranging from extreme to moderate were set. As shown in Table 1, a total of 20 different prices were set. In other words, this study comprised 20 questionnaire configurations, from Questionnaire A to Questionnaire T. Following the hypothetical questions, Item 8 was designed to eliminate those who rejected agricultural water. The final part of the questionnaire asked for the manufacturer's basic information and water usage.

Table 1. Hypothetical scenarios and price list

Hypothetical Scenarios Questionnaire Form	*Hypothetical Scenarios (Unit: dollars/ton)*
A	2
B	3
C	4
D	5
E	6
F	7
G	8
H	9
I	10
J	11
K	12
L	13
M	14
N	15
O	16
P	18
Q	20
R	24
S	26
T	30

Source: this study.

3.3 Sampling

This study focused on industries and science parks with heavier water usage, including industries from the Hsinchu Industrial Park, Jhongli Industrial Park, Taichung Industrial Park, Linyuan Industrial Park, Hsinchu Science Park, Central Taiwan Science Park and Tainan Technology Industrial Park. The above industrial and science parks were sampled. Using the directory purchased from the Ministry of Economic Affairs, questionnaires were sent via mail to all registered manufacturers. The mailings were followed by a telephone reminder. Manufacturers unwilling to participate in the survey were excluded from the study. After verification, 1085 manufacturers participated in the survey. The questionnaires were sent out on June 21, 2010 and the response deadline was August 6, 2010. As shown in Table 2, 135 questionnaires were collected, representing a return rate of 12.44%. To facilitate understanding of the aforementioned hypothetical scenarios, the questionnaire response rate to the price list of each given scenario is shown in Table 2.

Table 2. Questionnaire response rate

Area	6/8Copies sent, A	Copies responded, B	Copies rejected, C	Invalid, D (6/21-8/6No. of calls)	Valid Sample, E=A-(C+D)	Return Rate, F=B/E
Hsinchu Industrial Park	491	29	46	276	169	17.16%
Jhongli Industrial Park	442	40	11	176	255	15.69%
Taichung Industrial Park	1,207	29	109	665	433	6.70%
Tainan Technology Industrial Park	138	10	6	79	53	18.87%
Linyuan Industrial Park	27	4	4	16	7	57.14%
Hsinchu Science Park	306	21	9	177	120	17.50%
Central Taiwan Science Park	94	2	10	36	48	4.17%
Total	2,705	135	195	1,425	1,085	12.44%

4. Empirical Results

In this study, the hypothetical scenarios assumed that the government promised agricultural water of equivalent quality (raw water), "water quality guarantee and compensation for loss due to water scarcity," "exclusive pipelines for transferring agricultural water with free pipeline connection" and use of agricultural water for non-agricultural sectors during drought induced water shortages. Table 3 shows the result of multinomial logit model analysis. The relationship between price list and WTP conformed to demand theory. The significance variables were WT2T (the maximum amount of water that can be transferred), and X78 (the amount of water taken in 2009 influenced the willingness of industries to assume the price of transferring agricultural water during drought season.) In the conditions set under the hypothetical scenarios, the acceptable price that non-agricultural sectors were willing to pay for transferring agricultural water was NT$28 per ton. The predictive value was 66% and above.

Table 3. Results of Multinomial Logit Model Analysis

Variable	English Code	Coeff.	Std.Err.	t-ratio	P-value
constant	ONE	-0.04736	0.876312	-0.05405	0.956898
price list	BIT2	-0.03027	0.030223	-1.00143	0.316618
acceptable amount of water for transferring	WT2T	0.032882	0.012232	2.68823	0.007183
amount of water taken in 2009	X78	-1.39E-05	4.92E-06	-2.83035	0.00465
type of company	FAC	0.845882	0.664886	1.27222	0.203294
location of factory	LOCATE	-7.33E-02	1.03E-01	-0.71373	0.475394
interrupted production due to water shortage	SOL1B	0.100238	0.098517	1.01746	0.308933
number of days of factory reserve water	SOL1C2	-0.05525	0.061541	-0.89772	0.369334

5. Conclusion and Suggestions

During droughts brought about by climate changes, non-agricultural sectors often transfer in agricultural water in order to ensure a stable water supply or maintain normal operations of production lines. With continuing climate changes, such water transferring has become increasingly frequent. Climate changes are very likely to impact current water management. Collecting a complete set of data is the foremost condition for designing an appropriate water transferring mechanism. This study focused on industries and science parks with heavier water usage, including industries in the Hsinchu Industrial Park, Jhongli Industrial Park, Taichung Industrial Park, Linyuan Industrial Park, Hsinchu Science Park, Central Taiwan Science Park and Tainan Technology Industrial Park.

Results showed that among the sampled industries, the WTP price for non-agricultural sectors to transfer agricultural water during droughts was NT$28 per ton. The survey showed that industrial water users were already aware of payments in exchange for using agricultural water. The Intergovernmental Panel on Climate Change (IPCC) has suggested that in times of climate changes, an appropriate water exchange mechanism should be established for equitable and reasonable allocation of scarce water resources, and for protecting the rights of agricultural sectors.

Acknowledgments

This article was partial funded and extracted from 2 detailed project: 1). " Evaluation the willingness to pay of non-agricultural sectors transferred agriculture water due to water-shortage risk caused by climate change. " under National Hsinchu University of Education, Taiwan and Agricultural Engineering Center, Taiwan , Science and Technology Program, Council of Agriculture, Taiwan (project code: 99 Agriculture science -7.4.1-water -b1). 2) "Evaluate the Economic Cost of Irrigation Water Supply Stability and Drought : Lost in Changing Environment and Society" under National Hsinchu University of Education, Taiwan, Ministry of Science and Technology, Taiwan (project code:NSC103-2625-M-134-001 Finally, I would like to thank Council of Agriculture, Taiwan and Ministry of Science and Technology, Taiwan for partial funding of this article.

Reference

Brewer, J., Glennon, R., Ker, A., & Libecap, G. (2008). 2006 Presidential Address Water Markets in the West: Prices, Trading, and Contractual Forms. *Economic Inquiry 46*(2), 91-102. http://dx.doi.org/10.1111/j.1465-7295.2007.00072.x

Chiueh, Y. W. (2008). Evaluation the compensation to farmers for paddy irrigation water transferring in Kaohsiung area. *Journal of Humanities and Social Sciences of NHCUE, 1*, 133-146. (In Chinese)

Chiueh, Y. W.(2007). Market failure or government failure? Market analysis of transferring agricultural water for alternative uses in Taiwan. *Journal of Humanities and Social Sciences, 14*, 309-338.(In Chinese)

Chiueh, Y. W., & Cheng, C. C. (2007). The evaluation of compensation for transferring irrigation water to alternative uses in Tao-Yuan area. J*ournal of Taiwan Agricultural Engineering, 53*(4), 35-43. (In Chinese)

Christophe, B., & Stephane, C. (2002). Irrigation water demand for the decision maker. *Environment and Development Economics, 7*(4) ,643-657.

Chu, J., Wang, C., Chen, J., & Wang, H. (2009). Agent-Based residential water use behavior simulation and policy implications: A case-study in Beijing City. *Water Resource Management, 23*, 3267-3295.http://dx.doi.org/10.1007/s11269-009-9433-2

Consuelo V. O., Sumpsi, J. M., Garrido, A., Blanco, M., & Iglesias, E. (1998). Water pricing policies, public decision making and farmer's response: Implications for water policy. *Agricultural Economics, 19(1998)*, 193-202. http://dx.doi.org/10.1016/S0169-5150(98)00048-6

Cortignani, R., & Severini, S. (2009). Modeling farm-level adoption of deficit irrigation using Positive Mathematical Programming. *Agricultural Water Management, 96(2009)*, 1785-1791. http://dx.doi.org/10.1016/j.agwat.2009.07.016

Danilov-Danilyan, V. I., & Khranovich, I. L. (2009). Dependable water use under market conditions. *Water Resources, 36*(2), 214-224. http://dx.doi.org/10.1134/S0097807809020109

Dixon L. S., MOORE, N. Y., & Schechter, S. W. (1993). *California's 1991 drought water bank, Rand Prepared for the California Department of Water Resources.*

Frederic, L. (2006). Managing water diversion from Canada to the United States: An old idea born again? *International Journal, 62*(1), 81-92. http://dx.doi.org/10.2307/40204247

Garrido, R. (2005). Price setting for water use charges in Brazil. *Water Resources Development, 21*(1), 99-117. http://dx.doi.org/10.1080/07900620042000316839

Hansen, L. G. (1996). Water and energy price impacts on residential water demand in copenhagen. *Land Economics 72*(1), 66-79. http://dx.doi.org/10.2307/3147158

Nauges, C., & Thomas, A. (2000). Privately Operated Water Utilities, Municipal Price Negotiation, and Estimation of Residential Water Demand: The Case of France. *Land Economics, 76*(1), 68-85. http://dx.doi.org/10.2307/3147258

Qureshi, M. E., Shi, T., Qureshi, S. E., & Proctor, W. (2009). Removing barriers to facilitate efficient water markets in the Murray-Darling Basin of Australia. *Agricultural Water Management, 96*(2009), 1641-1651. http://dx.doi.org/10.1016/j.agwat.2009.06.019

Renzetti, S. (1992). Estimating the Structure of Industrial Water Demands: The Case of Canadian Manufacturing. *Land Economics, 68*(4), 396-404. http://dx.doi.org/10.2307/3146696

Strand, J., & Walker, I. (2005).Water markets and demand in Central American Cities. *Environment and Development Economics 10*(3), 313-335. http://dx.doi.org/10.1017/S1355770X05002093

Taylor, R. G., & Young, R. A. (1995). Rural-to-Urban water transfers: Measuring direct foregone benefits of irrigation water under uncertain water supplies. *Journal of Agricultural and Resource Economics, 20*(2), 247-262.

Walid, K., & Kameel, V. (2006). Water Demand Management Measures: Analysis of Water Tariffs and Metering in Barbados. *Journal of Eastern Caribbean Studies, 31*(3), 1-26.

Willis D. B., & Whittlesey, N. K. (1998). The effect of stochastic irrigation demands and surface water supplies on On-farm water management. *Journal of Agricultural and Resource Economics, 23*(1), 206-224.

Willis, D. B., Caldas, J., Frasier, M., & Norman, K. (1998). The effects of water rights and irrigation technology on streamflow augmentation cost in the Snake River Basin. *Journal of Agricultural and Resource Economics, 23*(1), 225-243.

15

Assessment of Trace Gas Emissions From Wild Fires in Different Vegetation Types in Northern Ghana: Implications for Global Warming

Nyadzi Emmanuel[1], Ezenwa I. S. Mathew[2], Nyarko K. Benjamin[3], A. A. Okhimamhe[1],
Bagamsah T. Thomas[4] & Okelola O. Francis [1]

[1] WASCAL CC&ALU, Federal University of Technology, Minna, Nigeria

[2] Department of Soil Science, Federal University of Technology, Minna, Nigeria

[3] Department of Geography and Regional Planning, University of Cape Coast, Cape Coast, Ghana

[4] Maryland Department of Agriculture: 50 Harry S Truman Parkway Annapolis, MD 21401, USA

Correspondence:Nyadzi Emmanuel , WASCAL Coordinating Unit, Federal University of Technology, Minna, Nigeria. E-mail: enyadzi@yahoo.com

Abstract

Biomass burning in Northern Ghana is a major cause for concern because of its potential contribution to global warming, hence climate change. This study assessed the emission of trace gases from human activities in the Guinea savanna of Northern Ghana using the guidelines of the Intergovernmental Panel on Climate Change. Carbon content of biomass was determined from four different vegetation covers in the study area; namely, widely open savanna woodland, grass/herb with scattered trees, open savanna woodland and closed savanna woodland. Under each vegetation cover, five plots (1 m x 1 m) were demarcated for the estimation of above-ground biomass density. Using the combustion furnace method, emitted carbon, methane and carbon monoxide were estimated. Results showed that the emitted methane (CH_4) and carbon monoxide (CO) differed significantly ($p<0.05$) under all the vegetation types. The gases were in perfect correlation ($r=1.00$) with the quantity of above-ground biomass density and carbon released, with more CO being emitted. Emission of CH4 and CO per hectare of burnt area in the open savanna woodland category was the highest with 0.001719 ton and 0.045119 ton respectively. Over time, emission of these gases may increase their atmospheric concentration, causing major health problems. The contribution to global warming, thus climate change, may also become quite significant. This underscores the fact that existing flaws in the wild fire management policy of Ghana must be effectively dealt with and appropriately implemented with regular reviews to reduce the annual wild fires that are very rampant in Northern Ghana, especially during the dry season.

Keywords: Trace gas emission, global warming, biomass burning, wild fires, Guinea savanna

1. Introduction

One of the most debated and a researched issue is the increase in concentration of greenhouse gases into the atmosphere and its effect on global warming (Schils et al., 2008). Typically major land-use changes, particularly through widespread use of wild fires, contribute significantly to atmospheric greenhouse gas emissions (Shimada et al., 2000). Two major causes of bush fires have been generally recognized: natural and anthropogenic (Jones, 1979; Langaas, 1995). Most of these fires, whether accidental or deliberate, are assumed to be generated by humans during dry periods, which vary between ecosystems and climate zones (Jones, 1979; Korem, 1985). Elsewhere around the globe, wild fires occur regardless of season. Australia, for example, is prone to bushfires irrespective of the season. Summer and autumn seasons are, however, considered to be the vulnerable periods in southern Australia, while in the Northern Territory experiences most of its fires in winter and spring (Middelmann, 2007). Studies on annual spatial distributions of burnt areas across the United States have shown the seasonal peak of biomass burning generally occurring during June to August (Zhang & Kondragunta, 2008). Africa has been called a 'fire continent' (Trollope & Trollope, 1996) because of pervasive anthropogenic fires that burn the savanna vegetation annually (Mbow et al., 2000; Reid et al., 2000; Laris, 2002; Danthu et al., 2003). Farmers and pastoralists in Africa have developed traditional ways of avoiding the overwhelming nature of fire. In Senegal they practise early season grass burning because they find it safest and most beneficial (Bucini

& Lambin, 2002). 'Green flush' from perennial grasses and landscape protection from destructive fires is also considered to be the reason for early season burning (Mbow et al., 2000). Meanwhile, in Mali, the earliest burning is carried out to suppress the growth of unpalatable grasses that have no use for grazing (Laris, 2002). In the savannas of Ghana, bushfires, which are mostly man-made, are very common. These fires are typically grass fires and their intensity is usually lower than that of the forest fires (Bagamsah, 2005).

In tropical savanna woodland regions, the frequency of bush fires is on the increase, with the rise in population pressures and intensive use of rangeland. These ecosystems naturally consist of layers of grass interspersed with trees and shrubs, with an estimated area of about 1900 million hectares (Bolin et al., 1979). Some of the extensive uses of fires in these tropical regions include shifting cultivation and deforestation, and clearances of agricultural residue are some of the extensive uses of fire in the tropical regions (Crutzen & Andreae, 1990; Hao et al., 1990). Biomass burning is a large source of atmospheric carbon and a variety of greenhouse and trace gases, aerosols and pollutants that may significantly influence climate and atmospheric chemistry, particularly in the tropics (Hao & Liu, 1994; Rudolph et al., 1995; Andreae, 1997; Korontzi et al., 2003; van der Werf et al., 2010). Bush fires in the savanna region of African result mostly from human activities, and may produce as much as a third of the total emissions from biomass burning across the globe (Hao et al., 1990; Cahoon et al., 1992; Stott, 1994). In recent decades, the implication of this has become apparent as studies began to establish a link between biomass emissions and the global budgets of many radioactively and chemically active gases such as carbon dioxide, carbon monoxide, methane, nitrous oxide, tropospheric ozone, methyl chloride and elemental carbon particulate (Andreae, 1990; Hao & Liu, 1994; Rudolph et al., 1995; Korontzi et al., 2003). Therefore, biomass burning is now recognized as a significant global source of emission contribution.

However, despite this recognition, uncertainties about emissions from non-energy sources (e.g. biomass burning, vegetation, soil, ocean, non-vehicle mobile sources) at the global level are considerable (IPCC, 2001). This makes the development of a complete emission inventory a vital contribution to the successful study of global atmospheric chemistry and climate change (Jain et al., 2006). At local, national and regional scales, these inventories are also critical as they contribute to the achievement of the ultimate objective of the United Nations Framework Convention on Climate Change (UNFCCC, 1992). In Ghana, an average of 68 ± 4 thousand km^2 of land is burnt annually; of this, 37 ± 2.6 thousand km^2 occur in the Northern region of Ghana. Approximately 53–56% of the total annual burnt land across Ghana occurs in the Northern region, which constitutes 29% of total dry land-cover of the country (Kugbe, 2012). This study aims at assessing the emission levels of CH_4 and CO from biomass burning in four different vegetation covers (based on the site level classification in Northern Ghana) and their implications for global warming.

2. Materials and Methods

2.1 Description and Location of the Study Area

This study was conducted in the Northern Region of Ghana (Figure 1), which has a population of about 2,468,557 (GSS, 2010). It lies within the Guinea Savanna Agro-Ecological Zone and forms part of the Volta Basin between the latitudes 8° 30'N and 10° 30'N and the longitudes 2° 30'W and 0° 00'W respectively. Characteristically, the area is characterized by two main seasons, wet and dry. The dry season starts in November and ends in April. The study site experiences a mono-modal rainfall pattern, which starts from May and ends in October. Within this period, rainfall peaks in August and September (Bagamsah, 2005). The mean monthly temperatures vary from about 36°C in March/April to 27°C in August. Relative humidity ranges between 20% and 85% (Cobbina et al., 2011). Approximately, 80% of the soils are upland soils developed in-situ from Voltaian sandstone and classified under the group of Lixisols (FAO, 1988). In general, the soils of the region have much lower organic matter content and nutrient status than those in the southern regions, thus the potential productivity of the soils of this zone may be regarded as being appreciably lower than that of the majority of the forest zone (Wills, 1962). The general vegetation of the study area is the mid-dry savanna type, with patches of dry woodland savanna and wet savanna (Menz and Bethke, 2000) that are classified as Guinea savanna (Lawson, 1985). Typically, the natural vegetation comprises a mix of tree (*Daniellia oliveri, Lophira spp, Terminalia glaucescens, Guiera senegalensis, Combretum glutinosum*, etc.) and grass (*Andropogon spp, Cymbopogon spp, Pennisetum spp*, and *Settaria spp, Aristida stipoides, Pennisetum spp*, and *Hyparrhenia spp)* species (Bagamsah, 2005). In terms of land use, northern Ghana is mostly utilized for agriculture and the typical crops cultivated include okra (*Abelemoschus esculentus*), groundnut (*Arachis hypogea*), tomato (*Lycopersicion esculantum)*, pepper (*Capsieum spp*), sweet potato (*Ipomoea batatas*), guinea corn (*Sorghum spp*), maize (*Zea Mays*), cowpeas (*Vigna spp*), cassava (*Manihot spp*) and yam (*Diascorea spp.*).

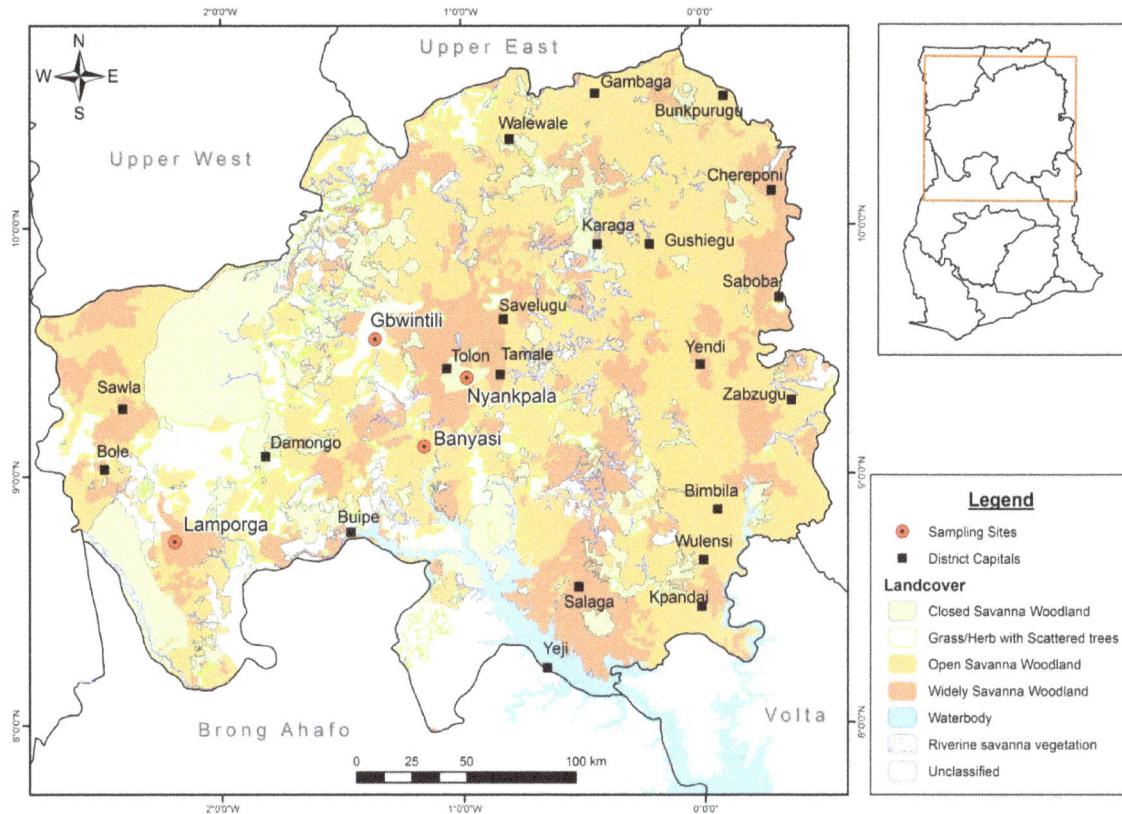

Figure 1. Map of the study area

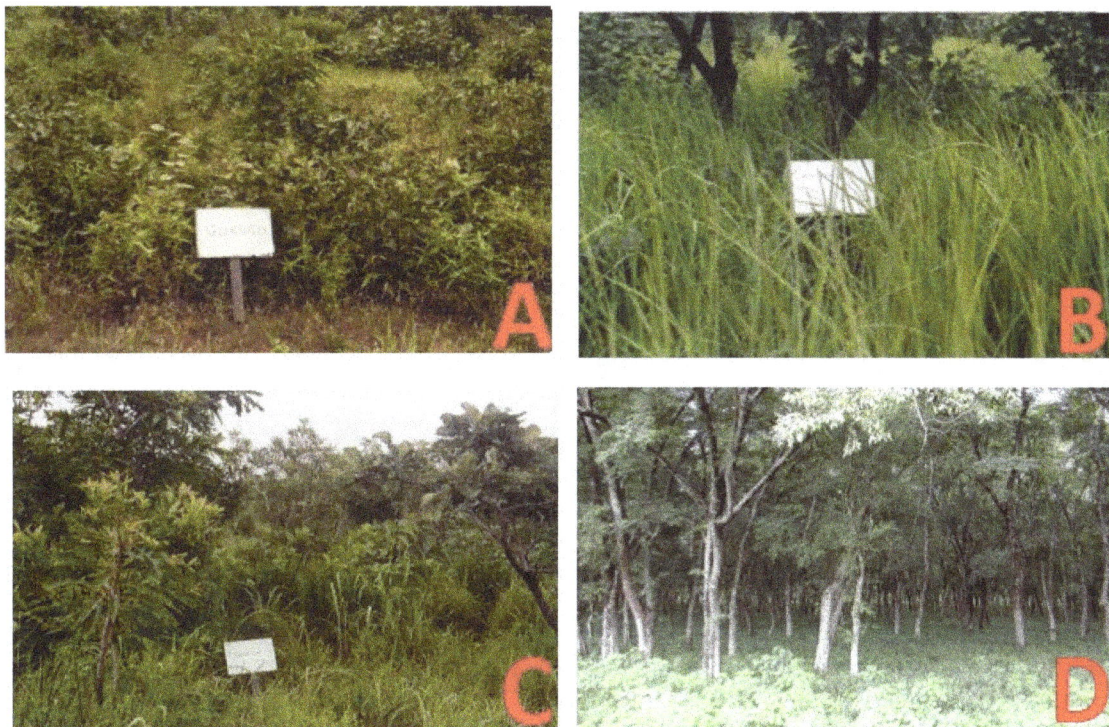

Figure 2. The four vegetation types as classified by Bagamsah (2005). (**A**) Widely Open Savanna Woodland. (**B**) Mixture of grass/herb fallow with scattered trees and shrubs. (**C**) Open savanna woodland vegetation dominated by shrubs (OSW). (**D**) Closed savanna woodland (CSW)

2.2 Site Selection and Experimental Layout

Sampling was carried out within four different vegetation types (Figure 1); namely, widely open savanna woodland (WOSW), grass/herb with scattered trees (GHST), open savanna woodland (OSW) and closed savanna woodland (CSW), as classified by Bagamsah (2005). This classification was based on Dansereau's (1951) methodology for describing and recording vegetation on a structural basis. Figure 2 indicates the different vegetation types within which sampling were done. Using a GPS, a community was selected from each of the locations covered by the different vegetation types and they were used as experimental sites. The communities selected were Gwintili (GHST), Nyankpala (CSW), Banyasi (OSW) and Lamporga (WOSW). From each of these locations, samples of above-ground biomass (AGB) were collected from five randomly selected plots, each covering an area of $1m^2$ (Figure 3).

2.3 Determination of Carbon Content of Biomass Using the Combustion Furnace Method

The combustion furnace method used in this study was adopted from Benscoter et al. (2011).The biomass was dried to a constant dry mass for four days at a temperature of 65°C and milled thoroughly in a Cyclone Mill. Each crucible was weighed (W_1), filled with 5g of the milled biomass and weighed again (W_2). For ashing, the milled biomass and the crucibles were placed in a combusted muffle furnace carbolite at a temperature of 550 °C for four hours and weighed again (W_3). Then the proportion (%) of organic carbon content was calculated using the following formulae:

$$\% \text{ Ash } = \frac{W_3 - W_1}{W_2 - W_1} \times 100$$

$$\% \text{ Organic matter} = 100 - \% \text{ Ash}$$

$$\% \text{ Organic carbon } = \frac{\% \text{ Organic matter}}{2}$$

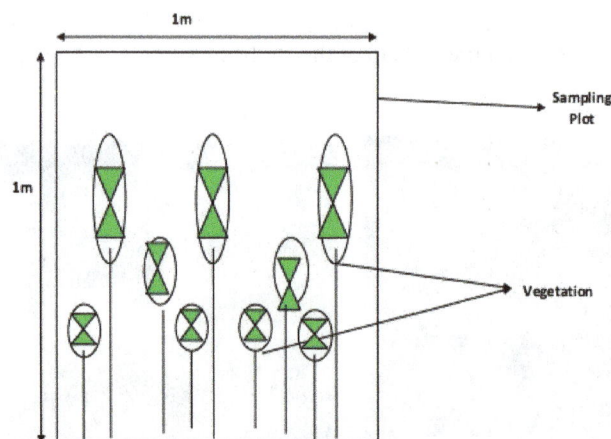

Figure 3. Schematic diagram showing experimental plot for sampling grass above-ground biomass

2.4 Estimating the Amount of Carbon Release, Methane and Carbon Monoxide Emitted by Biomass Burning

The carbon released from biomass burning in the study area was estimated using equations from the IPCC (1996) which involved the estimation of above-ground biomass density and carbon content in live and dead biomass. However, it must be noted that the estimated above-ground biomass density did not include standing trees; rather, the emphasis was on dried grass, litter, weeds and shrubs (Crutzen & Andreae, 1990). The default values used in the calculations were adopted from Hao et al. (1990) and Menaut et al. (1991). After quantifying the released carbon from the burnt savanna vegetation, emissions of CH_4 and CO were calculated and the emission ratios of CH_4 (0.004) and CO (0.06) from Lacaux et al. (1993) were adopted. The estimated emission values were subjected to statistical analysis at a confidence level of 95%. It is worthy of mention that the default values used in this study introduced some level of uncertainties into the estimation. However, the field and laboratory experiments carried out in the study reduced these uncertainties to an acceptable level. Additionally, the differences in biomass distribution across the study area made it difficult to accurately estimate the emitted carbon CH_4 and CO, so it is expected that there may be some degree of underestimation or overestimation of these values.

3. Result and Discussion

3.1 Grass Above-ground Biomass Density, Carbon Content and Total Carbon Emission

The estimated above-ground biomass density for the selected vegetation classes varied significantly at P<0.05. On the average, biomass density was highest on OSW with a density of ~4.8t/ha. This result is comparable to those obtained by other authors. For example, Saarnak et al., (2003) and Bagamsah (2005) reported a range of 3.36–7.80 t/ha and 2–3 t/ha respectively in northern Ghana; Shea et al. (1996) measured 3.7 t/ha in the savannas of South Africa, while Bourliére and Hardly (1983) reported 3.2–4.4 t / ha in the savannas of Cote d'Ivoire. Figure 4 shows the estimated above-ground biomass density for the different vegetation types.

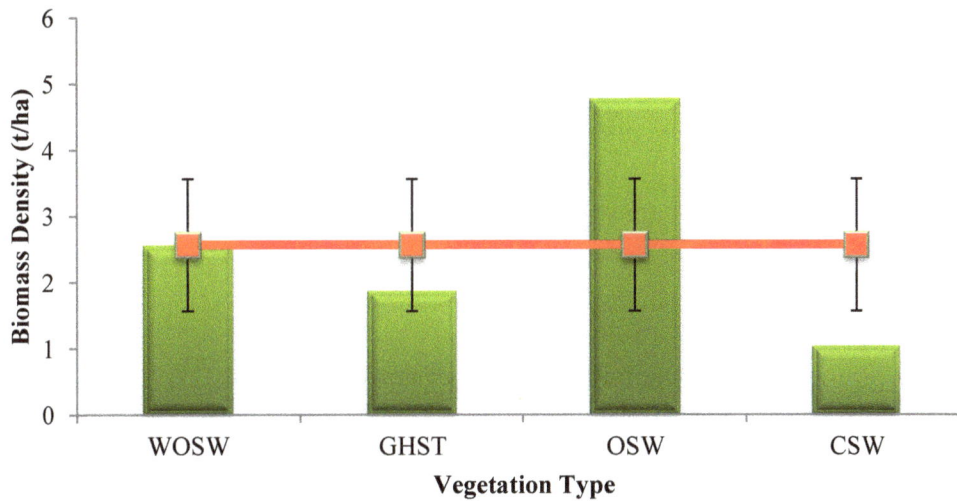

Figure 4. Estimated above-ground biomass density of grasses within the vegetation covers

Figure 5 shows that organic carbon content across the different vegetation types generally ranged from 29.3 % to 45.5%. On the average, GHST had the highest organic carbon content of 45.46%, while the least was CSW with 29.28% organic carbon content. The difference in the measured carbon content of the vegetation types was significant at P<0.05. Schlesinger (1991) noted that the carbon content of biomass is almost always found to be between 45% and 50%. With the exception of GHST, the carbon content of the other vegetation types fell below this range. Recent studies have shown that these assumptions introduce some degree of error of ~5% in forest carbon stock estimates (Martin & Thomas 2011; Thomas & Malczewski, 2007; Saner et al., 2012; Melson et al., 2011).

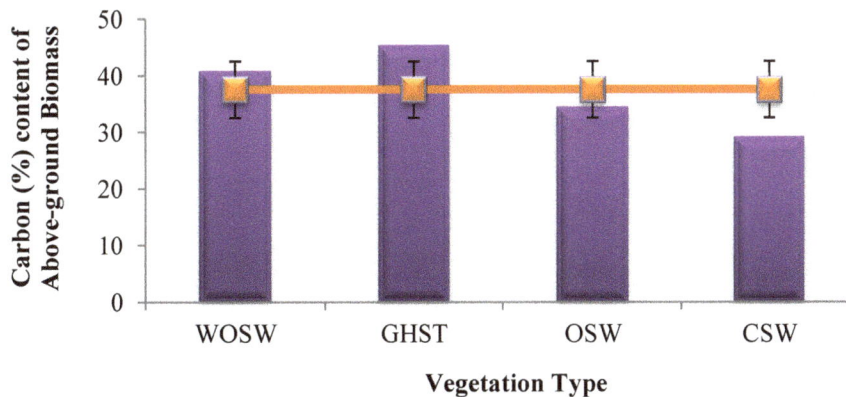

Figure 5. Estimated carbon content of grass above-ground biomass

The total carbon released per hectare for the different vegetation types ranged from ~0.000624 tC/ ha to ~0.398 tC/ha. Figure 6 shows that carbon emissions from OSW, WOSW, GHST and CSW were ~0.322 tC/ha, ~0.238 tC/ha, ~0.216 tC/ha and ~0.050 tC/ha respectively. The carbons released from the four types of vegetation during burning were significantly different at $p < 0.05$.

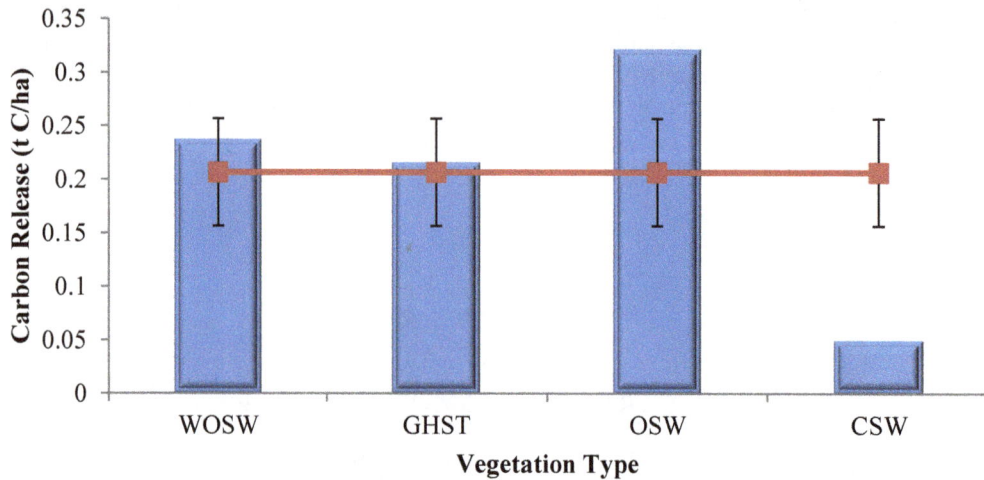

Figure 6. Total carbon emitted per hectare

Figures 5 and 6 clearly show that the carbon content of biomass varies across different vegetation types and this could be attributed to the prevailing conditions such as soil type and climate, as well as the degree of maturity of plants. It was also observed that vegetation types with high above-ground biomass density emitted high amounts of carbon into the atmosphere. This could be attributed to the fact that carbon emission is highly dependent on the fraction of dead and live biomass at the time of burning. Additionally, the class of vegetation with high biomass density contained a higher percentage of dead biomass per hectare than those with low biomass density. Also, Figure 6 shows that a higher level of carbon emission was recorded from one hectare of burnt OSW vegetation when compared with the other vegetation types.

3.2 Relationship Between Biomass Density and Total Carbon Released

According to Chuvieco et al. (2008), gaseous emission from the burning biomass of a given vegetation type is directly proportional to the area of vegetation burnt. This observation also applies to this study, given that biomass density and total carbon emitted have a significantly strong positive correlation of $r = 0.70$ and $p = 0.001$ (Figure 7). This implies that biomass density increases with an increase in total carbon and vice versa.

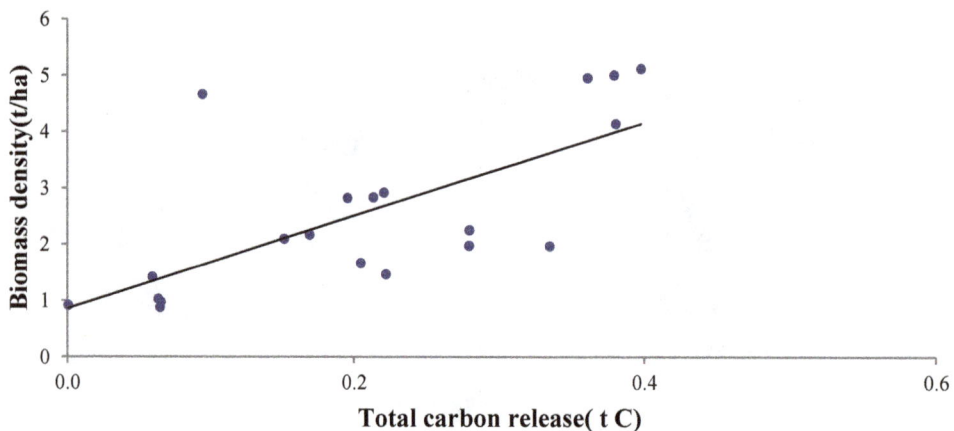

Figure 7. Correlation between above-ground biomass density and total carbon release

3.3 Trace Gas Emission From Biomas Burning

Table 1 shows that, on the average, more CO is emitted compared with CH_4 during savanna wild fires. This is consistent with the results of a similar study by Saarnak et al. (2003) and Bagamsah (2005). This difference in the emission of CH_4 and CO is a result of the difference in the emission ratios applied to CH_4 and CO, which were significantly ($P<0.05$) different across the four vegetation types, with OSW recording the highest emissions per hectare of burnt land, amounting to 0.001719 t/ha and 0.045119 t/ha respectively, while CSW recorded the least emission of 0.000268 t/ha and 0.007043 t/ha of CH_4 and CO respectively. The emissions from CH_4 and CO differ significantly ($p<0.05$) for all vegetation cover types and were perfectly correlated ($R=1.00$) with the quantity of above-ground biomass density and carbon emitted.

The results obtained on estimated emission values (Table 1) from the selected types of vegetation cover in northern Ghana were lower than the results obtained by Saarnak et al. (2003) and Bagamsah (2005). The difference could be linked to variation in above-ground biomass of the different test plots used in the studies, and this may be attributed to a reduction in biomass density within the study area in the previous two and half decades. There is a possibility that, over time, the accumulation of these emissions could increase their atmospheric concentration. In the short term, this could reduce visibility and cause health problems such as an increase in respiratory ailments and human mortality (Levine, 1991; Beer & Meyer 2000; Johnston et al., 2002; Bagamsah, 2005; Shindell et al. 2009). In the long term, it could contribute to global warming and climate change, since CO and CH_4 have a higher global warming potential compared with CO_2 (Rufus & Kirk, 2002; IPCC, 2007).

Table 1. CH_4 and CO gases emitted per hectare of burnt area *(Mean values with Variance in Parenthesis)*

Vegetation Cover	Emitted CH_4 (tons/ha)	Emitted CO (tons/ha)
WOSW	0.001154(1.47E-07)	0.030303(0.000101)
GHST	0.001269(4.39E-08)	0.033304(3.02E-05)
OSW	0.001719(4.7E-07)	0.045119(0.000324)
CSW	0.000268(2.21E-08)	0.007043(1.52E-05)

3.4 Implications of CH_4 and CO Emissions for Global Warming

Biomass burning is the third largest source of CH_4, contributing about 40 Tg/year to the global methane emission rate (IPCC, 2001) and it is increasing at a rate of about 30 to 40 million tonnes per year (Duxbury & Mosier 1993). Additionally, global CO emissions from biomass burning estimated using remote sensing techniques were 511 Tg, 565 Tg and 429 Tg for 1997, 1998 and 1999 respectively (Duncan et al., 2003). Research has proven that each of these gases has a different effect on the atmosphere (MacCarty et al., 2007) because they have different absorptive properties and atmospheric lifespan. The global warming potential for CH_4 and CO is 21 and 3 times respectively the global warming potential of CO_2 (Rufus et al., 2002; IPCC, 2007), implying that the emission of 1 unit mass of CH_4 or CO has a climate impact equivalent to that of the emission of 21 and 3 units of mass of CO_2 over the 100-year period following these emissions. It is projected that by the year 2030 the world is likely to be 1–2 °C warmer than today. Given the full range of uncertainties, the range could be from 0.5 °C to 2.5 °C (Moss et al., 2000). A number of studies have emphasized the interactive nature of CH4 and reactive GHGs, and the effects these interactions can have on overall climate change (Fuglestvedt et al., 1996; Daniel & Solomon, 1998; Kheshgi & Jain, 1999; Kheshgi et al., 1999; Hayhoe et al., 2000).

Against this background, it can be safely assumed that the increasing atmospheric concentration of CH_4 and CO from sources such as the study area may be a contributing factor to the continued rise in the average global temperature. Therefore, the need for concerted and collective efforts aimed at reducing these sources to the barest minimum through the implementation of concrete measures backed by appropriate policies cannot be overemphasized. It involves the bottom up participation of major stakeholders from community level to policy and decision makers in the government, which can be achieved through massive awareness creation and education. This entails, first and foremost, the design and implementation of relevant national policies. Ghana, in 2006, developed a practical and comprehensive National Wildfire Policy which is expected to encourage the practice of agro-forestry, natural regrowth and reforestation. Also, it emphasized the development of capacities of all stakeholders in bushfire management and the establishment of a district, regional or national bushfire management fund or insurance scheme to ensure effective and sustainable participation of all stakeholders in the bushfire management activities. Furthermore, it provides research institutions and other relevant bodies with the required resources to develop appropriate adaptive measures to bushfire management and to support commercial charcoal producers with improved and sustainable methods of wood carbonization (NWP, 2006). These policy strategies are similar to those in other developing countries like Nigeria, Zambia and The Gambia, and have not been fully exploited. Consequently, the intended target has not been met due to implementation flaws.

In contrast, developed countries like the USA, Canada and Australia have made headway in the implementation of their wild fire policies. For example, in the Victoria state of Australia, the many devastating bushfire scenarios led to the introduction of a Bushfire Safety Policy Framework. It consists of strategies that will regulate bushfire management activities. The recommended approaches adopted are effective awareness and education activities through broad-based media campaigns, government and fire agency websites and publications, locally delivered community education programmes, events and activities, a 'bottom up' community capacity building method aimed at skill and network building to enhance communities' ability to plan for and execute their own bushfire safety options, integration of local community fire plans with other municipal fire and emergency management plans, and the use of a national fire danger rating system that informs communities about different levels of forecast fire danger (Lapsley, 2012). Other initiatives by the Australian Government include the passage of the 1999 Environment Protection and Biodiversity Conservation Act, which provide guidelines to state and territory governments, local councils and other authorities such as fire and emergency services, and individuals on the regulations covering bushfire management activities.

Unlike the developing countries, the developed ones are able to make progress in their bushfire policy implementation as a result of the strict legislation and implementation procedures they adhere to. Their policies are mostly dynamic documents that are reactive or receptive to the experiences of stakeholders as they change with time. In most cases, these policies are usually subjected to reviews which are carried out annually. This best practice is advocated in this paper for implementation in developing countries like Ghana, as a national obligation to reduce the country's greenhouse gas footprint, hence global warming.

4. Conclusions

This study concludes that the burning of different vegetation types contributes in different ways to the release of carbon and emission of CH_4 and CO gases. The results of this study showed that the emissions of CH_4 and CO differed significantly ($p<0.05$) for all vegetation types and were perfectly correlated ($r=1.00$) with the quantity of above-ground biomass density and carbon released. More CO was emitted than CH_4 during savanna wild fires across all the vegetation cover, and the main reason for such discrepancies was the emission ratio used. OSW recorded the highest emission per hectare with 0.001719 t/ha and 0.045119 t/ha for CH_4 and CO respectively, while CSW recorded the least emission of 0.000268 t/ha and 0.007043 t/ha of CH_4 and CO respectively. The estimated above-ground biomass density for the selected vegetation classes ranged from ~1.0t/ha to ~4.8t/ha but was highest on OSW. The organic carbon content of the vegetation types also ranged from 29.3% in CSW to 45.5% in GHST. The total carbon released, of which CH_4 and CO were estimated, ranged from ~0.000624 t C/ha in OSW to ~0.398t C/ha in CSW. The findings confirmed the results of similar studies conducted within the savanna regions of Africa. By implication, there is the need to project beyond the immediate environment the consequences of such sources of potent greenhouse gas emissions, in order to ensure that countries like Ghana continue to maintain very minimal contributions to global warming. In order to achieve this objective, the existing flaws in the wild fire management policy of Ghana must be effectively dealt with and appropriately implemented with regular reviews to reduce the annual wild fires, which are rampant in Northern Ghana.

Acknowledgements

This work was fully funded by the German Federal Ministry of Education and Research (BMBF) through the West Africa Science Center of Climate Change and Adapted Land Use (WASCAL) Master's Research Programme in the Federal University of Technology, Minna, Niger state, Nigeria. We appreciate the technical contributions made by William Attakora (Savanna Agriculture Research Institute, Nyankpala-Tamale), Emmanuel Boakye (Climate Change & Biodiversity – Universite Felix Houphouet Boigny Ufr Biosciences, Abidjan – Cote d'Ivoire) and Hardi Mohammed (Nyankpala).

References

Andreae, M. O. (1990). Biomass burning: Its history, use, and distribution and its impact on environmental quality and global climate. In J. Levine (Ed.), *Global biomass burning: Atmospheric, climatic, and biospheric implications* (pp. 3–21). Cambridge: MIT Press.

Andreae, M.O. (1997). Emissions of trace gases and aerosols from savanna fires. In B. W. van Wilgen, M. O. Andreae, J. G. Goldammer, & J. A. Lindesay, (Eds.), *Fire in the Southern African Savanna: Ecological and Environmental Perspectives* (pp. 161–183). Johannesburg: Witwaters and University Press. Retract from http://www.researchgate.net/publication/44158698_Emissions_of_trace_gases_and_aerosols_from_southern_African_savanna_fires

Bagamsah, T. T. (2005). The impact of bushfire on carbon and nutrient stocks as well as albedo in the savanna of northern Ghana. *Ecology and Development Series,* 25. Retrieved from http://books.google.com.gh/books/about/The_Impact_of_Bushfire_on_Carbon_and_Nut.html?id=Tow-_cMnn48C&redir_esc=y

Bagamsah, T. T. (2005). *The impact of bushfire on carbon and nutrient stocks as well as albedo in the savanna of northern Ghana. Ecology and Development Series* (No. 25), 2005. http://books.google.com.gh/books/about/The_Impact_of_Bushfire_on_Carbon_and_Nut.html?id=Tow-_cMnn48C&redir_esc=y

Beer, T., & Meyer, C. P. (2000). The impact on the environment – The atmosphere. In *Fire! The Australian experience: Proceedings from the National Academies Forum Seminar* (pp. 59–77). National Academies Forum, Canberra, ACT, Australia.

Benscoter, B. W., Thompson, D. F., Waddington, J. M., Flannigan, M. D., Wotton, M., DeGroot, W., & Turetsky, M. R. (2011). Interactive effects of vegetation, soil moisture, and bulk density on the depth of burning of thick organic soils. *International Journal of Wild Land Fire, 20,* 418–429. http://dx.doi.org/10.1071/WF08183

Bolin, B., Degens, E. T., Duvigneaud, P., & Kempe, S. (Eds.) (1979). The global carbon cycle. *SCOPE Report, 13,* 1–56. Wiley, Chichester, England.

Bourliére, F., & Hadley, M. (1983). Present-day savannas: An overview. In F. Bourliére (Ed), *Ecosystems of the world* (vol. 13, pp. 1–17). *Tropical Savannas.* Amsterdam: Elsevier.

Bucini, G., & Lambin, E. F. (2002). Fire impacts on vegetation in Central Africa: a remote-sensing-based statistical analysis. *Applied Geography, 22*(1), 27-48. http://dx.doi.org/10.1016/S0143-6228(01)00020-0

Cahoon, D. R., Stocks, B. J., Levine, J. S., Coffer III, W. R., & O' Neil, K. P. (1992). Seasonal distribution of African savanna fires. *Nature, 359,* 812–815. http://dx.doi.org/10.1038/359812a0

Chuvieco, E., Giglio, L., & Justice, C. (2008). Global characterization of fire activity: toward defining fire regimes from earth observation data. *Glob. Chang. Biol., 14,* 1488–1502. Retrieved from http://onlinelibrary.wiley.com/doi/10.1111/j.1365-2486.2008.01585.x/pdf

Cobbina, S. J., Armah, F. A., & Obiri, S. (2011). Multivariate Statistical and Spatial Assessment of Groundwater Quality in the Tolon-Kumbungu District, Ghana. *Research Journal of Environmental and Earth Sciences, 4*(1), 88–98. Retrieved from www.sciencepub.net/newyork/ny0611/006_21356ny0611_38_48.pdf

Crutzen, P. J., & Andreae, M. O. (1990). Biomass Burning in the Tropics: Impact on Atmospheric Chemistry and Biogeochemical Cycles. *Science,* 250(4988), 1669–1678. http://dx.doi.org/10.1126/science.250.4988.1669

Daniel, J. S., & Solomon, S. (1998). On the climate forcing of carbon monoxide. *Journal of Geophysical Research: Atmospheres (1984–2012), 103*(D11), 13249-13260. http://onlinelibrary.wiley.com/doi/10.1029/98JD00822/abstract

Dansereau, P. (1951). Description and recording of vegetation upon a structural basis. *Ecology, 32,* 172–229. http://dx.doi.org/10.2307/1930415

Danthu, P., Ndongo, M., Diaou, M., Thiam, O., Sarr, A., Dedhiou, B., & Vall, A. O. M. (2003). Impact of bush fire on germination of some West African Acacias. *Forest Ecology and Management, 173,* 1–10. http://dx.doi.org/10.1016/S0378-1127(01)00822-2

Duncan, B. N., Martin, R. V., Staudt, A. C., Yevich, R., & Logan, J. A. (2003). Inter-annual and 25 seasonal variability of biomass burning emissions constrained by satellite observations. *J. Geophys. Res., 108*(D12), 4100. http://dx.doi.org/10.1029/2002JD002378

Duxbury J. M., & Mosier, A. R. (1993). Status and issues concerning agricultural emissions of greenhouse gases. In: Kaiser H.M., Drennen, T.W. (Eds), *Agricultural Dimensions of Global Climate Change* (pp. 229–258). Delray Beach, FL: St. Lucie Press. http://onlinelibrary.wiley.com/doi/10.1002/sd.3460020305/abstract

FAO. (1988). Soil Map of the World. Revised Legend. *World Soil Resources Report* 60. FAO-UNESCO, Rome, 119 pp.

Fuglestvedt, J. S., Isaksen, I. S. A., & Wang, W. C. (1996). Estimates of indirect global warming potentials from CH_4, CO, and NO_x. *Climate Change, 34*, 405–437. http://dx.doi.org/10.1007/BF00139300

GSS (Ghana Statistical Service). (2010). *Ghana Population and Housing Census, 2010.* http://www.statsghana.gov.gh/

Hao, W. M., & Liu, M. H. (1994). Spatial and temporal distribution of tropical biomass burning. *Global Biogeochem. Cy., 8*, 495–503. http://onlinelibrary.wiley.com/doi/10.1029/94GB02086/abstract

Hao, W. M., Liu, M., & Crutzen, P. J. (1990). Estimates of annual and regional releases of CO_2 and other trace gases to the atmosphere from fires in the tropics, based on the FAO statistics for the period 1975–1980. In: Goldammer, J.G. (Ed), *Fire in the tropical biota: ecosystem processes and global challenges* (pp. 440–462). Springer- Verlag, Berlin. http://link.springer.com/book/10.1007%2F978-3-642-75395-4

Hayhoe, K., Jain, A., Kheshgi, H. S., & Wuebbles, D. (2000). Contribution of CH_4 to multi-gas reduction targets: The impact of atmospheric chemistry on GWPs. In J. van Ham (Ed.), *Non-CO₂ Greenhouse Gases: Scientific Understanding, Control and Implementation* (pp. 425–432). New York: Springer. http://dx.doi.org/10.1007/978-94-015-9343-4_67

Intergovernmental Panel on Climate Change (IPCC). (2001). Climate Change 2001, Working Group I: The Scientific Basis. In J. T. Houghton, Y. Ding, & D. J. Griggs (Eds.). Cambridge: Cambridge University Press.

IPCC. (1996). *Greenhouse Gas Inventory. Reference Manual, Workbook. Revised 1996 IPCC Guidelines for National Greenhouse Gas Inventories.*

Jain, A. K., Tao, Z., Yang, X., & Gillespie, C. (2006). Estimates of global biomass burning emissions for reactive greenhouse gases (CO, NMHCs, and NO_x) and CO_2. *J. Geophys. Res., 111.* http://dx.doi.org/10.1029/2005JD006237

Johnston, F. H., Kavanagh, A. M., Bowman, D. M. J. S., & Scott, R. K. (2002). Exposure to bushfire smoke and asthma: an ecological study. *Medical Journal of Australia, 176*, 535–538. Retrieved from http://www.climate.atmos.uiuc.edu/atuljain/publications/2005JD006237.pdf

Jones, R. (1979). *Annu. Rev. Anthropol., 8*,445. http://dx.doi.org/10.1146/annurev.an.08.100179.002305

Kheshgi, H. S., Jain, A. K. (1999). Reduction of the atmospheric concentration of methane as a strategic response option to global climate change. In P. Reimer, B. Eliasson, & A. Wakaun (Eds.), *Greenhouse Gas Control Technologies* (pp. 775–780). New York: Elsevier.

Kheshgi, H. S., Jain, A. K., Kotamarthi, V. R., & Wuebbles, D. J. (1999). Future atmospheric methane concentrations in the context of the stabilization of greenhouse gas concentrations. *Journal of Geophysical Research: Atmospheres (1984–2012), 104*(D16), 19183-19190. Retrieved from http://www.climate.atmos.uiuc.edu/atuljain/publications/1999JD900367.pdf

Korem, A. (1985). Bushfire and Agriculture Development in Ghana. Tema, Ghana Publishing Cooperation, 220.

Korontzi, S., Justice, C. O., & Scholes, R. J. (2003). Influence of timing and spatial extent of savanna fires in southern Africa on atmospheric emissions. *J Arid Environ, 54*, 395–404. http://dx.doi.org/10.1006/jare.2002. 1098

Kugbe, J. (2012). Spatio-temporal dynamics of bush-fire nutrient losses and atmospheric depositional gains across the northern savanna region of Ghana. PhD Dissertation, Faculty of Agriculture, University of Bonn. Issue 90 of *Ecology and Development Series*. Retrieved from http://www.hss.ulb.uni-bonn.de/2012/3006/3006.pdf

Lacaux, J. P., Cachier, H., & Delmas, R. (1993). Biomass burning in Africa: An overview of its impact on atmospheric chemistry. In *Fire in the Environment*.

Langaas, S. (1995). *Night-Time Observations of West Africa Bushfires from Space*. Studies on Methods and Application of Thermal NOAA/AVHRR Satellite Data from Senegal and the Gambia. Thesis / Dissertation, Department of Geography, University of Oslo, Norway.

Lapsley, C. (2012). Bushfire Safety Policy Framework. Fire Services Commission, Victoria State - Australia. Retrieved from http://fire-com-live-wp.s3.amazonaws.com/wp-content/uploads/2013-Bushfire-Safety-Policy-Framework.pdf

Laris, P. (2002). Burning the seasonal mosaic: preventive burning strategies in the wooded savanna of southern Mali. *Human Ecology, 30,* 155–186. http://dx.doi.org/10.1023/A:1015685529180

Lawson, G. W. (1985). *Plant Life in West Africa.* Accra: Ghana Universities Press.

Levine, J. S. (1991). *Global Biomass Burning: atmospheric, climatic, and biospheric implications.* Massachusetts: MIT Press. Retrieved from http://onlinelibrary.wiley.com/doi/10.1029/90EO00289/pdf

MacCarty, N., Ogle, D., Still, D., Bond, T., Roden, C., & Willson, B. (2007). Laboratory comparison of the global-warming potential of six categories of biomass cooking stoves. Aprovecho Research Center, *Advanced study in appropriate technology laboratory.* Retrieved from http://scscertified.com/lcs/docs/Global_warming_full_9-6-07.pdf

Martin, A. R., & Thomas, S. (2011). A reassessment of carbon content in tropical trees, *PLoS One, 6,* e23533:1–e23533:9. http://journals.plos.org/plosone/article?id=10.1371/journal.pone.0023533.

Mbow, C., Nielsen, T. T., & Rasmussen, K. (2000). Savanna fires in east-central Senegal: Distribution patterns, resource management and perceptions. *Human Ecology, 28*(4), 561-583. http://dx.doi.org/10.1023/A:102648 7730947

Mbow, C., Nielsen, T. T., & Rasmussen, K. (2000). Savannah fires in east-central Senegal: distribution patterns, resource management and perceptions. *Human Ecology, 28*(4), 561–583. http://dx.doi.org/10.1023/A:102 6487730947

Melson, S. L., Harmon, M. E., Fried, J. S., & Domingo, J. B. (2011). Estimates of live-tree carbon stores in the Pacific Northwest is sensitive to model selection. *Carbon Balance Manag., 6,* 1–16. http://dx.doi.org/10.1186/1750-0680-6-2

Menaut, J. C., Abbadie, L., Lavenu, F., Loudjani, P., & Podaire, A. (1991). Biomass burning in West African savannas. In S. J. Levine (Ed.), *Global Biomass Burning: Atmospheric, Climatic, and Biospheric Implications* (pp. 133-142). Cambridge: MIT Press.

Menz, M., & Bethke, M. (2000). Vegetation map of Ghana. Regionalization of the IGBP Global Land Cover Map for Western Africa (Ghana, Togo and Benin). In *Proceedings of the 20th EARSeL-Symposium.* June 2000, Dresden, in press. Remote Sensing Research Group. Institute of Geography. University of Bonn, Germany.

Middelmann, M. H. (Ed.) (2007). *Natural Hazards in Australia. Identifying Risk Analysis Requirements.* Geoscience Australia, Canberra.

Moss, A. R., Jouany, J., & Newbold, J. (2000). Methane Production By Ruminants: Its Contribution To Global Warming. *Ann. Zootech., 49,* 231–253. INRA, EDP Sciences. Retrieved from http://animres.edpsciences.org/articles/animres/abs/2000/03/z0305/z0305.html

NWP. (2006). *National wildfire policy, Ghana.* Retrieved from http://www.rspo-in-ghana.org/sitescene/custom/userfiles/file/national_wildlife_management_policy.pdf_on

Reid, R. S., Kruska, R. L., Muthui, N., Taye, A., Wotton, S., Wilson, C. J., & Mulatu, W. (2000). Land-use and Land-cover dynamics in response to changes in climatic, biological and socio-political forces: the case of South western Ethiopia. *Landscape Ecology, 15,* 339–355. http://dx.doi.org/10.1023/A:1008177712995

Rudolph, J., Khedim, A., Koppmann, R., & Bonsang, B. (1995). Field study of the emissions of methyl chloride and other halocarbons from biomass burning in western Africa, *J. Atmos. Chem., 22,* 67–80. http://dx.doi.org/10.1007/BF00708182

Rufus, D. E., & Kirk, R. S. (2002). *Carbon Balances, Global Warming Commitments, and Health Implications of Avoidable Emissions from Residential Energy Use in China: Evidence from an Emissions Database.* http://www.giss.nasa.gov/meetings/pollution2002/d3_edwards.html

Saarnak, C. F., Nielsen, T. T., & Mbow, C. (2003). Local study on the Trace Gas Emissions from Vegetation Burning Around the Village of Dalun-Ghana with respect to Seasonal Vegetation Changes and Burning Practices. *Climatic Change, 56,* 321–338. http://dx.doi.org/10.1023/A:1021788509191

Saner, P., Loh, Y. Y., Ong, R. C., & Hector, A. (2012). Carbon stocks and fluxes in tropical lowland Dipterocarp rainforests in Sabah, Malaysian Borneo. *PLoS ONE, 7,* e29642:1–e29642:11. http://journals.plos.org/plosone/article?id=10.1371/journal.pone.0029642

Schils, R., Kuikman, P., Liski, J., van Oijen, M., Smith, P., Webb, J., ... Hiederer, R. (2008). Review of existing information on the interrelations between soil and climate change. *CLIMSOIL*. European Communities. http://ec.europa.eu/environment/soil/review_en.htm

Schlesinger, W. H. (1991). *Biogeochemistry, an Analysis of Global Change*. New York: Academic Press. http://www.sciencedirect.com/science/book/9780126251579

Shea, R. W., Shea, B. W., Kauffman, J. B., Ward, D. E., Haskins, C. I., & Scholes, M. C. (1996). Fuel biomass and combustion factors associated with fires in savanna ecosystems of South Africa and Zambia. *Journal of Geophysical Research: Atmospheres (1984–2012)*, *101*(D19), 23551-23568. http://onlinelibrary.wiley.com/doi/10.1029/95JD02047/abstract

Shimada, S., Takahashi, H., Kaneko, M., & Haraguchi, A. (1999, November). The estimation of carbon resource in a tropical peatland: a case study in Central Kalimantan, Indonesia. In *Proceedings of the International Symposium on Tropical Peatlands, Bogor, Indonesia* (pp. 9-18).

Shindell, D. T., Faluvegi, G., Koch, D. M., Schmidt, G. A., Unger, N., & Bauer, S. E. (2009). Improved attribution of climate forcing to emissions. *Science, 326*, 716–718. http://dx.doi.org/10.1126/science.1174760

Stott, P. (1994). Savanna landscapes and global environmental change. In N. Roberts (Ed), *the Changing Global Environment* (pp. 287–303). Cambridge: Blackwell.

Thomas, S. C., & Malczewski, G., (2007). Wood carbon content of tree species in eastern China: Interspecific variability and the importance of the volatile fraction. *J. Environmental Management, 85*, 659–662. http://dx.doi.org/10.1016/j.jenvman.2006.04.022

Trollope, W. S. W., & Trollope, L. A. (1996). Fire in African savanna and other grazing ecosystems. Paper presented at the seminar on '*Forest fire and Global Change*', held in Shushenkoye in the Russian Federation from 4th–10th August 1996.

UNFCCC. (1992). *United Nations Framework Convention on Climate Change*. Climate Change Secretariat, Geneva, June.

van der Werf, G. R., Randerson, J. T., Giglio, L., Collatz, G. J., Mu, M., Kasibhatla, P. S., ... & van Leeuwen, T. T. (2010). Global fire emissions and the contribution of deforestation, savanna, forest, agricultural, and peat fires (1997–2009). *Atmospheric Chemistry and Physics, 10*(23), 11707-11735. http://dx.doi.org/10.5194/acp-10-11707-2010

Wills, J. B. (1962). *Agriculture and land use in Ghana, Ghana Ministry of Food and Agriculture*. London: Oxford University Press. http://www.mofa.gov.gh/site/wp.../10/AGRICULTURE-IN-GHANA-FF-2010.pdf

Wuver A. M., Attuquayefio, D. K., & Enu-Kwesi, L. (2003). A study of bushfires in Ghanaian coastal wetland. II. Impact on floral diversity and soil seed bank. *West African Journal of Applied Ecology, 4*. 2003. Retrieved from www.ajol.info/index.php/wajae/article/viewFile/45583/29066

Zhang, X., & Kondragunta, S. (2008). Temporal and spatial variability in biomass burned areas across the USA derived from the GOES fire product. *Remote Sensing of Environment, 112*, 2886–2897. Elsevier Inc. Retrieved from http://www.goes-r.gov/resources/.../1-s2.0-S0034425708000576-main.pdf

16

When Rhetoric Meets Reality: Attitudinal Change and Coastal Zone Management in Ghana

Elaine T. Lawson[1]

[1] Institute for Environment and Sanitation Studies (IESS), University of Ghana, Legon, Ghana

Correspondence: Elaine T. Lawson, Institute for Environment and Sanitation Studies (IESS), University of Ghana, Legon, Ghana. E-mail: elaine_t@ug.edu.gh

Abstract

The current poor state of coastal natural resources in Ghana has been attributed to pressures largely from anthropogenic sources, as well as to the negative attitudes of resource users. In order to facilitate attitudinal change educational programmes have focused on the linear model of behaviour, where an awareness of environmental problems is thought to lead to positive environmental behaviour. This paper presents the results of a study of the environmental attitudes of some coastal residents and the socio-economic milieu in which these attitudes are expressed. The results indicated that (1) majority of the respondents lacked access to basic infrastructure, (2) their main environmental concerns were linked to their desire for better living conditions, (3) they have generally positive environmental attitudes and (4) their positive environmental attitudes did not translate to good environmental behaviour because of factors mentioned in (1) and (2). The paper recommends the consideration of environmental and socio-economic concerns of resource users, which influence behavioural intentions during the policy-making processes.

Keywords: attitudinal change, coastal zone, environmental attitudes, environmental behaviour, environmental concerns, Ghana, intentions, natural resources

1. Introduction

For thousands of years humans have had a close relationship with the sea, with the vast resources of the ocean are contributing to the survival of coastal communities (English, 2003). The high concentration of human populations along coasts makes coastal ecosystems some of the most impacted and altered worldwide (Adger et al., 2005). For example it is estimated that about 40% of the human population is compressed into 5% of the inhabited land-space along the margins of ocean, seas and great lakes (Olsen, 2009). In addition, a United Nations Environment Programme (UNEP) report in 2009 confirmed that key habitats supporting coastal ecosystems, such as mangroves, are declining in area (UNEP, 2012).

The current poor state of Ghana's coastal zone is as result of extensive pressures largely from anthropogenic sources, resulting in environmental and socio-economic impacts on the functioning of the coastal ecosystems. Drivers of degradation include population increase, poverty, over-exploitation of fisheries resources, farming, industrial and extractive activities. For example the decrease in mangroves and coastal shrubs could be attributed to economic and agricultural activities in the coastal zone (Coleman et al., 2005) (Tables 1 and 2).

Table 1. Evidence of coastal degradation in some coastal habitats in Ghana

Classes	Area (ha) in 1990	Area (ha) in 2000	% change (1990 to 2000)
Water	116,188	117,829	1.41
Mangrove	2,605	1,905	-26.87
Settlement	1,117	1,792	37.67
Coastal shrubs	27,884	21,181	-24.04
Agriculture	15,363	20,450	24.88

Source: Coleman et al. (2005).

Table 2. Mangrove area estimates in Ghana

Year	1980	1990	1997	2000	2005	2006
Area (km²)	181	168	214	138	124	137

Source: UNEP, 2007.

Another pressing consideration for Ghana is the impact of global environmental change on its coasts and coastal resources. Projected increase of flooding in low and coastal areas might further impact marine ecosystems and coastal livelihoods (EPA, 2011). Damage to the coastal zone in the form of flooding, land loss, and forced migration is estimated to reach €4 million per annum by the 2020s, rising to €4.75 million per annum by the 2030s (World Bank, 2010). The severity of marine and coastal ecosystem degradation also affects the access of coastal communities to goods and services necessary for life.

Community participation has become an important component of the management of coastal natural resources. In most cases those most affected by resource loss and depletion are the poor people because they directly rely on these resources, with limited alternatives available to them (Nkemnyi et al., 2011; Lawson et al., 2012). Yet there are increasing media reports attributing the current state of degradation in Ghana's coastal zone largely to the negative attitudes of coastal resource users. Attitudes are thought to reflect how natural resources are perceived. According to Schultz et al. (2004), environmental attitudes can be defined as a collection of beliefs, affect and behavioural intentions a person holds regarding environmentally related activities and issues. One of the difficulties of researching into environmental attitudes is identifying the relationship between attitudes and actual behaviour. Yet attitudes are thought to be predictors of behaviour. They also help to estimate support for planned and implemented natural resource management strategies (Sesabo, 2006). Attitudes have also been assessed to help determine future management options (Lepp, 2007; Lawson et al., 2010). In order to facilitate an attitudinal change in the coastal zone some implementing agencies have concentrated on designing educational and awareness programmes, thus focusing on a linear model of behaviour. In this model awareness of environmental problems and knowledge of how to tackle them is thought to lead to positive environmental behaviour. Such a behaviour change is to also lead increased engagement on the part of local communities (Nelson & Agrawal, 2008; Lawson & Bentil, 2014). The traditional thinking especially in the field of environmental education has been that knowledge about the environment and its associated issues lead to an increase in awareness and an environmentally responsible behaviour. Attitudes of local resource users can also help shape different forms of environmental management and policies positively (d'Aquino & Bah, 2013). Although still widely used, most current research on environmental behaviour have criticised this linear model as being simplistic. For example Hines et al. (1987) in discussing his model of responsible environmental behaviour linked a person's desire to act to a host of personality factors, determined by locus of control, attitudes (toward the environment and toward taking action), and personal responsibility (toward the environment). Situation factors, such as economic constraints, social pressures and opportunities to choose different actions could either counteract or to strengthen the variables in the model. The Theory of Reasoned Action (Fishbein & Ajzen, 1975) and the Theory of Planned Behaviour (Ajzen & Driver, 1992) both introduced the crucial intention-behaviour relationship (Barr et al., 2007). Hence many of the factors that predict behaviour do so indirectly by first influencing intentions. Other researchers such Kaiser et al. (2005) and Levine and Strube (2012) have made similar findings.

Using 304 women from two coastal communities this paper examines the linkages between environmental attitudes and behaviour in two coastal communities in Ghana. It answers the questions:

1) What is the socio-economic milieu in which environmental decisions are made?

2) What are the main environmental concerns?

3) What are the current trend of environmental attitudes and behaviour in the coastal zone of Ghana?

4) What key elements are required in the translation of attitudes into behaviour?

The significance of this study lies in the fact that it attempts to bridge the gap between research and policy recommending the completion of policy aimed at managing coastal natural resources in Ghana effectively. From the review of studies on coastal resource management in Ghana, it was noted that assessment of community attitudes was not adequately addressed, yet it is one of the key components in effective coastal zone management. Current research has also shown that the factors that may influence attitudes and behaviour towards natural resources are many. They include demographic and socio-economic factors (Pomeroy & Carlos, 1996; Wright & Shindler, 2001) such as gender (Hill, 1998; Mehta & Heinen 2001), age, educational and income levels

(Fiallo & Jacobson, 1995). Many studies have shown that people often hold positive attitudes towards their environment (Ramos et al., 2007; Mehta & Heinen, 2001; Mehta & Kellert, 1998). Another important consideration is how local people's support for natural resource management can be influenced by benefits (and costs) obtained from these resources as against socio-economic and demographic consideration (Sesabo, 2006; Dolisca et al., 2006). The results of this study also reveal the importance of context in research as well as the difficulty in developing a "one-size-fits-all" solution to changing local resource users' attitudes towards natural resources. Attempts to impose solutions that have worked in developed countries into a developing setting such as prevails in Ghana is a recipe for disaster. Hence the importance of local case studies as is reported in this paper.

2. Methodological Approach

The research reported in this paper was undertaken in two communities in the coastal zone of Ghana. A purposive non-probability sampling technique was used to select the communities and the respondents. This method was used because the aim of the study was to reach a specialised population, namely women living in coastal communities. Because the case study approach was used, the generalisations of findings are theoretical rather than statistical. Eisenhardt (1989) further explains that in case studies random selection is neither necessary nor even preferable.

2.1 Study Areas

Ghana, a tropical country on the west coast of Africa, is divided into ten administrative regions. The country has an estimated population of about 23.4 million (GSS, 2009) with a population density varying from 897 per km^2 in Greater Accra Region to 31 per km^2 in the Northern Region. Ghana's population is predominately rural and the urban population is skewed towards the south with Accra the capital city having 17% of the total population (ADB, 2011), while adult literacy rate (age 15 and above) stands at 65%. In addition, Ghana's life expectancy at birth increased by 11.5 years to 64.6 (UNDP, 2013). The country is divided into five distinct geographical regions.

The Ghanaian coastline is 550 km from the border with Cote d'Ivoire to the border with Togo, and generally covers a low lying area of 30metres above sea level. The offshore zone in Ghana is about 26,000 km^2. The continental shelf (200 meters deep) is narrow and generally extends seaward between 20 to 35 kilometres, except in Takoradi where it reaches 90 kilometres. The coastal zone covers areas in the Western, Central, Greater Accra and Volta regions (Figure 1).

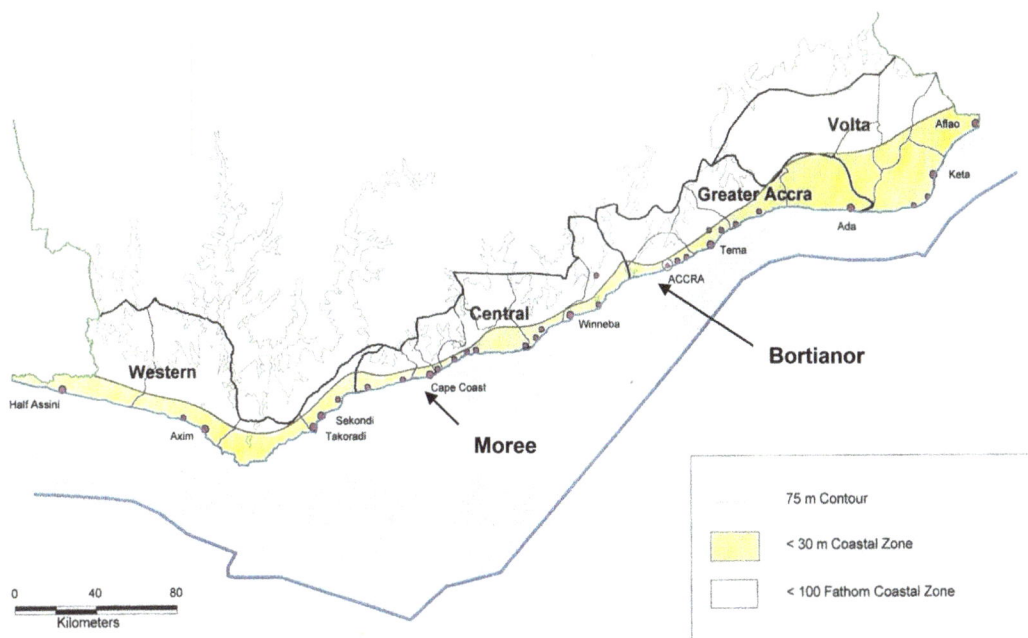

Figure 1. Map of the coastal zone of Ghana

The study communities were chosen because of the high interaction between residents and the coastal natural resources. The first community, Bortianor is found in the Ga South Municipal in Greater Accra Region of Ghana.

It is a fishing community with a total population estimated in the 2000 Census to be 5,446. Out of this 2,683 are males whilst 2,763 are females. The main economic activities in the Ga South Municipal include fishing in the coastal areas and farming in the inland areas. The structure of the local economy is predominantly agriculture, followed by the industrial and the service sectors.

The second community, Moree is also a traditional fishing community found on a rocky headland overlooking the Atlantic Ocean in the Abura-Asebu-Kwamankese (A-A-K) District. According to the 2000 population census, Moree has a total population of 17,761. Out of this 8,577 are males whilst 9,184 are females. Marine fishing and fish processing are the main occupations in the community, though as the town has grown the services industry including retailing and informal artisanal activities, have expanded considerably (Marquette et al., 2002).

2.2 Data Collection

Primary data was collected by means of a number of participatory methods, which included visualizations, poverty profiles, mappings, focus group discussions, personal interviews and participant observations. The focus group discussions were held at the beginning of the study. Two focus groups were organised in each community, an extra one some representatives of government at the local level. The questions for the focus group discussions were generated based on extensive study of secondary data such as newspapers and other reports on the study areas, as well as through observations. Some of the questions were developed or altered as the discussions progressed. Questions for the focus group discussions examined perceptions of poverty, socio-economic and environmental concerns, dependence on coastal natural resources, knowledge on institutions involved in managing natural resources in the communities as well as extent of community involvement. The discussions and audio recordings generated a lot of data which was analysed qualitatively using thematic analysis. This method involves "identifying, analyzing and reporting patterns (themes) within data" (Braun & Clarke, 2006:79). Information from the focus group discussions was also important in creating questions for the Likert Scales.

2.2.1 Personal Interviews

In all 304 women; 151 in Bortianor and 153 in Moree were involved in the personal interviews. The target population was women living in the coastal zone of Ghana above sixteen years (16) years. This lower limit was chosen because there was the need to interview women who could appreciate the issues at stake. The actual dwelling compounds that were visited for the interviews were identified through purposive sampling, in an attempt to reach and to interview women with different backgrounds. The final questionnaire, which also doubled as an interview guide depending on whether the respondent could read and write contained both open ended and closed questions and was made up of nine sections.

Likert scales were used to assess respondents' environmental attitudes. This scaling technique used statements that required the respondents to indicate the extent to which they agreed or disagreed with the statements. The measurement scale was explained to the respondents to avoid any confusion. The questions for this section were developed using information from the focus group discussions and an extensive review of literature which resulted in the selection of three statements from the New Ecological Paradigm (Dunlop et al., 2000), which were found to be relevant to the study.

The study prioritised respondents' environmental concerns through the judgment of severity placed by respondents on certain environmental issues (Quah et al., 2003). The paired comparison methodology measures environmental values and concerns of respondents by providing a clear order of the relative priority they place on coastal natural resources. An added advantage is that apparent preference intransitivities or inconsistent choice patterns are easily observable (Chuenpagdee, 1998). The pairs were presented to the respondents who were asked to choose the environmental concern they saw as most critical.

2.2.2 Participant Observations

The method used was passive participation, which means that the researcher was present at the area of activity but was more of a bystander or spectator. From the beginning of the study, throughout the study period many visits were made to both Bortianor and Moree. This enabled the researcher to identify who the key players of the community were and determine how to gain access into the communities. In order to understand the context in which to interpret people's comments, the researcher also attended some local events, visited local markets and public places and also met with some local government employees. Behaviours of randomly selected respondents were observed to assess the extent to which their actual behaviours agreed with their professed behaviours.

2.3 Analyses

Simple frequencies and cross-tabulations dependent and independent variables were analysed quantitatively using Statistical Package for the Social Sciences (SPSS) version 16. Data from the factual questions were presented in frequency tables showing the raw numbers and in terms of percentages of the total. Descriptive information was also presented in tables.

In developing Likert-style questions, each question must have a similar psychological weight and direction in the respondent's mind. This method has been used quite extensively in assessing attitudes, including environmental attitudes. The challenge however, is developing the right questions that can be easily understood by the respondents and that can also effectively assess their environmental attitudes. For this study the respondents made a variety of responses in the form of a five point scale ranging from strongly agree (SA), agree (A), indifferent (I), disagree (D), strongly disagree (SD). A numeric score was given to each item in order to reflect the degree to which the respondent agreed and disagreed with the item. The scores were totalled to measure the attitude of the respondent. The maximum score was 75 (5X15) and the minimum 15 (5X1), since no statements were left blank. The scores were then grouped into two categories. Scores ranging from 15 to 45 represented a negative attitude whilst scores from 46 to 75 a positive attitude. On the other hand a four point Likert scale was used to measure behaviour. The middle point was removed since these are actual behaviours and respondents had to have either participated in the activities or not.

The total score gave some idea of the strength of a respondent's attitude. The items of statements all had approximately the same level of importance (size) to the respondent, and are all more or less talking about the same concept (direction), which concept the scale is trying to measure. The "direction" of the statements was mixed up, to make sure statements were carefully read, and the informant had to consider their answers carefully. A simple way of evaluating paired comparison data is to use the preference score for each item which is the number of times the respondent prefers that item over other items in the choice set (Peterson & Brown, 1998). The paired comparison questions were thus analysed in a straight-forward fashion by examining a matrix of preferences and calculating an arithmetic average of the preference for each value across all other values with which it was paired. This is because scaling the scores would have not produced different results (Neuman & Watson, 1993). Where no circular triads were produced, the result showed a clear ordering of environmental values of women towards coastal natural resources.

3. Results

3.1 Socio-Economic Background of Respondents

About half of the respondents interviewed were below the age of 40 years (Table 3). More respondents from Bortianor were married. Marriage is an inherent part of the fisheries industry, which is reflected in the gender division of labour: men fish and women trade. The right of a wife to buy fish through a husband or son is thus an important institution in the fishing economy (Overå, 1998; Marquette et al., 2002).

As is expected, fishing industry employed the highest number of women, although the percentage (67.2%) was higher in Moree. The vital role played by women can be especially seen during the lean fishing season. When there are problems with the effective processing of fish, there is a remarkable decrease in the availability of fish in the market and an increase in prices. About 18.3% of respondents in Bortianor as compared to 12.5% in Moree considered themselves unemployed. Many of these used to be involved in the post harvest fishing industry and had lost their initial investment as a result of dwindling fish catch. Whilst petty trading used to be a secondary occupation for most, it is becoming increasingly profitable to sell along the main roads, to visitors and tourists than to depend on fish for their livelihood.

3.1.1 Respondents' Perceptions on Poverty and Vulnerability

Poverty levels among the post-harvest fisheries sector in Ghana is estimated to be 29%. Although this figure is lower than the food and export crops sectors it is still significant (NDPC, 2003). The poor rely heavily on the fisheries post-harvest sector because it is an easy sector to enter with few barriers to stop the unskilled engaging in activities such as labouring and petty- trading (IMM, 2004). Activities include carrying fish around from place to another, sorting by-catch from canoes and trawlers, selling small quantities of fish to other poor consumers. These often make up the majority of people in the post-harvest sector (IMM, 2004). In order to understand their perceptions of poverty, the respondents were first asked if they thought themselves as being poor. More than half of the respondents described themselves as poor, 67.1% in Bortianor and 72% in Moree, which could again be a reflection of the growing economic instability in the post-harvest fisheries sector. The study also asked respondents to cite the four contributory factors of poverty. Interestingly, although issues relating to environment and natural resources did not seem to be priority issues, it was mentioned in both areas. Most of their concerns had to do with their ability to take care of themselves and their families. Decreasing fish catch and food insecurity were ranked highest.

Table 3. Age, marital status and educational levels of respondents

	Bortianor		Moree	
Age	N	%	N	%
16-20 years	12	7.9	20	13.1
21-25 years	20	13.2	25	16.3
26-30 years	15	9.9	21	13.7
31-35 years	19	12.5	12	7.8
36-40 years	21	13.9	7	4.6
41-45 years	19	12.6	17	11.1
46-50 years	14	9.3	18	11.8
51-55 years	8	5.3	11	7.2
>56 years	11	7.3	19	12.4
Do not know	12	7.9	3	2.0
Total	**151**	**100**	**153**	**100**
Marital Status				
Single	19	12.6	37	24.2
Married	114	75.5	93	60.8
Separated/Divorced	11	7.3	16	10.5
Widowed	7	4.6	7	4.6
Total	**151**	**100**	**153**	**100**
Education				
None	84	55.6	94	61.4
Primary	44	29.1	32	20.9
Middle school	8	5.3	6	3.9
JHS	8	5.3	16	10.5
SHS	2	1.3	4	2.6
Vocational/tech	3	2.0	1	0.7
University	2	1.3	-	-
Total	**151**	**100**	**153**	**100**

3.1.2 Access to Basic Amenities

Poor access to some basic amenities like potable water and good sanitation all influence the environmental attitudes of respondents. For example structures for rubbish collection in the communities were virtually non existent. In addition only 6% of respondents in Bortianor and 15% of respondents in Moree had access to toilets in their homes or compounds. Majority of the respondents were aware of the fact that beach defecation was wrong and had some awareness of its impacts but insisted they did it because if a lack of other options. Most of them expected the government through the District Assemblies to provide them with places of convenience as well as other basic amenities.

3.2 Environmental Concerns

Individual expression of environmental concern is often influenced by their perception of the biophysical environment. Respondent's local environmental concern was high. The choices made by respondents were analysed in a straight-forward fashion. 20% of the results in Bortianor as compared to 16% of the results in Moree were inconsistent. The results could thus be used (Figure 2).

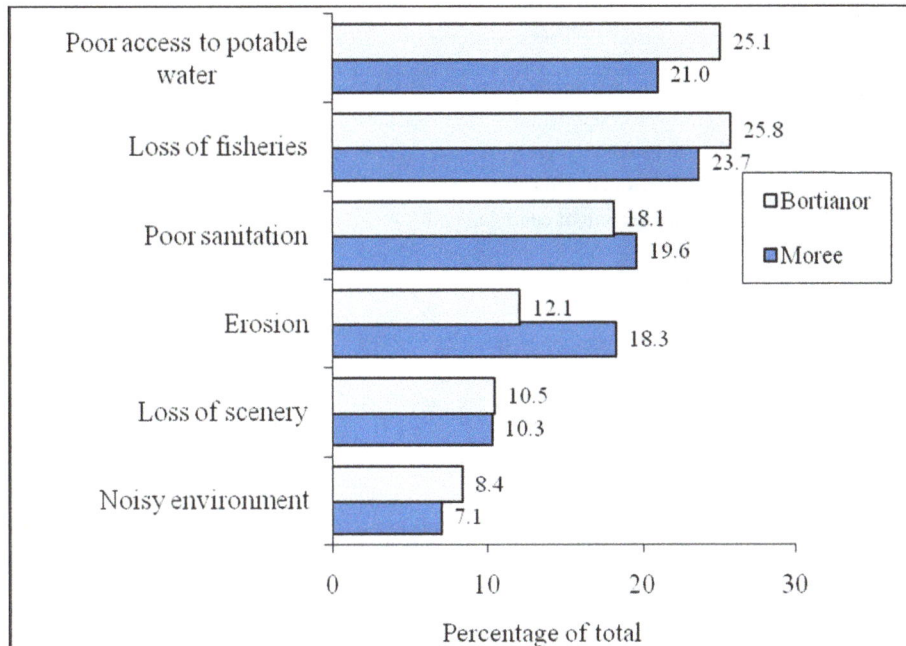

Figure 2. Environmental concerns in study areas

Poor access to potable water and the loss of fisheries were high priority areas in both communities. These were followed by poor sanitation. Coastal Erosion was judged as a bigger environmental problem in Moree as compared to Bortianor.

3.3 Attitudes and Behaviour of Respondents Towards Trends of Natural Resource Degradation

The constructions of the attitudinal questions were based on specific, identified problems affecting natural resources in the area. In addition, Three of the 15 statements were adapted from the New Ecological Paradigm (Dunlop et al., 2000) to suit the local setting (Tables 4 and 5). The statements assess attitudes towards natural resources through:

- Sanitation, pollution, loss of fish and natural resource use (Questions 1, 2, 3, 6, 8, 9, 10, 11, and 13).
- Anti anthropocentric (note 1) attitudes (Questions 5, 12, and 15).
- Institutions for managing coastal natural resources (Questions 4, 7, and 14).

The scale achieved adequate internal consistency with a Cronbach's (1951) Alpha reliability coefficients 0.729 in Bortianor and 0.735 in Moree. This showed that the scales developed were reliable and valid. There were few exceptions however. For example in Bortianor more than half of respondents agreed with the statement that it is okay to throw rubbish on the beach if you are poor (Tables 4 and 5). Their strong disagreement to the statement that plants and animals have as much right to exist as humans is also noteworthy. More than 20% were undecided concerning the questions humans have a right to modify the natural environment to suit their needs, the natural resources cannot get depleted, people who pollute the beaches and the surrounding mangroves should pay to clean it.

Table 4. Environmental attitudes of respondents in Bortianor

Items in scale	SA (%)	A (%)	I (%)	D (%)	SD (%)
1. The coastal zone is being degraded by indiscriminate dumping of refuse and defecation.	39.7	45.0	5.3	9.3	0.7
2. It is okay to throw rubbish on the beach if you are poor.	10.9	34.4	9.9	29.5	15.3
3. All households in the communities should have toilets and bathrooms.	56.3	35.1	3.3	4.6	0.7
4. Resource users should be involved in the management of coastal natural resources.	41.1	31.8	7.3	18.5	1.3
5. Humans have a right to modify the natural environment to suit their needs.	26.5	42.4	22.5	6.6	2.0
6. Outdoor defecation and indiscriminate waste disposal along the beach cause diseases.	59.6	27.2	4.6	7.3	1.3
7. Management of coastal natural resources is the sole responsibility of the government	29.8	39.1	15.9	14.6	0.7
8. Only unconcerned and uncommitted citizens dump garbage and defecate along the beaches.	5.6	30.4	7.3	39.5	17.3
9. People who pollute the beaches and the surrounding mangroves should pay to clean it.	6.0	31.1	21.9	34.4	6.6
10. Fishing with dynamite and nets with small mesh size is not good	58.3	24.5	2.0	13.9	1.3
11. The natural resources cannot get depleted	4.0	41.7	26.5	27.8	0
12. Plants and animals have as much right to exist as humans.	2.0	23.2	7.3	53.6	13.9
13. If the current degradation of natural resources continues we will soon experience a major ecological catastrophe.	29.8	54.3	10.6	4.6	0.7
14. Priority in resource allocation should be given to activities that create greatest economic returns even if it may harm the environment.	9.9	47.5	9.3	32.5	0.8
15. Humans were meant to rule over the rest of nature.	4.0	53.6	11.3	28.8	1.3

SA-Strongly agree, A-agree, I-indifferent, D-disagree, SD-strongly disagree.

Table 5. Environmental attitudes of respondents in Moree

Items in scale	SA (%)	A (%)	I (%)	D (%)	SD (%)
1. The coastal zone is being degraded by indiscriminate dumping of refuse and defecation.	40.5	43.1	5.2	10.5	0.5
2. It is okay to throw rubbish on the beach if you are poor.	13.7	35.8	10.4	25.5	14.6
3. All households in the communities should have toilets and bathrooms.	16.3	35.1	13.3	34.6	0.7
4. Resource users should be involved in the management of coastal natural resources.	38.6	32.7	7.2	20.3	1.3
5. Humans have a right to modify the natural environment to suit their needs.	24.2	42.5	23.5	7.8	2.0
6. Outdoor defecation and indiscriminate waste disposal along the beach cause diseases.	58.8	27.5	4.6	7.8	1.3
7. Management of coastal natural resources is the sole responsibility of the government	27.5	41.2	15.0	15.7	0.7
8. Only unconcerned and uncommitted citizens dump garbage and defecate along the beaches.	11.1	47.1	7.2	24.2	10.5
9. People who pollute the beaches and the surrounding mangroves should pay to clean it.	5.2	35.3	19.6	33.3	6.5
10. Fishing with dynamite and nets with small mesh size is not good	56.2	28.1	2.0	12.4	1.3
11. The natural resources cannot get depleted	5.2	39.9	26.1	28.8	0.0
12. Plants and animals have as much right to exist as humans.	10.7	29.0	9.8	39.8	10.7
13. If the current degradation of natural resources continues we will soon experience a major ecological catastrophe.	28.1	54.9	11.1	5.2	0.7
14. Priority in resource allocation should be given to activities that create greatest economic returns even if it may harm the environment.	1.3	23.5	17.8	44.2	13.2
15. Humans were meant to rule over the rest of nature.	4.6	54.9	10.5	28.1	2.0

SA- Strongly agree, A- agree, I- indifferent, D- disagree, SD- strongly disagree.

The scales reflect negative and positive environmental attitudes. The higher the number, the stronger is the intention to perform the behaviour. Hence in this study a person with a negative attitude has low score and is least ready to support natural resources management initiatives. The environmental attitudes of respondents in both areas towards natural resource degradation were overwhelmingly positive with 94.7% of respondents in Bortianor compared to 90.8% in Moree exhibiting positive attitudes. From the chi-squared computed p-value obtained was 0.271. Thus there was no statistical significant difference between the environmental attitudes of the respondents in Bortianor and Moree.

3.4.1 Actual Participation in Some Environmental Behaviours

Actual behaviours were measured with respondents indicating their how often they participated in five activities (Table 6). To make this as practical as possible, the selected behaviours were simple everyday activities easily understood by the respondents. Some of these were also rephrased from the questions in Tables 4 and 5. In addition behaviours of randomly selected respondents were observed to assess the extent to which their actual behaviours agreed with their professed behaviours.

Table 6. Respondents' participation in actual environmental behaviours

Behaviour	Bortianor				Moree			
Do you...	Y	S	R	N	Y	S	R	N
1. Throw rubbish in gutters, bushes and on the beach?	56.3	35.7	4.9	3.1	64.8	18.5	5.2	11.5
2. Defecate in on the beach and its surroundings?	58.6	29.3	6.2	5.9	61.0	27.1	10.2	1.7
3. Encourage friends and family to keep their surroundings clean?	20.4	21.3	26.5	31.8	17.5	35.0	32.6	14.9
Participate in tree planting and clean up exercises?	22.7	34.5	17.0	25.8	27.2	40.8	20.7	11.3
Buy fish you know is caught with small nets or dynamite?	31.1	20.3	31.0	17.6	37.1	23.2	11.4	28.3
6. Cut mangroves and other trees along the beaches for fuel wood?	22.1	28.2	19.1	30.6	15.1	20.0	28.3	36.6

Y: Yes; S: Sometimes; R: Rarely; N: No.

More than 80% of respondents in both areas disposed of their domestic and human waste on the beaches. In addition more than half had at one time or the other bought fish that they knew was caught by unapproved means. Similarly, most respondents admitted they used the beaches as places of convenience and rubbish disposal although they knew it was wrong. However they still did it, sometimes because of the lack of other alternatives. Others strongly disagreed that those who dump rubbish indiscriminately should be punished, but not until there were suitable facilities provided by the District Assembly. Some disagreed that sand winning should be illegal because they argued there were no jobs. There were also some who said that it was wrong to fish with dynamite or other illegal means and yet still bought fish caught this way because they still had to feed their families. Such respondents would generally exhibit positive environmental attitudes yet negative environmental behaviour.

Majority of the respondents also believed that the management of coastal natural resources is the sole responsibility of the government and its agencies such as the District Assemblies and the area/town councils. They believed that since they paid some form of tax directly or indirectly these monies should be used in cleaning the communities, for management programmes and to provide basic infrastructure.

3.5 Key Elements Required in the Translation of Attitudes Into Behaviour

An analysis of the relationship between education and environmental attitudes showed that education affected the attitudes of respondents towards coastal natural resources positively. Hence knowledge and education (formal, non-formal and informal) remain effective routes to attitude change (Mehta & Heinen 2001; Dolisca et al., 2007). Environmental awareness which may be created through various programmes at various levels, influences how people perceive and interact with coastal natural resources. It also helps to create environmentally positive behaviour in people. However environmental education programmes cannot be implemented in isolation. For example there is substantial evidence to show that poor people will increase their dependence on natural sources to survive (Lawson et al., 2012; Lawson & Bentil, 2014), which will lead to further degradation irrespective of the levels of environmental education received. Hence addressing environmental as well as socio-economic concerns such as the provision of sanitation facilities, potable water, alternative livelihoods as well as the enforcement of existing laws and equipping relevant institutions to work effectively all influence behavioural intensions and must be addressed simultaneously. This will help natural resource users to translate environmentally positive attitudes into environmentally positive behaviour (Figure 3).

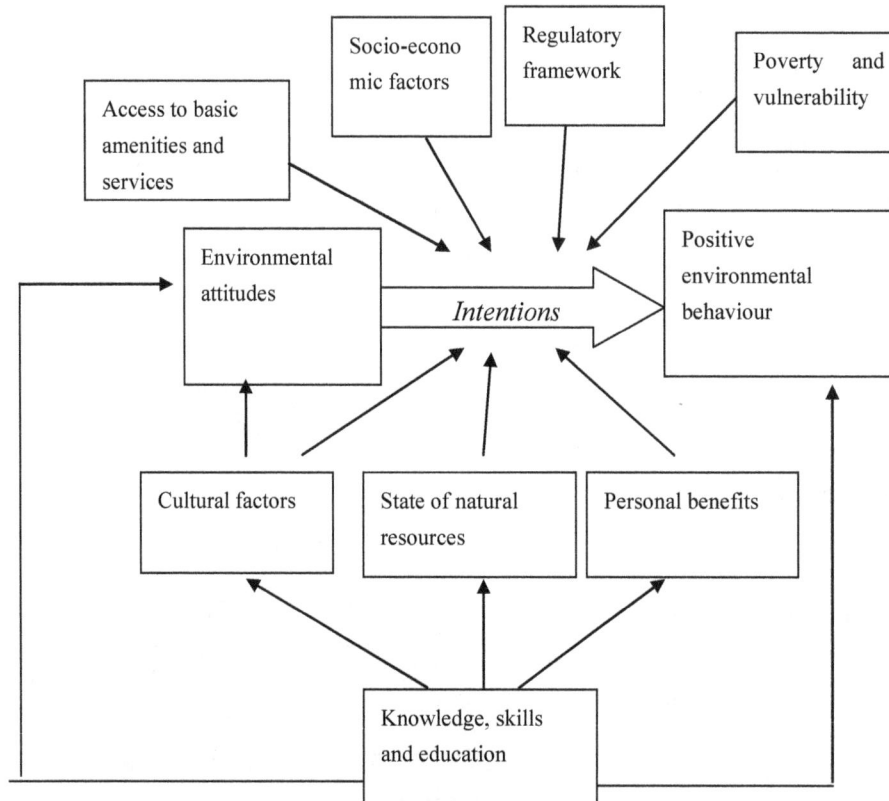

Figure 3. Relationship between attitudes and behaviour

4. Discussion

Whilst the results presented in this paper cannot be generalized they have provided a snapshot of the nature of environmental attitudes among some selected residents in two coastal communities in Ghana. The results presented have also critically examined the often expressed assumption that positive environmental attitudes result in positive behaviour. This paper has examined the socio-economic milieu in which environmental decisions are made. It has assessed the main environmental concerns of respondents as well their environmental attitudes and behaviours. Based on the findings, it can be concluded that efforts to manage the coastal zone may be more effective when the socio-economic background of resource users (Cinner & Pollnac, 2004) as well as how they prioritise their environmental concerns are understood. Environmental education remains important. However the reality is that whilst some level of awareness could lead to positive attitudes it does not always lead to positive environmental behaviour. Hence concentrating *only* on the creation of positive attitudes is not the solution to the challenges facing the effective management of coastal natural resources. Equal effort must also be places on the factors that influence the translation of attitudes into behaviour, which in this paper were identified to include improved access to basic amenities and infrastructure, livelihood options and the right governance regimes.

The Sustainable Livelihoods Approach (SLA) offers a way of addressing coastal livelihood complexities whilst providing an entry point to supporting coastal poverty and livelihood change (Campbell et al., 2006; Badjeck, 2010). The approach embraces the complexity of rural livelihoods from the perspective of the poor. It takes the poor as the centre of the development process because they are the ones most in need of support and it provides a mechanism for enhancing empowerment of the poor (Campbell et al., 2006). The approach provides a framework for understanding and guiding policy-making in coastal zone whilst improving the well-being of community members, enabling them to expend more energy in other activities such being actively involved in community initiatives to manage coastal natural resources (Schulte et al., 2013)

Another crucial yet often understated requirement for the successful translation of positive attitudes into behaviour is an effective governance regime. This can be seen from the percentage of respondents who believe that the management of coastal zone is the responsibility of the government and its agencies. Access to, and use of natural resources by individuals, households and communities are influenced by policies, institutions and the context in which individuals, households and communities live (Prowse & Chimhowu, 2007). The lack of a comprehensive national coastal zone management policy means that the management of the coastal environment depends on laws and policies from other sectors as well as traditional governance systems. Local governments often do not have enough capacity, funds and authority to implement an integrated coastal management. This has led to a gap between the development of the laws and policies and their enforcement (Lawson et al., 2010). Enforcement of exixtsing laws is critical for the effective management of coastal resources. When governance regimes are effective, they challenge resource users to take responsibility for their actions and includes more of their knowledge and perceptions (Mahona et al., 2008).

Respondents' behaviours were in conflict with their professed attitudes (Kollmuss & Agyeman, 2002). The findings of this study are similar with Kotey (1998) who studied the attitudes of residents of Cape Coast towards beach sanitation. She also found out that that generally people expressed positive attitudes. These positive attitudes did not necessarily translate into good conservation practices; people's beliefs are sometimes in conflict with their actions. According to Schultz & Zelezny (1999), one viable explanation for why positive environmental attitudes exist in a "developing" setting is that culture (defined as knowledge, attitudes, values, beliefs, and behaviours shared by a group of people that are communicated from one generation to the next), and held values play an important role in determining environmental attitudes. This could cause a significant discrepancy between people's attitudes and their actual behaviour (Chan, 1996). As already mentioned, an understanding of the complexities of the interrelationships between human activity and the natural environment is therefore necessary for the improvement of environmental quality.

In conclusion positive environmental attitudes do not necessarily result in positive behaviour as this study has shown. Factors that cause environmental degradation in the coastal environment need to be addressed simultaneously. Education and awareness on environmental issues should go hand-in-hand with the provision of infrastructure, livelihood options and the right governance regimes. Since coastal natural resources serve as a source of well-being and livelihood for many of the coastal residents its management must be carried out in partnership with major groups such as local communities. These communities are not homogenous and as such the needs of different groups such as women and the poor must be taken into consideration. The strategies that will be most effective in managing coastal natural resources are those that seek to reduce poverty, create environmental awareness and provide improved access to social amenities and infrastructure and provide effective governance systems that ensure that those who pollute the coastal environment are penalised.

References

Adger, W. N., Hughes, T. P., Folke, C., Carpenter, S. R., & Rockström, J. (2005). Social-ecological resilience to costal disasters. *Science, 309*(5737), 1036-1039. http://dx.doi.org/10.1126/science.1112122

African Development Bank. (2011). Gender, Poverty and Environmental Indicators on African Countries. Economic and Social Statistics Division Statistics Department, African Development Bank. Belvédère Tunis, Tunisia. 312 pages

Ajzen, I., & Driver, B. L. (1992). Contingent value measurement: on the nature and meaning of willingness to pay. *Journal of Consumer Psychology, 1*(4), 297-316. http://dx.doi.org/10.1016/S1057-7408(08)80057-5

Badjeck, M. C., Allison, E. H., Halls, A. S., & Dulvy, N. K. (2010). Impacts of climate variability and change on fishery-based livelihoods. *Marine Policy, 34*(3), 375–383. http://dx.doi.org/10.1016/j.marpol.2009.08.007

Barr, S., & Gilg, A.W. (2007). A conceptual framework for understanding and analyzing attitudes towards environmental behaviour. *Geografiska Annaler Series B, 8*(4), 361-379. http://dx.doi.org/10.1111/j.1468-04 67.2007.00266.x

Braun, V., & Clarke, V. (2006). Using thematic analysis in psychology. *Qualitative Research in Psychology, 3*, 77-101. http://dx.doi.org/10.1191/1478088706qp063oa

Brian, J. (2003). Integrated Coastal Resource Management: A Prescription for Sustainable development. *Electronic Green Journal, 1*(19).

Campbell, J., Whittingham, E., & Townsley, P. (2006). Responding to coastal poverty: should we be doing things differently or doing different things? In C. T. Hoanh, T. P. Tuong, J .W. Gowing, & B. Hardy (Eds.), *Environment and Livelihoods in Tropical Coastal Zones* (pp. 230-257). CABI Publishing, Wallingford UK and Cambridge MA USA.

Chan, K. W. (1996). Environmental attitudes and behaviour of secondary school students in Hong Kong. *The Environmentalist, 16*, 297-306. http://dx.doi.org/10.1007/BF02239656

Chuenpagdee, R. (1998). *Scales of relative importance and damage schedules: non-monetary valuation approach for Natural Resource Management.* Ph.D. Dissertation submitted to the Faculty of Graduate Studies (Resource Management and Environmental Studies), University of British Columbia. 269 pages.

Cinner, J. E., & Pollnac, R. B. (2004). Poverty, perceptions and planning: why socio-economic matter in the management of Mexican Reefs. *Ocean*s and *Coastal Management, 47*, 479-493. http://dx.doi.org/10.10 16/j.ocecoaman.2004.09.002

Coleman, T. L., Manu, A., & Twumasi, Y. A. (2005). Application of landsat data to a base-line ecological study of mangrove vegetation along the coast of Ghana. In *Proceedings of the 31st International Symposium on Remote Sensing of Environment (ISRSE) – Global Monitoring for Sustainability and Security Conference.* The International Society of Photogrammetry and Remote Sensing (ISPRS). June 20-21. Saint Petersburg, Russia Federation.

d'Aquino, P., & Bah, A. (2013). A Participatory Modeling Process to Capture Indigenous Ways of Adaptability to Uncertainty: Outputs From an Experiment in West African Drylands. *Ecology and Society, 18*(4). http://dx.doi.org/10.5751/ES-05876-180416

Directorate of Fisheries. (2003). *Ghana: post-harvest fisheries overview.* Directorate of Fisheries, Ministry of Food and Agriculture, Ghana. Brightsea Press Ltd, Exeter, UK. 81 pages.

Dolisca, F., Carter, D. R., McDaniel, J. M., Shannon, D. A., & Jolly, C. A. (2006). Factors influencing farmers' participation in forestry management programs: A case study from Haiti. *Forest Ecology and Management, 236*, 324-331. http://dx.doi.org/10.1016/j.foreco.2006.09.017

Dunlap, R. E., Van Liere, K., Mertig, A., & Jones, R. E. (2000). Measuring endorsement of the New Ecologial Paradigm: a revived NEP Scale. *Journal of Social Issues, 56*, 425-442. http://dx.doi.org/10.1111/002 2-4537.00176.

Eisenhardt, K. M. (1989). Building theories from case study research. *Acad*emy *Management Review, 14*(4), 532-550. http://escholarship.org/uc/item/6kp8g491.

English, B. (2003), Integrated coastal resource management: A prescription for sustainable development. *Electronic Green Journal, 1* (19). Retrieved August 27, 2008, from http://escholarship.org/uc/item/6k p8g491.

EPA. (2011). *Ghana's Second National Communication, under the United Nations Framework Convention on Climate Change,* Environmental Protection Agency, Accra. 168pp.

Ferrol-Schulte, D., Wolff, M., Ferse, S., & Glaser, M. (2013). Sustainable Livelihoods Approach in tropical coastal and marine social–ecological systems: A review *Marine Policy, 42*, 253-258 http://dx.doi.org/10.1016/j.marpol.2013.03.007

Fiallo, E., & Jacobson, S. (1995). Local communities and protected areas: attitudes of rural residents towards conservation and Machalilla National Park, Ecuador. *Environmental Conservation, 22*, 241-249. http://dx.doi.org/10.1017/S037689290001064X

Fishbein, M., & Ajzen, I. (1975). *Belief, attitude, intention and behaviour: an introduction to theory and research.* Addison-Wesley Publishing, Reading, MA.

Ghana Statistical Service (GSS). (2009). *Ghana at a Glance 2009.* GSS, Accra.

Government of Ghana (GoG). (2010). *Medium Term National Development Policy Framework: Ghana Shared Growth and Development Agenda (GSGDA I), 2010-2013 Volume I: Policy Framework.* NDPC, Accra.

Hill, C. (1998). Conflicting attitudes towards elephants around the Budongo Forest Reserve, Uganda. *Environmental Conservation, 25*(3), 244-250. http://dx.doi.org/10.1017/S0376892998000307

Hines, J. M., Hungerford, H. R., & Tomera, A. N. (1987). Analysis and synthesis of research on responsible environmental behaviour: A meta-analysis. *Journal of Environmental Education, 18*(2), 1-8. http://dx.doi.org/10.1080/00958964.1987.9943482

IMM. (2004). *A Framework for Linking Poverty to Policy in the Post-Harvest Fisheries Sector.* Output from the Post-Harvest Fisheries Research Programme Project R8111. IMM Ltd, Exeter, UK. 4 pages

Kaiser, F. G., Hübner, G., & Bogner, F. X. (2005). Contrasting the theory of planned behavior withthe value-belief-norm model in explaining conservation behavior. *Jornal of Applied Social. Psychology, 35,* 2150-70. http://dx.doi.org/10.1 111/j.1559-1816.2005.tb02213.x

Kollmuss, A., & Agyeman, J. (2002). Mind the gap: why do people act environmentally and what are the barriers to pro-environmental behaviour? *Environmental Education Research, 8*(3), 239-260. http://dx.doi.org/10.108 0/13504620220145401

Kotey, V. D. (1998). *Attitude towards environmental degradation of the Cape Coast Beach.* M. Ed. Thesis. University of Cape Coast, Ghana. 150 pages

Lawson, E. T., & Bentil, G. (2014). Shifting sands: changes in community perceptions of mining in Ghana. *Journal of Environment, Development and Sustainability, 16*(1), 217-238. http://dx.doi.org/10.1007/s10 668-013-9472-y

Lawson, E. T., Gordon, C., & Schluchter, W. (2012). The Dynamics of Poverty Environment Linkages in the Coastal Zone of Ghana. *Oceans and Coastal Management, 67,* 30-38. http://dx.doi.org/10.1016/j.oce coaman.2012.05.023

Lawson, E. T., Schluchter, W., & Gordon, C. (2010). Using the paired comparison methodology to assess environmental values in the coastal zone of Ghana. *Journal of Coastal Conservation, 14*(3), 231-238. http://dx.doi.org/10.1007/s11852-010-0096-1

Lepp, A. (2007). Residents' attitudes towards tourism in Bigodi village, Uganda. *Tourism Management, 28,* 876-885. http://dx.doi.org/10.1016/j.tourman.2006.03.004

Levine, D. S., & Strube, M. J. (2012). Environmental attitudes, knowledge, intentions and behaviors among college students. *Journal of Social Psychology, 152*(3), 308-326. http://dx.doi.org/10.1080/00224 545.2011.604363

Mahona, R., McConneya, P., & Royb, R. N. (2008). Governing fisheries as complex adaptive systems *Marine Policy, 32,* 104-112. http://dx.doi.org/10.1016/j.marpol.2007.04.011

Marquette, C. M., Koranteng, K. A., Overå, R., & Bortei-Doku Aryeetey, E. (2002). Small-scale fisheries, population dynamics and resource use in Africa: The Case of Moree, Ghana. *AMBIO, 31*(4), 324-336.

Mehta, J. N., & Kellert, S. R. (1998). Local attitudes toward community-based conservation policy and programmes in Nepal: a case study in the Makalu-Barun Conservation Area. *Environmental Conservation, 25*(4), 320-333. http://dx.doi.org/10.1017/S037689299800040X

Mehta, J., & Heinen, J. (2001). Does community-based conservation shape favorable attitudes among locals? An empirical study from Nepal. *Environmental Management, 8*(2), 165-177. http://dx.doi.org/10.1007/s002 670010215

National Development Planning Commission. (2003). *Ghana Poverty Reduction Strategy 2003-2005. An agenda for growth and prosperity. Volume I: analysis and policy statement.* Government of Ghana and National Development Planning Commission, Accra, Ghana.

Nelson, F., & Agrawal, A. (2008). Patronage or participation? Community-Based natural resource management reform in Sub-Saharan Africa. *Development and Change, 39,* 557-585. http://dx.doi.org/10.1111/j.1467-7 660.2008.00496.x

Neuman, K., & Watson, B. G. (1993) Application of paired comparison methodology in measuring Canadian's forest values. *Proceedings of the Survey Research Methods Section. American Statistical Association* (pp. 1091-1094).

Nkemnyi, M. F., Koedam, N., & De Vreese, R. (2011). *Livelihood and Conservation: Reconciling Communities' livelihood needs and Conservation Strategies in the Bechati forest area, Western Cameroon. Saarbrücken.* Germany: LAMBERT Academic Publishing.

Olsen, S. B. (2009). Building Capacity for the Adaptive Governance of Coastal Ecosystems Priority for the 21st Century. Retrieved September 9, 2012, from http://www.ferrybox.eu/imperia/md/content/dahlem/wg4_ paper_-_building_capacity_for_the_adaptive_governance_of_coastal_ecosystems.pdf

Overå, R. (1998). *Partners and competitors. Dr.Polit Dissertation.* Norway: University of Bergen.

Pomeroy, R. S., & Carlos, M. B. (1996). *A review and evaluation of community-based coastal resources management projects in the Philippines* (pp. 1984-1994). ICLARM RR# 6.

Prowse, M., & Chimhowu, A. (2007). Making agriculture work for the poor. Natural Resource Perspectives 111. ODI, London. Retrieved July 7, 2011, from http://www.odi.org/sites/odi.org.uk/files/odi-assets/publicatio ns-opinion-files/584.pdf

Quah, E., Tan, K. C., & Choa, E. (2003). *Environmental valuation: damage schedules. Paper to be presented at the Economics and Environment Network National Workshop.* May 2-3, 2003 at the Australian National University, Canberra, Australia.

Ramos, J., Santos, M., Whitemarsh, D., & Monteiro, C. (2007). Stakeholder perceptions regarding the environmental and socio-economic impacts of the Algarve artificial reefs. *Hydrobiology, 580,* 181-191. http://dx.doi.org/10.1007/s10750-006-0454-z

Rodary, E. (2009). Mobilizing for nature in southern African community-based conservation policies, or the death of the local. *Biodiversity Conservation, 18*(10), 2585-2600. http://dx.doi.org/10.1007/s10531-009-9666-7

Savory, A. (1988). *Holistic resource management.* Washington, D.C.: Island Press.

Schlaepfer, R. (1997). *Ecosystem-based management of natural resources: a step towards sustainable development.* IUFRO Occasional Paper No. 6.

Schultz, P. W., & Zelezny, L. (1999). Values as predictors of environmental attitudes: Evidence for consistency across cultures. *Journal of Environmental Psychology, 19,* 255-265. http://dx.doi.org/10.1006/jevp.199 9.0129

Schultz, P. W., Shriver, C., Tabanico, J. J., & Khazian, A. M. (2004). Implicit connections with nature. *Journal of Environmental Psychology, 24,* 31-42. http://dx.doi.org/10.1016/S0272-4944(03)00022-7

Sesabo, J. K., Lang, H., & Tol, R. S. J. (2006). *Perceived Attitude and Marine Protected Areas (MPAs) establishment: Why households' characteristics matters in Coastal resources conservation initiatives in Tanzania.* FNU-99.

UNDP. (2013). *The Rise of the South: Human Progress in a Diverse World. Explanatory note on 2013 HDR composite indices Ghana.* UNDP. Retrieved December 17, 2013, from http://hdr.undp.org/sites/default/files/Country-Profiles/GHA.pdf

United Nations Environment Programme. (2012). 21 Issues for the 21st Century: Result of the UNEP Foresight Process on Emerging Environmental Issues. *United Nations Environment Programme (UNEP).* Nairobi, Kenya.

World Bank. (2010). *Economics of adaptation to climate change: social synthesis report.* The International Bank for Reconstruction and Development.

Wright, A. S., & Shindler, B. (2001). The role of information sources in watershed management. *Fisheries, 26*(11), 16-23. http://dx.doi.org/10.1577/1548-8446(2001)026<0016:TROISI>2.0.CO;2

Note

Note 1. This encompasses the view that destructiveness is rooted in anthropocentrism, an arrogant view that we are separate from and superior to nature, which exists to serve our needs (Barnhill et al. 2006).

17

Fatal Elephant Encounters on Humans in Bangladesh: Context and Incidences

A.H.M. Raihan Sarker[1], Amir Hossen[2] & Eivin Røskaft[2]

[1] Institute of Forestry and Environmental Sciences, University of Chittagong, Chittagong 4331, Bangladesh

[2] Department of Biology, Norwegian University of Science and Technology, NTNU, Realfagbygget 7491 Trondheim, Norway

Correspondence: Eivin Røskaft, Department of Biology, Norwegian University of Science and Technology, NTNU, Realfagbygget 7491 Trondheim, Norway. E-mail: roskaft@bio.ntnu.no

Abstract

Here we report the context encounters of elephant attacks on humans in Bangladesh, during the period 1989 to 2012. Attack rates significantly increased over this study period. The proportion of encounters that caused deaths or injuries differed statistically significant between the two sexes (men more deaths), age groups (elder more deaths), time of the day (more deaths during night), place of casualty (more deaths outside forests), weapon used by elephants (more deaths when elephants were using both trunk and leg) and study sites. No difference was found between seasons, elephant group size, or financial status, occupation and household size of victims. Elephant family groups were mostly responsible for attacks in the north, while single bulls were more responsible in the southeast. The place of casualty (inside or outside forests), time of the day, gender and regions were all significant in explaining the variation in encounters which resulted in human deaths or injuries. Conflict mitigation approaches including incentive-, awareness-or training programs from the forest department could help to reduce the conflict between humans and elephants in Bangladesh.

Keywords: human-elephant conflict, human injuries and deaths, Bangladesh

1. Introduction

Humans are considered to be users of ecosystem services as forests and natural resources and conflict intensity between humans and wildlife is dependent on the level of resource utilisation, access and control priority (Hossain, 2008; Jones & Carswell, 2004; Robbins, 2012). The potential cost of living near a protected area is particularly high when the concerned animals are large and dangerous, such as the Asian (*Elephas maximus)* and African elephants (*Loxodonta africana)*. One of the calamities of the interaction between elephants and people is that elephants kill people even without any direct provocation (Røskaft, Larsen, Mojaphoko, Sarker & Jackson, 2014).

In Bangladesh, the Asian elephant population is restricted to the south-eastern and northern forest areas. In addition, a trans-border movement of elephants into Bangladesh from Meghalaya and Assam in India and Myanmar has been observed. According to Sarker and Røskaft (2011), the size of the wild elephant population in Bangladesh varies between 150 and 200 animals. Wild elephants in Bangladesh can thus be divided into two categories, (i) a local migratory- and (ii) a trans-border migratory population. Local migratory elephants frequently visit different habitat patches in their home range at specific time intervals.

Wild elephants which have been considered protected in Bangladesh, often venture into villages, mostly at night, to look for food, thereby frequently damaging crops and attack villagers. In 2014 (December 17), three people were killed and another one was injured as raiding elephants invaded several villages at Satkania upazila of Chittagong district in south-eastern Bangladesh (PA, 2014). In November 2001, a herd of some 15 elephants stormed the villages at Rangunia in Chittagong district. They pulled down a number of bamboo and straw houses, damaged crops and killed four people in addition to damaging 15 villagers (Reuters, 2001). A middle-aged man was killed by an elephant attack at Kamarkhali village in Durgapur upazila of the Netrokona district, which is located near the political border between Bangladesh and India (New-Nation, 2015). Three victims died when herds of elephants stormed two farming villages near a forest at Ukhia, Cox's Bazar (Personal communication with Matiur Rahman, ex-deputy commissioner of Cox's Bazar district). Elephants destroyed nearly 15 straw and bamboo-made houses, leaving some 100 people homeless (Reuters, 2002). In India, during the years 1980-2000, about 150-200 people lost their lives yearly due to attacks by wild elephants --- a total of 3,000 to 4,000 people over these two decades (Sukumar, 2004). Information from Sri Lanka indicates that 30-50 people are killed annually (Bandara & Tisdell, 2002). Similar figures are found in Kenya during the past decades caused by African elephants (Sukumar, 2004).

In Bangladesh, elephants have gradually been confined to 'pocketed herds' in small patches of forests, which in most cases are surrounded by human settlements. Such 'pocketed herds' represent an extreme stage of human-elephant conflict in the south-eastern and northern regions of Bangladesh and make the wild elephants almost a seasonal invader into farm fields and dwelling places resulting in many human fatalities (Sarker & Røskaft, 2010).

Manslaughter by elephants is common during two kinds of contexts: (i) under "normal" circumstances in the wilderness or in settlements within a large, natural elephant habitat and (ii) under "abnormal" circumstances, such as when a large herd or clan of mostly crop-raiding elephants disperse to a new habitat (Sukumar, 2004). Human deaths in the forest are usually chance encounters when people are walking alone or in small groups from one settlement to another, visiting a small shrine on a hill top, grazing livestock, collecting firewood and other forest products, or even when humans are sleeping under the shade of a tree (Ramakrishnan & Ramkumar, 2007). While most incidents have been taking place during daytime, it is noteworthy that several of these incidents occurred at dark / dusk very close to human settlements where bull elephants were waiting for the cover of darkness to enter cultivated land (Sukumar, 2004). For example, forest dependant resource collectors and farmers are mostly at risk of wild elephant attacks when they are working inside or near the forest (Sarker & Røskaft, 2010). Victims have mostly been men guarding crop fields at night from simple structures at ground level, although on occasion's women or children have been attacked when a riding bull broke down a hut (Ramakrishnan & Ramkumar, 2007).

Elephant biology also attributes to its periodical aggressive behaviour leading to conflict and human fatalities (Sukumar, 1989). Attacks resulting in fatalities in the forest are frequently due to solitary bulls but occasionally also by members of family groups (Sukumar, 2004). The elephants are often killed together with the human victim (Sarker & Røskaft, 2010). In contrast, practically all human killings within cultivated land are caused by sub-adult or adult bulls (Sarker & Røskaft, 2011). A flashlight shined at a bull or the sound of a dog barking often evokes an aggressive reaction (Sukumar, 1989). Farmers sometimes use dogs to warn them of the presence of raiding elephants. If an elephant chase the dog, the latter would naturally tend to run back to its master, bringing behind it an elephant which might redirect its aggression to the person (Sukumar, 1989). Another common pattern seen is that a few notorious bulls may be responsible for multiple killings. Both in the forest and within fields, elephants kill people by lashing out with the trunk, grasping and flinging, trampling or goring with the tusks. There have been occasional incidents involving photographers approaching elephants and even a curious foreigner possibly unaware of the danger from a wild elephant (Sukumar, 2004).

Management of the human-elephant conflict has become a topic of national, regional and international significance (Røskaft, Larsen, Mojaphoko, Sarker & Jackson, 2014; Sarker, 2010; Skarpe, du Toit & Moe, 2014). The high intensity of the conflict level between humans and elephants in Bangladesh has interrupted the smooth and long co-existence. People who live close to elephant habitats are now regarding them as an agricultural pest, an invader that damage property, and as a life-threatening animal (Sarker & Røskaft, 2011). Therefore, there is a need to formulate preventive strategies in order to protect the vulnerable human population from wild elephant attacks and hence to save many valuable human as well as elephant lives. Thus, the aim of this study was to investigate the proportion of encounters intensity and consequences of elephant attacks on humans - in Bangladesh - as well as to describe the circumstances when those incidents commonly occur.

2. Materials and Methods

2.1 Study Areas

The study was performed in the south-eastern and northern regions of Bangladesh where the conflict between humans and wild elephants are common. The south-eastern region of Bangladesh consists of 11 of the 34 protected areas in the country and is managed by the Chittagong and Rangamati Forest Circle. This study was mainly confined to the regions near the Rangamati Forest Reserve (RFR), Banshkhali Forest Reserve (BFR), Chaunti Wildlife Sanctuary (CWS) and Teknaf Wildlife Sanctuary (TWS) based on their higher level of conflict vulnerability compared to other sites in the south-eastern region (Figure 1). The south-eastern region is associated with the Arkan forest area of Myanmar via a cross border corridor near the north bank of the Naf

River between Teknaf, Bangladesh and Arkan province, Myanmar. Several corridors of elephants are networked over the entire south-eastern region.

The northern region only consists of the Sherpur forest reserve (SFR) (Figure 1), which includes Madhutila Eco-park. This area embraces a cross border corridor between Korigram, Mymensigh and the Sylhet forest regions of Bangladesh and West Bengal and Assam forest in India (Choudhury, 2007). Trans-border migratory wild elephants frequently visit the cross border corridor during movements from the Maghalaya forest region of Assam, India, to the Nalitabari and Zinaighati forest regions of Bangladesh.

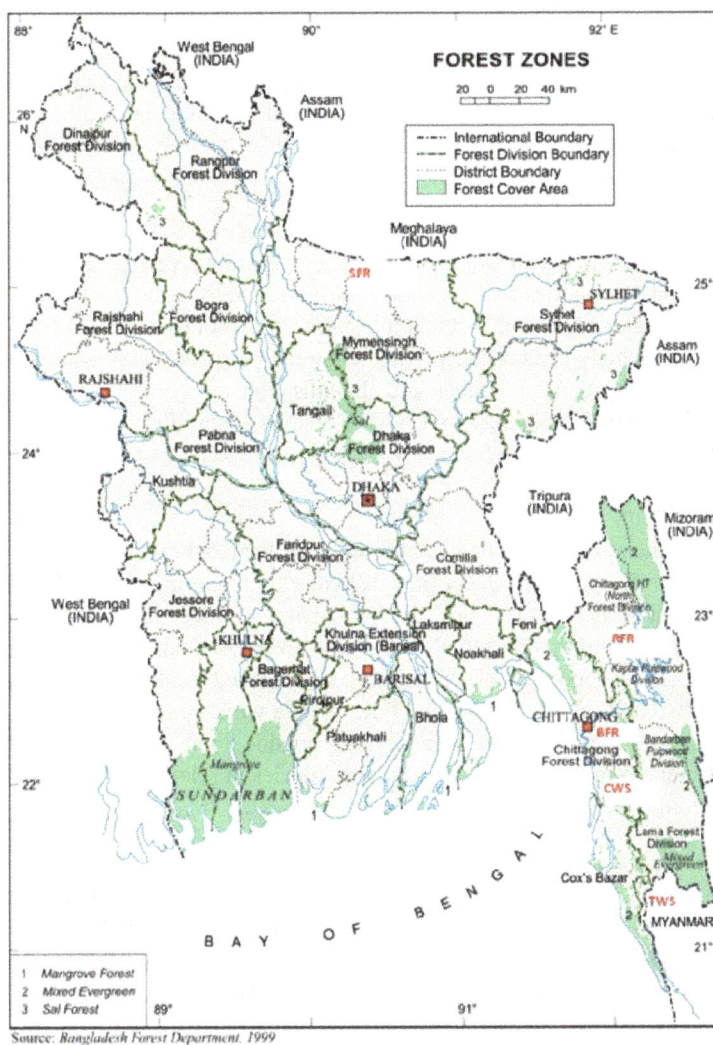

Figure 1. Location of study sites (SFR = Sherpur forest reserve, RFR = Rangamati Forest Reserve, BFR = Banshkhali Forest Reserve, CWS = Chaunti Wildlife Sanctuary and TWS = Teknaf Wildlife Sanctuary)

2.2 Data Collection and Analyses

Prior to the fieldwork, secondary resources, including published and unpublished research studies, conflict-related forest departmental documents and daily local and national newspapers, were studied to obtain the number and type of identified cases. We collected data on encounters of wild elephants, which caused either human death or injury during the period 1989-2012 by contacting Bangladesh Forest Department and requesting copies of investigation reports and other such information on elephant attacks. Injured victims were interviewed directly. In case of death of the victim, information was collected either by interviewing family or witnesses to the incident. Information on the victim's sex (male, female), age (young below 30 years, old above 30 years), occupation (work at home or outside the home), financial status (poor income below Tk 5000 per month, solve income above 5000 Tk per month); time of the day (night between 6pm and 6 am; day between 6am and 6 pm), season (dry October to May, wet June to September), year (1989-2000, 2001-2012), place of casualty (inside a protected forest, outside s protected forest in human dominated areas), elephant group size (single bulls, family groups) and activity of the victim during the attack by the wild elephant. Characteristics of elephant attacks were collected by using a semi-structured questionnaire. We applied a cross-examination process through checking records available in local police stations as well as in District Commissioner Offices to ensure data reliability. In case of a victim's death, we also checked the post-mortem report from the local hospital. Sometimes, the victim's family had left their settlement after an incident. In such a case, interviews were obtained from the victim's close relatives. A total of 224 recorded incidents of encounters between humans and wild elephants which resulted in deaths or injuries were recorded.

For simplicity, if at least one person died due to encountering wild elephant, we considered it as a death incidence, or if at least one person was injured it was considered as an injury incidence. Unless the attack was witnessed or confirmed by the local police they were not included in our data. The collected primary raw data were sorted on the basis of importance and usability for the purpose of easy, meaningful and high quality quantitative analyses.

2.3 Statistical Analyses

Most tests were performed using simple chi-square tests. A binary logistic regression analysis was performed to examine variations in human deaths or injuries as a dependent variable, and by using a set of independent variables such as sex, age group, financial status of victims, occupation status, household size, time of the day, season, year, place, study region, elephant group, body parts used as weapons by wild elephants (*i.e.,* such as used the trunk, the legs, the tusk, or both trunk and legs). After coding and digitalisation the collected data, data analyses were performed by using SPSS version 20.0 (SPSS, Chicago, USA). The level of significance was set at $P = 0.05$.

3.Results

3.1 Socioeconomic Characteristics of Victims

The proportion of the 224 encounters between humans and wild elephants which resulted in human deaths over the study period, was 76.8% (n = 172), while 23.2% (n = 52) caused human injuries. More males than females were attacked by elephants (Table 1). Furthermore, the proportion of incidents causing human deaths was statistically significantly higher in males than in females (Table 1). More encounters were registered in the older than the younger age group. The proportion of encounters that caused deaths or injuries was statistically significant different between two age groups (Table 1). Older people were more frequently killed than younger people. Finally, although more poor people and people working outside home were victims of elephant attacks, there were no significant differences in frequencies of deaths or injuries in relation to household size, occupation level or financial status of victims (Table 1).

3.2 Study Periods

A significant increase in frequency of encounters between humans and wild elephants was observed between the two periods 1989 - 2000 to 2001 - 2012 (Table 1). The majority of incidents between humans and wild elephants were recorded during 2001 to 2012. A significant higher frequency of deaths with more than 80.0 % of the incidents causing human deaths occurred during this last period (Table 1).

3.3 Period of the Day and Season

The proportion of encounters resulting in deaths and injuries varied significantly between the two day periods (day or night) but insignificant between the two different seasons (Table 1). More attacks occurred during night with a much higher frequency of deaths compared to daylight period (Table 1). Although the dry season was much longer than the wet season, relatively more attacks occurred during the dry season.

3.4 Place (Inside or Outside the Forest)

Although the number of attacks did not differ inside or outside the protected forests, the proportion of deaths or injuries varied statistically significantly in relation to whether it occurred inside or outside the protected forests (Table 1). Most of the encounters, which caused human deaths, were reported inside the forest, while human injuries were considerably higher outside the forests (Table 1).

3.5 Study Sites

The highest proportion of encounters between humans and wild elephants was recorded in RFR (n = 91, 40.6%), while the lowest was recorded in CWS (n = 23, 10.3%). The number of encounters in the other study sites were; TWS (N = 43, 19.2%), BFR (n = 42, 18.8%) and SFR (n = 25, 11.2%). The proportion of encounters causing human deaths or injuries varied statistically significantly between the four study sites (χ^2 = 10.72, df = 4, P = 0.030). Although most incidents occurred in the south-eastern region the frequencies of injuries and deaths were insignificant between two regions (Table 1).

Table 1. Numbers of encounters between humans and wild elephants that resulted in human deaths or injuries in relation to socio-economic status, year and period of the day of incidents and place of casualty across the study sites and χ^2 tests of independence between victims' status in the form of human deaths or injuries

Variables			Victim's status		Total	Statistics		
			Death	Injury				
			(n=172)	(n=52)	(n=224)	χ^2	df	P
Socioeconomic characteristics	Gender	Male	139	23	162	26.7	1	0.001
		Female	33	29	62			
	Age group	Young (0 to 30 years)	44	22	66	5.37	1	0.020
		Old (> 30 years)	128	30	158			
	Financial status	Poor (<Tk 5000)	137	38	175	1.01	1	0.315
		Solvent (>Tk 5000)	35	14	49			
	Household size	Small HHs (up to 4 members)	79	21	100	0.49	1	0.481
		Large HHs (> 4 members)	93	31	124			
	Occupation status	Work outside home	114	30	144	1.28	1	0.257
		Work at home	58	22	80			
Period of incidents	Study periods	1989 to 2000	31	17	48	5.10	1	0.024
		2001 to 2012	141	35	176			
	Time of the day	Daylight (6 am to 6 pm)	60	34	94	15.3	1	0.001
		Night(6 pm to 6am)	112	18	130			
	Seasons	Wet (June to September)	32	12	44	0.51	1	0.477
		Dry (October to May)	140	40	180			
Place of casualty	Forest reserve	Inside the reserve (within the reserve boundary)	98	10	108	22.8	1	0.001
		Outside the reserve (up to 1 km away from the reserve boundary)	74	42	116			
	Region	Northern (i.e., SRF)	16	9	25	2.58	1	0.108
		Southeastern (i.e., RFR, CWS, BFR, TWS)	156	43	199			

3.6 Elephant Social Groups

Single elephants were bulls (n = 108, 48.2%), while groups normally counted family herds (n = 116, 51.8%). There was no statistically significant difference in the encounters of single bulls or group attacks in relation to the victim's gender (χ^2 = 0.001, df = 1, P = 0.974), age group (χ^2 = 0.68, df = 1, P = 0.088), financial status (χ^2 = 3.32, df = 1, P = 0.190), or between the two study periods (χ^2 = 0.87, df = 1, P = 0.352). These encounters were not significantly different with time of day (χ^2 = 0.39, df = 1, P = 0.529), different seasons (χ^2 = 2.39, df = 1, P = 0.496). Consistent with these findings, there was no statistically significant variation in the elephant social group encountering attack rate, independent of whether the encounter occurred inside or outside of the protected forests (χ^2 = 0.05, df = 1, P = 0.826).

The incidents of single and group encountering attacks varied statistically significant between the two regions (χ^2 = 8.97, df = 1, P = 0.003); family groups (80.0 %) were mostly responsible in the north, while single bulls (51.8 %) were more responsible in the southeast.

Table 2. Results of logistic regression analyses to examine the effects of 12 independent variables (see text for details) on variations in human death or injury as a dependent variable

Independent variables	B	SE	Wald	df	P	Odds ratio[1]
Place (inside/outside)	2.577	0.528	23.865	1	0.0001	13.162
Time of the day	-1.726	0.444	15.080	1	0.0001	0.178
Gender	1.871	0.507	13.601	1	0.0001	6.496
Region	-1.224	0.638	3.675	1	0.055	0.294
Status of attacked elephant	0.736	0.429	2.946	1	0.086	2.087
Age	-0.533	0.453	1.386	1	0.239	0.587
Occupation	-0.587	0.517	1.290	1	0.256	0.556
Body parts as weapon	0.668	0.757	0.780	1	0.377	1.950
Season	0.451	0.520	0.755	1	0.385	1.571
Monthly income (Tk)	0.355	0.472	0.564	1	0.453	1.426
Household size	0.267	0.429	0.387	1	0.534	1.306
Year of incidents	-0.026	0.523	0.002	1	0.961	0.975

[1] A measure of association between each independent and dependant variables. When B is negative, the odds ratio must be inverted (1/odds ratio) to indicate the relevant odds. Cox & Snell r^2 = 0.297, Nagelkerke r^2 = 0.449.

3.7 Body Parts Used as Weapons by Wild Elephants

Elephants usually used their trunk (n = 92, 41.1%), leg (n = 15, 6.7%), tusk (n = 19, 8.5%), or both their trunk and legs (n = 98, 43.8%) as a weapon when encountering a human being (χ^2 = 34.35, df = 3, P = 0.0001). If an elephant used the trunk and leg combined, then the proportion of encounters causing human deaths (n = 92, 93.9%) dramatically increased compared to other types of weapons, such as by the trunk only (n = 56, 60.9%), by the legs only (n = 8, 53.3%), or by the tusk only (n = 16, 84.2%). However, there was no difference between the single bull or family groups in the use of these different body parts when encountering humans (χ^2 = 3.09, df = 3, P = 0.378).

A binary logistic regression analysis was performed to examine variations in human deaths or injuries as a dependent variable using twelve independent variables [*i.e.,* study sites, year of incidents, season, time of the day, inside/outside forest, group size of wild elephants, weapon used by wild elephants and socio-demographic characteristics of victims (such as monthly income, age, gender, occupation and family size)]. The analysis shows that place of casualty (*i.e.,* inside or outside forest reserve), time of the day (*i.e.,*daylight and night), gender (*i.e.,* male and female) and regions (*i.e.,* northern and south-eastern) were significant in explaining the variation in encounters which resulted in human deaths or injuries (Table 2).

4. Discussion

4.1 Socio-Economic Features of Victims

As previously discussed by Bandara and Tisdell (2003), of a total number of 536 causalities due to elephant's encounter attacks between 1992 and 2001 in Sri Lanka, 75.0% were males, 13.0% females and 12.0% infants. In our study, the binary logistic regression analysis also revealed that the highest proportion of encounters causing human deaths was reported in males. Males are more involved in forest-related activities as well as farming near forest reserves, whereas females are more involved in non-forest activities (Aziz, 2002; Røskaft, Hagen, Hagen & Moksnes, 2004). During male-dominated activities, men are encountered by elephants in most forest areas, while women are attacked in their residences during the night (see also, (Røskaft, Bjerke, Kaltenborn, Linnell & Andersen, 2003; Treves & Naugthon-Treves, 1999)). When agricultural crops are not available in the field, wild elephant herds may attack in the residential areas where grains are stored. Under such circumstances, most males escaped easily from their homes due to having better physical fitness, while females are busy to protect household assets as well as their children from the encountering of wild elephants.

In our study, we found that the proportion of encounter rate of wild elephants which resulted in human deaths was higher among older people, most likely due to many factors, such as lack of physical fitness of older people to escape from the wild elephant, or it may be that this is the most common age group (*i.e.,* adults) that collects natural resources in the forests (IUCN, 2004b). When old people are entrapped by an attacking elephant, they are unable to escape from this dangerous place. Furthermore, the old age group is the most frequent group engaged in forest-related activities. For this reason, people from this group are more easily affected by elephant's encounters.

4.2 Study Period, Periods of the Day, Season and Place (Inside or Outside of the Forest) of Elephant Encounters

There was significant increase in attacks by elephants during the study period, probably due to many factors, the most important being lack of space for elephants. The number of people encountered by wild elephants gradually increased over the last 24 years (1989 to 2012), which was partly due to the population increase in Bangladesh as well as the dramatic reduction in elephant habitats due to several anthropogenic factors. The encountering rates between wild elephants and human increased from 1989-2000 to 2001-2012 as in our study more than two-third of all incidents were recorded during 2001 to 2012. The first human death was reported in the Rangamati region in 1989. During the years before the turn of the century, the intensity of encountering wild elephants was negligible due to less migration of local people towards forested areas. However, after the year 2000, a huge number of humans have been migrating to the forested zones due to the employment crisis and poverty. They have encroached the elephant's habitats, blockaded corridors and transformed forest land into agricultural land at extreme levels and continuing the overexploitation of forest resources for their subsistence or livelihood purposes. Moreover, weak forest management systems are considered to be the key causes for the deterioration or degradation of elephants' habitats across study areas, which forces wild elephants come to close contact with humans and ultimately increases the intensity of human-elephant conflict over recent years (Khan, Rashid & Khan, 1983; Sarker, 2010; Sarker & Røskaft, 2011).Thus, encounters of wild elephants on human are increasing in and around the protected areas in recent times in both the northern and south-eastern part of Bangladesh.

Human encounters by wild elephants also varied during the daytime. As previously discussed by Sukumar (2004), the highest encounters of wild elephants occurred during dark/dusk during the harvesting period, which was consistent with our findings. A few attacks by single bulls might also occur inside the forest during noon, afternoon and early evening when victims are alone inside the forest and are busy collecting forest resources (Feeroz, Aziz, Islam & Islam, 2004). However, at noon and in the afternoon, elephants are mostly resting in a quiet shady place inside the forest or near a body of water. During the evening, the elephants come out from these resting areas, while humans are still active inside the forest and in the cropping fields. During night, when most fatalities occurred, wild elephants are coming close to human settlements where humans are disturbed. Thus they are attacked when they try defending their homes and properties. During the cropping season, wild elephants frequently raid forests close to residential areas. When moving between such areas, they frequently use roads during, resulting in an increased rate of attacks in areas outside of, but close to the forests. The encounter rates increased during the peak cropping season in areas outside the forests due to the availability of a ripen paddy, which attracts wild elephants. Wild elephants prefer the paddy because it provides a nutritious source for grazing. The availability of agricultural crops is lower in the agricultural fields during the spring and summer, but farmers home garden are rich in horticultural crops during that time. Wild elephants encounter home gardens and more causalities occurred in the settlements during the non-pick seasons (Sarker & Røskaft, 2010; Sukumar, 2006). Elephants enter human settlement zones only after dark, and elephants usually move close to the forest–village boundary in the early evening during the cropping seasons (Sukumar, 1989). Settlement zones including crop fields and home gardens are not however, generally guarded on a 24-hour basis and are tended only during the daylight hours. Most of the encounters, therefore, take place at night when the crop fields or homestead gardens are unattended. Farmers construct thatched houses (huts) during the harvest season (especially the rice harvest) to guard their fields and home gardens at night. Such guarding activities are always carried out in groups (Sarker & Røskaft, 2010). Therefore, in the early evening, large guard groups make it possible to guard against crop raiding and drive elephants from the field. However, farmers are unable to maintain large groups throughout the late evening due to lack of sleep. Moreover, they are financially unable to hire night guards. On the other hand, illicit fellers entered into the forest at night to extract timber and other resources for their livelihood and encountered wild elephants in there. Under such condition, they failed to trace out the presence of wild elephants and in most of the cases they failed to escape from the elephant's attacks resulting in human deaths. The logistic regression analysis also shows similar results in our study. Thus, the highest proportion of elephant's attacks occurred during night and resulted in a much higher frequency of deaths compared to daylight period the inside the forest.

4.3 Study Sites (Regions)

Currently, wild elephants are only found in south-eastern and northern hilly forest areas (Islam, Mohsanin, Chowdhury, Choudhury, Aziz, et al., 2012). In the northern parts of Bangladesh, elephants are only found in and around the Sherpur forest reserve (SFR), whereas in the south-eastern part, they are found in several locations, more specifically in and around TWS, CWS, BFR and RFR under Chittagong and Rangamati Forest Circles. The south-eastern region, particularly the Chittagong, Cox's Bazar and Rangamati districts in Chittagong Hill Tracts (CHTs), was the region with the highest number of wild elephant's encounters on humans. This region is also a refuge of diversified indigenous and non-indigenous people. A large number of illiterate people have settled down in this region from other parts of Bangladesh because the region is rich in forest-based natural resources (Sarker & Røskaft, 2011).

We found that the encounter rates of wild elephants were higher in RFR while lowest in CWS. The RFR has large areas with elephant habitats and the availability of both local and trans-border migratory wild elephants from India and Myanmar. Unsustainable tribal and non-tribal human activities inside the forests are thus responsible for more encounters of wild elephants in the south-eastern region of Bangladesh (IUCN, 2004a). The findings of the study revealed that the proportion of encounter which resulted in human deaths or injuries was considerably higher in the RFR than other study sites. The reason is that wild elephants in this region have gradually been confined to 'pocketed herds' in small patches of forest due to construction of the Kaptai dam and highway, which has caused isolation of wild elephants from their mainstream population. Furthermore, most farmers inhibited around RFR are involved in *jhum* (shifting cultivation), which makes the wild elephants almost a seasonal aggressive invader into farm fields and densely dwelling places located in the RFR and resulting in many human fatalities during these seasons. Moreover, small size of reserves increases the frequent contact of humans and wild elephants, thus ultimately increases encountering rates between wild elephants and human which causes more human deaths or injuries. Probably, due to this reason, we found that the higher rate of encounters of wild elephants causing human deaths or injuries was higher in RFR. On the other hand, the CWS has high human densities with settlements that surround the forest areas. Thus, only a low number of elephant herds can potentially intrude into these due to the blockade of the corridor and enormous human interfering activities (Sarker & Røskaft, 2011). The high rate of human causalities might hinder the elephant conservation strategy due to a dramatic increase in the negative attitudes of people (Sarker & Røskaft, 2010; Sarker & Røskaft, 2011).

4.4 Elephant Social Groups

No significant difference between single males or family groups encountering attacks was found with respect, but single or group encountering attacks varied according to the time of day. During the daylight, the forest walking people are mostly affected by single males, whereas at night, the number of attacks by family groups increased. Encounters were amplified during the peak crop season, but the off-season group encountering attacks increased compared to single encountering attacks, which might be due to differences in the behaviour between single bulls and family groups.

4.5 Body Parts Used as Weapons by Wild Elephants

The fatality rate (80.3% human death rate) might be considered as a consequence of the elephant's encountering attack behaviour, while the intensity of fatalities was dependent on the style of elephant encountering attack. During an encountering attack, the wild elephant either used the leg, trunk, or tusk as a weapon or a combination of the trunk and leg with different results. We would expect that wild bull elephants most frequently practiced offensive attacks (leg and trunk) due to their changing hormone levels, whereas herds practice more defensive attacks (a single weapon) as a less offensive strategy (IUCN, 2004b). However, when herds are carrying infants in the group or when they are passing through during the breeding period, they showed a more aggressive behaviour and normally used an offensive encountering attack style which is trunk and leg (Sukumar, 1991). During an aggressive offensive encountering attack, both bulls and herds used legs and trunks combined at similar frequencies and consequently increased the fatality rate. Therefore, family groups were mostly responsible for human deaths in the north, while single bulls were more responsible in the southeast. However, when elephants are less aggressive, they mostly used the leg or sometimes only the trunk to avoid a conflict. During frontal encountering attacks, they used the tusk, trunk or even the leg and trunk combined during an encounter and attack raid. As a consequence of such encounters, fewer fatalities in terms of deaths were observed, while a medium number of fatalities occurred after using the trunk or tusk while a high fatality rate occurred after using the leg and trunk combined.

5. Concluding Remarks

The findings of this study indicate that the rate of encountering wild elephants has gradually increased into recent times, also being more serious because they more frequently are causing human deaths. All types of forest-related illegal activities, such as settlement in forests, illegal forest entrances and forest resource exploitation, weaken the forest management system and are primary causes for the high rate of human-elephant conflict. Human and elephant conflict has introduced a new dimension of conflict between the local people and forest authorities. However, most causality occurs during illegal forest-related activities and in illegal settlements. Indeed, the new wildlife act allows compensation for people injured or killed even during illegal activities and in illegal settlements. However, this new dimension of conflict is limited by the lack of mutual cooperation between local people and forest authorities. Wild elephants play a significant role in protecting natural forests, which offer different types of ecosystem services that are essential for human life, the conservation of elephants should be a human moral, ethical and mandatory task to ensure the survivability of this threatened species in the wild. If it is possible to reduce the human and elephant conflict level by terminating the revenge killing of elephants, then the forest will be protected and human life would be safer. The conflict mitigation approaches and welfare activities, including incentive programs, awareness programs and training programs from the forest department, could help to reduce the conflict between humans and wildlife. Unless active conservation measures are taken immediately, the elephant will most probably go extinct in Bangladesh in a few years to come.

Acknowledgements

AH was supported by a grant through the Norwegian University of Science and Technology (Quota).We are very grateful to our five research gatekeepers in the five study areas in Bangladesh for their assistance and support during our field work. We are also grateful to the staff and officers of the Bangladesh Forest Division for their assistance and cooperation.

References

Aziz, M. A. (2002). *Ecology of Asian elephants, Elephas maximus and its interaction with man in the Chittagong Hill Tracts*. M. Sc. thesis, Department of Zoology, Jahanghirnagar University.

Bandara, R., & Tisdell, C. (2002). Asian elephants as agricultural pests: damages, economics of control and compensation in Sri Lanka. *Natural Resources Journal, 42,* 491-519.

Bandara, R., & Tisdell, C. (2003). Comparison of rural and urban attitudes to the conservation of Asian elephants in Sri Lanka: empirical evidence. *Biological Conservation, 110,* 327-342.

Choudhury, A. (2007). Impact of border fence along India-Bangladesh border on elephant movement. *Gajah, 26,* 27-30.

Feeroz, M. M., Aziz, M. A., Islam, M. T., & Islam, M. A. (2004). Human-elephant conflict in southeastern hilly areas of Bangladesh. In J. Jayewardene (Ed.), *Endangered elephants: past, present and future* (pp. 98-102). Colombo, Sri Lanka: Biodiversity & Elephant Conservation Trust.

Hossain, J. (2008). *Review of CBD programme work in Bangladesh*. Forest People's Programme.

Islam, M. A., Mohsanin, S., Chowdhury, G. W., Choudhury, S. U., Aziz, M. A., Uddin, M., & Azam, I. (2012). Current status of Asian elephant in Bangladesh. *Gajah, 35,* 21-24.

IUCN. (2004a). Assessment of elephant habitat and local capacity and awareness enhancement programme at Chunati Wildlife Sanctuary. *The World Conservation Union, Bangladesh Country Office.*

IUCN. (2004b). Human-elephant conflicts in Bangladesh and assessment of financial losses. *The World Conservation Union, Bangladesh Country Office.*

Jones, S., & Carswell, G. (2004). *The earthscan reader in environmental, development and livelihoods.* Trowbridge, UK: Cromwell press.

Khan, M. A., .R., Rashid, S. M. A., & Khan, A. Z. (1983). *Development of a management plan for the elephants of Cox's Bazar Forest Division (South), Bangladesh.* Department of Zoology, University of Dhaka.

New-Nation. (2015). One person killed in elephant attack in Netrokoma. Retrieved 17 December, 2014, from http://thedailynewnation.com/news/37696/one-killed-in-elephant-attack-in-netrakona.html

PA. (2014). *Two humans killed in Chittagong elephant attck.* Retrieved 17 December, 2014, from http://en.prothom-alo.com/bangladesh/news/57169/2-killed-in-Ctg-elephant-attack

Ramakrishnan, B., & Ramkumar, K. (2007). *Land acquisition perspectives of vital elephant corridors in the Coimbatore and Sathyamangalam Forest Divisions, Tamil Nadu, South India.* Wildlife Trust of India.

Reuters. (2001). *15 people killed in an elephant attck.* Retrieved from http://factsanddetails.com/asian/cat68/sub431/item2466.html

Reuters. (2002). *Elephants trample 3 humans.* Retrieved from http://igorilla.com/gorilla/animal/2002/elephants_trample_3.html

Robbins, P. (2012). *Political ecology: A critical introduction* (2nd ed.). Wiley-Blackwell.

Røskaft, E., Bjerke, T., Kaltenborn, B. P., Linnell, J. D. C., & Andersen, R. (2003). Patterns of self-reported fear towards large carnivores among the Norwegian public. *Evolution and Human Behavior, 24*(3), 184-198. http://dx.doi.org/10.1016/S1090-5138(03)00011-4

Røskaft, E., Hagen, M. L., Hagen, T. L., & Moksnes, A. (2004). Patterns of outdoor recreation activities among Norwegians: an evolutionary approach. *Annales Zoologici Fennici, 41*(5), 609-618.

Røskaft, E., Larsen, T., Mojaphoko, R., Sarker, A. H. M. R., & Jackson, C. (2014). Human dimensions of elephant ecology. In C. Skarpe, J. Du Toit, & S. Moe (Eds), *Elephants and Savanna Woodland Ecosystems: A study from Chobe National Park, Botswana* (pp. 271-288). Oxford: Wiley Blackwell.

Sarker, A. H. M. R. (2010). Human-wildlife conflict: A comparison between Asia and Africa with special reference to elephants. In E. Gereta, & E. Røskaft (Eds.), Conservation of natural resources; some African & Asian examples (pp. 186-210). Trondheim: Tapir academic press.

Sarker, A. H. M. R., & Røskaft, E. (2010). Human attitudes towards conservation of Asian elephants (Elephas maximus) in Bangladesh. *International Journal of Biodiversity and Conservation, 2*(10), 316-327.

Sarker, A. H. M. R., & Røskaft, E. (2011). Human attitudes towards the conservation of protected areas: a case study from four protected areas in Bangladesh. *Oryx, 45*(3), 391-400. http://dx.doi.org/10.1017/S0030605310001067

Skarpe, C., du Toit, J. T., & Moe, S. R., eds. (2014). *Elephants and savanna woodland ecosystems, A study from Chobe National Park, Botswana.* Volume 14. Cambridge: Wiley Blackwell.

Sukumar, R. (1989). *The Asian elephant - ecology and management. Cambridge.* UK: Cambridge University Press.

Sukumar, R. (1991). The management of large mammals in relation to male strategies and conflict with people. *Biological Conservation, 55*(1), 93-102. http://dx.doi.org/10.1016/0006-3207(91)90007-V

Sukumar, R. (2004). *The living elephants: Evolutionary ecology, behavior and conservation.* New York, USA: Oxford University Press.

Sukumar, R. (2006). A brief review of the status, distribution and biology of wild Asian elephants. *Int Zoo Yearbook, 40*, 1-8.

Treves, A., & Naugthon-Treves, L. (1999). Risk and opportunity for humans coexisting with large Carnivores. *Journal of Human Evolution, 36*(3), 275-282. http://dx.doi.org/10.1006/jhev.1998.0268

Permissions

All chapters in this book were first published in ENRR, by Canadian Center of Science and Education; hereby published with permission under the Creative Commons Attribution License or equivalent. Every chapter published in this book has been scrutinized by our experts. Their significance has been extensively debated. The topics covered herein carry significant findings which will fuel the growth of the discipline. They may even be implemented as practical applications or may be referred to as a beginning point for another development.

The contributors of this book come from diverse backgrounds, making this book a truly international effort. This book will bring forth new frontiers with its revolutionizing research information and detailed analysis of the nascent developments around the world.

We would like to thank all the contributing authors for lending their expertise to make the book truly unique. They have played a crucial role in the development of this book. Without their invaluable contributions this book wouldn't have been possible. They have made vital efforts to compile up to date information on the varied aspects of this subject to make this book a valuable addition to the collection of many professionals and students.

This book was conceptualized with the vision of imparting up-to-date information and advanced data in this field. To ensure the same, a matchless editorial board was set up. Every individual on the board went through rigorous rounds of assessment to prove their worth. After which they invested a large part of their time researching and compiling the most relevant data for our readers.

The editorial board has been involved in producing this book since its inception. They have spent rigorous hours researching and exploring the diverse topics which have resulted in the successful publishing of this book. They have passed on their knowledge of decades through this book. To expedite this challenging task, the publisher supported the team at every step. A small team of assistant editors was also appointed to further simplify the editing procedure and attain best results for the readers.

Apart from the editorial board, the designing team has also invested a significant amount of their time in understanding the subject and creating the most relevant covers. They scrutinized every image to scout for the most suitable representation of the subject and create an appropriate cover for the book.

The publishing team has been an ardent support to the editorial, designing and production team. Their endless efforts to recruit the best for this project, has resulted in the accomplishment of this book. They are a veteran in the field of academics and their pool of knowledge is as vast as their experience in printing. Their expertise and guidance has proved useful at every step. Their uncompromising quality standards have made this book an exceptional effort. Their encouragement from time to time has been an inspiration for everyone.

The publisher and the editorial board hope that this book will prove to be a valuable piece of knowledge for researchers, students, practitioners and scholars across the globe.

List of Contributors

Vasileios Markantonis
Joint Research Center of the European Commission (JRC), Ispra, Italy
Research Institute of Urban Environment and Human Resources (UEHR), Panteion University, Athens, Greece

Kostas Bithas
Faculty of Economics and Regional Development, Panteion University, Athens, Greece

A. I. Akintola
Department of Earth Sciences, Olabisi Onabanjo University, Nigeria

P. R. Ikhane
Department of Earth Sciences, Olabisi Onabanjo University, Nigeria

S. I. Bankole
Geosciences Department University of Lagos, Akoka Lagos, Nigeria

O. A. Mosebolatan
Department of Earth Sciences, Olabisi Onabanjo University, Nigeria

Chen Zhang
School of Engineering, University of Guelph, Canada

Edward A. McBean
School of Engineering, University of Guelph, Canada

Ernest L. Molua
Department of Agricultural Economics and Agribusiness, Faculty of Agriculture & Veterinary Medicine, University of Buea, Cameroon

John J Milledge
Faculty of Engineering & Science, University of Greenwich, UK

Alan Staple
Faculty of Engineering & Science, University of Greenwich, UK

Patricia J Harvey
Faculty of Engineering & Science, University of Greenwich, UK

Drissa Traore
Chang An University, Xi'an, China

Qian Hui
Chang An University, Xi'an, China
Department Environmental Sciences and Engineering, Chang An University, Xi'an, China

O. A. Okunlola
Department of Geology University of Ibadan, Ibadan, Nigeria

A. A. Afolabi
Department of Geology University of Ibadan, Ibadan, Nigeria

Zhen-guang Yan
State Key Laboratory of Environmental Criteria and Risk Assessment, Chinese Research Academy of Environmental Sciences, Beijing, P. R. China

Wei-li Wang
State Key Laboratory of Environmental Criteria and Risk Assessment, Chinese Research Academy of Environmental Sciences, Beijing, P. R. China

Xin Zheng
State Key Laboratory of Environmental Criteria and Risk Assessment, Chinese Research Academy of Environmental Sciences, Beijing, P. R. China

Zheng-tao Liu
State Key Laboratory of Environmental Criteria and Risk Assessment, Chinese Research Academy of Environmental Sciences, Beijing, P. R. China

Vincent R. Nyirenda
Department of Zoology and Aquatic Services, School of Natural Resources, Copperbelt University, Kitwe, Zambia

Peter A. Lindsey
Department of Zoology and Entomology, Mammal Research Institute, University of Pretoria, Pretoria, South Africa

Edward Phiri
Lusaka Agreement Task Force, Nairobi, Kenya; Directorate of Conservation and Management, Zambia Wildlife Authority, Chilanga, Zambia

Ian Stevenson
Conservation Lower Zambezi, Chirundu, Zambia

Chansa Chomba
Disaster Management Training Centre, School of Agriculture and Natural Resources, Mulungushi University, Kabwe, Zambia

Ngawo Namukonde
Department of Zoology and Aquatic Services, School of Natural Resources, Copperbelt University, Kitwe, Zambia

Willem J. Myburgh
Department of Nature Conservation, Faculty of Science, Tshwane University of Technology, Pretoria, South Africa

Brian K. Reilly
Department of Nature Conservation, Faculty of Science, Tshwane University of Technology, Pretoria, South Africa

Indika R. Palihakkara
Faculty of Agriculture, University of Ruhuna, Sri Lanka
Graduate School of Agricultural and Life Sciences, The University of Tokyo, Japan

Abrar J. Mohammed
Graduate School of Agricultural and Life Sciences, The University of Tokyo, Japan

Makoto Inoue
Graduate School of Agricultural and Life Sciences, The University of Tokyo, Japan

Robert Summers
Department of Agriculture and Food, Western Australia, WAROONA, Western Australia, Australia

David Weaver
Department of Agriculture and Food, Western Australia, ALBANY, Western Australia, Australia

Nardia Keipert
Peabody Energy Australia, 14/259 Queen Street Brisbane 4000, Australia

Jesse Steele
Newmont Mining Corporation Elko, Nevada 89801, Australia

Michael Ahlheim
University of Hohenheim, Institute of Economics, Stuttgart 70593, Germany

Oliver Fror
University of Koblenz-Landau, Institute for Environmental Sciences, Landau 76829, Germany

Jing Luo
Research Center for China's Borderland History and Geography, Chinese Academy of Social Sciences, Beijing 100732, China

Sonna Pelz
University of Hohenheim, Institute of Economics, Stuttgart 70593, Germany

Tong Jiang
National Climate Center, China Meteorological Administration, Beijing 100081, China
School of Geography and Remote Sensing / Collaborative Innovation Center on Forecast and Evaluation of Meteorological Disasters, Nanjing University of Information Science & Technology, Nanjing 210044, China

Yiliminuer
China Academy of Forestry Sciences in Xinjiang, Urumqi 830011, China
Institute of Applied Physical Geography, Catholic University of Eichstatt-Ingoldstadt, 85072 Eichstatt, Germany

Brett A. Buckland
Via Department of Civil and Environmental Engineering, Virginia Tech, Blacksburg, Virginia, United States

Randel L. Dymond
Via Department of Civil and Environmental Engineering, Virginia Tech, Blacksburg, Virginia, United States

Clayton C. Hodges
Via Department of Civil and Environmental Engineering, Virginia Tech, Blacksburg, Virginia, United States

Ya-Wen Chiueh
Professor, National Hsinchu University of Education, Taiwan

Cheng Chang Huang
Associate Researcher, Agricultural Engineering Research Center, Taiwan

Nyadzi Emmanuel
WASCAL CC&ALU, Federal University of Technology, Minna, Nigeria

Ezenwa I. S. Mathew
Department of Soil Science, Federal University of Technology, Minna, Nigeria

Nyarko K. Benjamin
Department of Geography and Regional Planning, University of Cape Coast, Cape Coast, Ghana

A. A. Okhimamhe
WASCAL CC&ALU, Federal University of Technology, Minna, Nigeria

Bagamsah T. Thomas
Maryland Department of Agriculture: 50 Harry S Truman Parkway Annapolis, MD 21401, USA

Okelola O. Francis
WASCAL CC&ALU, Federal University of Technology,
Minna, Nigeria

Elaine T. Lawson
Institute for Environment and Sanitation Studies (IESS),
University of Ghana, Legon, Ghana

A. H. M. Raihan Sarker
Institute of Forestry and Environmental Sciences,
University of Chittagong, Chittagong 4331, Bangladesh

Amir Hossen
Department of Biology, Norwegian University of Science
and Technology, NTNU, Realfagbygget 7491 Trondheim,
Norway

Eivin Røskaft
Department of Biology, Norwegian University of Science
and Technology, NTNU, Realfagbygget 7491 Trondheim,
Norway

www.ingramcontent.com/pod-product-compliance
Lightning Source LLC
Chambersburg PA
CBHW080255230326
41458CB00097B/4985